ATOMIC ADVENTURES

ALSO BY THE AUTHOR

Atomic Accidents
Atomic Awakenings

ATOMIC ADVENTURES

Secret Islands, Forgotten N-Rays, and Isotopic Murder—A Journey into the Wild World of Nuclear Science

JAMES MAHAFFEY

PEGASUS BOOKS
NEW YORK LONDON

ATOMIC ADVENTURES

Pegasus Books Ltd.
148 W. 37th Street, 13th Floor
New York, NY 10018

First Pegasus Books edition June 2017

Interior design by Maria Fernandez

Library of Congress Cataloging-in-Publication Data is available.

ISBN: 978-1-68177-421-3

10 9 8 7 6 5 4 3 2 1

Printed in the United States of America
Distributed by W. W. Norton & Company, Inc.
www.pegasusbooks.us

Written in memory of

Dr. Monte V. Davis

Adviser, mentor, and friend.

He would have liked it.

Contents

AUTHOR'S NOTE

Stories Told at Night around the Glow of the Reactor

AS THE TWENTIETH CENTURY DAWNED in Georgia, there remained two sure sources of income left over from the antebellum days: growing cotton and draining the sap out of pine trees. These two commodities, cotton fiber and turpentine, sold well to the industrial northern states, but to dress them up for sale required steel straps to bind the bales and steel hoops to hold the barrels together. The closest place to buy these necessary items was in Pennsylvania.

Finding this situation unacceptable, a mortgage broker in Atlanta, George Washington Connors, sold the concept of building a steel mill to a group of enthusiastic investors, and in 1901 the Atlanta Steel Hoop Company was formed. An old farm north of the city was the perfect place for it. The land, which had been fought over one afternoon in the summer of 1864 in the Battle of Peachtree Creek, was purchased from the brothers Kontz, A. L. and E. C., and Captain James W. English.

The site was cleared; a sprawling complex of buildings, steam engines, and open-hearth furnaces was erected; and a railroad spur was laid. A workforce of 120 men was hired. Soon 50 tons of steel, formed into the desired articles, were rolling out of the mill per year, and Atlanta rejoined the industrial world after thirty-seven years of sleep following the Late Unpleasantness.

By 1907 the company was in explosive expansion, and the name was changed to Atlanta Steel Company. There were now more than one thousand people drawing pay at the plant. Good times rolled, and steel hoops, bands, nails, rivets, welding rods, rebars, and Dixisteel barbed wire piled high at the loading platform. Most hirelings lived within walking distance in the Home Park neighborhood that had sprung up south of the plant, from Fourteenth Street down to Sixth, with the main north-south venue being Kontz Street. It became the northernmost suburb of Atlanta, and as automobiles started being used for transportation, the streets were even paved. North of the plant, the deer still roamed.

Downtown was hustling and jiving, with a new vaudeville theater, the Forsyth, built at the corner of Forsyth and Luckie in 1911. Always striving to beat the Atlanta Theater over on Edgewood Avenue, the Forsyth Theater was pleased to snag a one-week engagement of the world-famous Harry Houdini, extraordinary magician and escape artist, starting April 19, 1915.[1]

Houdini was a truly rare event. The Handcuff King worked the Orpheus Circuit, which usually ranged west all the way to San Francisco, north to Boston, and south as far as Richmond, when he wasn't entertaining Europe or Australia. To reach the masses in suburban Atlanta, some face time was in order. He hired a mule-drawn flatbed wagon on which he and his assistants could stand and perform. On Sunday, April 18, after church let out and lunch had been eaten, he parked it on the east side of Kontz Street, halfway between Sixth and Tenth Streets, where the steep hill was briefly interrupted by a flat, vacant lot. Across the street was the big, normally muddy backyard of

1 This was actually Houdini's second show at the Forsyth Theater. The first was announced in late December 1911, and his engagement was January 1–6, 1912. To have him appear in Atlanta was still a world-class novelty in 1915. Working back from the original story, I believe that the 1915 show is the one depicted here.

the house at Sixth and Kontz. An artesian well in back of the house, fed by an underground tentacle of Peachtree Creek running north, kept the area in semi-swamp mode.

Harry Houdini stood on the wagon, rolled up his sleeves, and started shuffling cards. Soon, a small wad of curious wanderers collected in front of the wagon. Word spread rapidly by speedy-foot telegraph through the Home Park community. Some Yankee was giving a free magic show up on Kontz! The crowd, hushed and spellbound, grew to hundreds, taking up the entire backyard across the street. Children were hoisted to shoulders so that they could see, and women were asked to remove their hats. There is no record of what tricks the great magician performed that day, but he was over his handcuff-escape phase and was probably showing off with cards and his new East Indian needle trick.

He first swallowed a handful of needles, one at a time, sometimes having dramatic trouble getting them down. There were at least thirty-five needles of various sizes, up to two and a quarter inches long. He then ate a length of thread, which was actually a five-foot-long piece of highly visible white twine, swallowing a couple of inches per gulp. He paused for a second to settle his stomach, then searched around in his mouth with a digit, seemed to discover something by feel, then slowly withdrew the thread, held delicately between his thumb and forefinger. The string was so long, a couple of assistants had to support it as it slowly emerged. The needles were spaced at random intervals, hanging on the string. He had apparently regurgitated the needles and threaded them onto the string with his tongue. The crowd was amazed, and Houdini enjoyed sellout crowds at the Forsyth.

That December, the name of the mill was changed to the Atlantic Steel Company, and Kontz was renamed Atlantic Drive. World Wars I and II came and went, and the fortunes of Atlantic Steel grew as the demand for Dixisteel products increased. By 1952, 2,100 people were working at the mill. At that time the Georgia School of Technology, headquartered down on North Avenue, was striving to become a research university and was buying up property north of Sixth Street.

The change of technology from simple to complex had been in effect since a person picked up the first rock and found it useful for cracking walnuts. In the 1950s, the movement seemed to accelerate. While the steel plant, built with the most advanced nineteenth-century mechanisms, faded into the heavy industry landscape, the feisty trade school on North Avenue loomed large. The name was changed to the Georgia Institute of Technology. In 1954, the governor of Georgia appointed a Tech graduate, Frank H. Neely, forward-looking industrialist and the head of Rich's department store, as chairman of the Georgia Nuclear Advisory Commission. Under aggressive prodding by Neely, Georgia Tech built the Radioisotopes Laboratory Facility at the southeast corner of the intersection of Sixth and Atlantic Drive in 1959, and plans for a large nuclear research reactor were laid down. The old house at the top of the hill on Atlantic was demolished, and the Electronics Research Building was erected with the parking lot paved directly over the artesian well, ensuring that there would always be a curious wet spot there. Sixth Street was renamed Ferst Street.

Behind the Electronics Research Building, right over the spot where the crowd had congregated to see Houdini on a fine April afternoon, a large hole was excavated, down to bedrock. Here was planted the Frank H. Neely Nuclear Research Reactor, a multipurpose neutron source, designed by the General Nuclear Engineering Corporation in Dunedin, Florida. The main building was a gleaming white steel cylindrical structure with a domed top, connected to a two-story annex having offices, a classroom, laboratories, and special shops for nuclear work. It was issued an operating license, no. R-97, by the Atomic Energy Commission on December 29, 1964, and proceeded to entertain with its own form of magic, doing tricks that would have baffled Houdini on a daily basis, doing everything from driving a LASER cavity with neutrons to investigating Legionnaire's Disease.[2]

By 1975, I was an eager PhD candidate at Georgia Tech wanting to do an experimental thesis dissertation using the impressive 5-megawatt, heavy-water moderated reactor. Funding for such work had

---

2 The Houdini story was first told by Georgia governor Lester G. Maddox. Upon his election in 1967, Maddox was treated to a tour of the Georgia Tech campus, and he was particularly intrigued by the nuclear reactor. He was invited to the control room, located high and overlooking the reactor bio-shield in the containment building. The operations crew encouraged him to sit at the control panel, where he put his feet up on the console and spun his tale about what had been on this spot back in the day. The governor didn't know anything about nuclear research, but he knew a lot about Atlantic Drive. He had grown up in a house right down the street, and he quit high school to work in the steel mill.

pretty much dried up by then, and I had watched as the last of the Era of Great Reactor Experiments disappeared over the horizon, but I was determined, as a candidate must be, to do something significant.

There was much to be learned. First of all, the operations staff at this facility was an underpaid, diversely tempered gang of refugees from the Georgia Nuclear Aircraft Laboratory and the nuclear navy, unappreciated by the Institute as a whole, and under constant bombardment by the Nuclear Regulatory Commission to do nothing wrong while generating neutron flux peaking at about trillion neutrons per square centimeter per second in the heart of a major metropolis. Thrown in were the nuclear scientists, ranging in personality from mildly eccentric to bat-shit crazy, and the ever watchful, ticking radiation counters of Health Physics. One thing these men had in common was that they all seemed to drive sports cars, from Fred Apple's bug-eyed Triumph to Bob Kirkland's Austin-Healy.

As an unfunded grad student with vague plans, I had little to stand on, but it became clear that if I was going to succeed as an experimentalist at this place, then the operations staff would have to *want* me to succeed. I would have to study them, connect with them, find their likes, their loves, fears, politics, and hatreds, and I would eventually have to convince the entire backbiting, claim-jumping, loosely coupled conglomeration to work together on my project, slaving away doing three shifts a day. They had to see me as basically competent, quick to learn, confident in my knife-edge narrow field to the point of cockiness, and one who would see and acknowledge the essential value of experience, something of which a grad student had vanishingly little. For my peculiar project, I had to run the reactor balls-to-the-wall for many hours to build up the iodine poisoning, then drop the power, level out, and map the axial neutron flux periodically as the iodine burned off, the net reactivity of the core increased, and the control-blades slowly sank into the reactor, maintaining precise criticality. It was harder than it sounds.

I spent many a day sitting in the control room, absorbing the atmosphere and a surprising influx of tales not told in the Real World on the other side of the air lock. Here, there was no sunlight, no passage of days or contact with anything happening outside. With the reactor stabilized and on autopilot, as it usually was, the only sounds were the ticking of a certain multi-pen chart recorder as it changed horizontal position on the paper roll or the squeaking of the deteriorating bearing in a swivel chair. The operations men, usually not a particularly talkative lot, under this condition of isolation would start to open up, spilling old government secrets or their guts. Orren Williams, a grad student working as a reactor operator, leaned back, put his feet up on the console, and described the beautiful, highly intelligent woman he was seeing and the house he was going to buy as soon as they were married.[3] There were tales of service on the *Nautilus* nuclear sub, the *Enterprise* nuclear aircraft carrier, and odd bits from the national labs, from Brookhaven to Los Alamos.

The control room of the Georgia Tech Research Reactor with Dave Cox at the console.

The most unusual individual of the lot was John Moon. He was one who did not own a sports car. Instead, he drove a 1965 Mercedes-Benz 190Dc diesel sedan, painted white-gray, flogging it fifty miles down from Dawsonville every morning. By pure chance, I owned an identical car, the same color and with a serial number that was

3 Williams's description of his beautiful, highly intelligent love interest, who was working downstairs in the reactor complex for a nuclear medical research foundation, sounded so irresistible, immediately upon graduation I married her, following a brief courtship. Our thirty-eighth anniversary is coming up in 2017. Orren Williams never talked to me again.

sequential with his. That was my immediate connection to this particular reactor operator.

The author, sitting on top of the reactor with his neutron counting equipment. The computer in the rack at right was built from scratch in the electronics shop downstairs. Notice the hole in his shoe.

Moon had been an operator at the Georgia Nuclear Aircraft Laboratory (GNAL), and his reputation ran in front of him, warning away the meek and sensitive. Moon was something of a wild man. He was the type who would eat a roach off the floor on a dare or decide to test the emergency escape tunnel after lunch.[4] One night

4 Moon's favorite stunt while idling at the control console was to ask a fellow operator, in this case Dean McDowell, did he know that bees can only sting you through open pores, and further that pores open only as you take a breath? If you simply hold your breath, a bee cannot sting you. McDowell found this claim difficult to believe, but Moon just happened to have a bee right here, under an inverted urine specimen cup. He took a deep breath, held it, and slipped his palm under the cup. The bee, mad as hell, tried repeatedly to stab him, to no success. See? "Let me try that!" McDowell enthused. He took his breath, slid his palm under the lip of the cup, screamed, and flung the bee-cup combination ceilingward. Moon collapsed with laughter. His palm was so heavily callused, there was no way for a bee to stick a stinger in him. The escape tunnel was at GNAL, not at Georgia Tech. It was a steel pipe, 3,600 feet long, just wide enough for a man to crawl through, leading from the underground control room of the Radiation Effects Reactor to the main gate at the "lethal fence."

we were at the console, running at one megawatt, and Moon was disclosing the secrets of his former place of employment as the reactor ran on autopilot while I took mental notes. A young student observer interrupted to ask an undergraduate question: "Mr. Moon, does a reactor glow when it's running?"

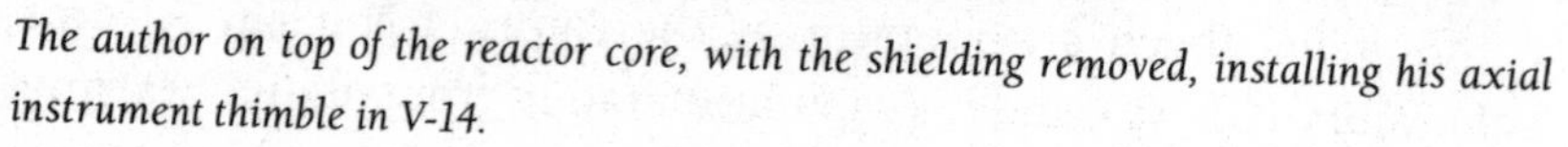

The author on top of the reactor core, with the shielding removed, installing his axial instrument thimble in V-14.

John Moon shifted slightly in his chair, cut a wicked smile, and touched his fingers together. "Why, yes, it does. Would you care to see it?"

"May I?"

Moon turned slowly in his seat and addressed his fellow reactor operator, David Cox, who was resting quietly. "Mr. Cox, go pull vertical fourteen so we can show this fellow what the core looks like." Dave roused himself and complied with due care, using the remotely controlled polar crane to lift the concrete plug on V-14 gently off the reactor face and put it aside. This was the glory-hole for my experiment, and the upper and lower shield plugs had already been removed.

I raised an eyebrow. Moon wasn't really going to send this kid to peer down the hole, was he?

No, he wasn't. Dave, being cautious not to stand over it, maneuvered a big, 45-degree mirror over the open hole and turned it to face the control room. The bright Cherenkov light coming from the naked reactor core, eleven feet down, washed through the glass wall of the control room and overdrove our retinas. The gamma rays went straight up out of the hole and through the steel roof of the containment building.

"Huh," mused the undergraduate. "I thought it would be red. You know, like a stove eye."

We chuckled. No, young man, the reactor glow is a lovely shade of blue.[5]

---

5 The Georgia Tech Research Reactor was shut down and de-fueled in 1996, for fear that international athletes would storm the building and steal the 97 percent enriched metallic uranium fuel during the 1996 Summer Olympics in Atlanta. The facility was decommissioned in 1999. It took a few years to knock it down and haul away the remains. Atlantic Steel was down to four hundred employees by 1997, but they were still turning out Dixisteel barbed wire. The mill was bought by Jacoby Development in 1998, erased from the old farmland, and replaced with a residential/commercial development named Atlantic Station. Neither facility, the reactor nor the steel mill, will ever be built again.

INTRODUCTION

The Curious Case of the N-Rays, a Dead End for All Times

I GOT A CALL FROM a friend who lives down the street, Dr. James Cooke, an anesthesiologist with Emory Hospital. Cooke grew up in Berrien Springs, Michigan, along with his friend, Gregory Stemm, founder and CEO of Odyssey Marine Exploration. Stemm needed a favor.

Odyssey Marine Exploration was in the habit of finding old shipwrecks that had gone down carrying a lot of gold, and Stemm was always on the lookout for new technology that claims to find precious metal at a distance, buried or under a lot of water. A fellow from Florida, to whom I will refer as Randy, had approached him with a new type of gold-finder instrument. Would I take a look at what he claims and make a preliminary evaluation? Be glad to. Send him up to Atlanta.

Randy arrived, not with his device, but with a photograph of it and some maps showing found booty. He looked like a walking American Legion post, decked out in a jacket and a hat crowded with Vietnam

War patches. We asked him to explain his device. It looked like a metal box, about a cubic foot, with a parabolic dish antenna on the front, some unlabeled knobs and switches, and a meter on the back. Randy explained that he was in the navy in the Vietnam conflict, working the radar set on a ship, when he noticed certain signals when the antenna was aimed at far-off metal objects. Using his knowledge of military radar, he had perfected his gold-finder using technology that could detect the special, undocumented radiation that is emitted by all matter, concentrating on the specific characteristics of radiations given off by gold and silver.

His description of how his instrument worked was vague and mysterious. I concluded that either his device was unprotected by a patent and he feared theft of the design, or that he had no idea how it worked. Claims of performance measures for it ranged all over the map, depending on how the question was worded, but I got the impression that this particular model gold-finder could locate an 80-ton block of pure gold at a distance of 40 miles. I paused to wonder how he had tested this capability.

My immediate impression of the device was not positive. For one thing, the probe in the center of the antenna dish was way too long, far outside the focal point of his parabola, but my opinion was clinched when he described its operation. Not just anyone could use the instrument correctly to find gold or silver. In fact, he was the only one who knew how to turn the knobs and read the subtle movements of the meter, and it was impossible to explain it to anyone else. He had, however, practiced with his children reading the dial, and he was confident that they would soon be able to use the instrument effectively.

That was all I needed to hear. He was describing something that in experimental science is called the Hieronymus Effect. This term is used to condemn a physical measurement in which a certain sensitivity of a human being is an essential component of the instrumentation. The problem with using the Hieronymus Effect is that no two people are exactly alike, and that can result in different outcomes of the same measurement. This complicates the problem of duplicating an experiment and compromises the integrity of results. The dependence on a Hieronymus Effect is therefore avoided in any scientific experiment or engineered data collection effort. Your measurement or detection must rely on the objectivity of physical equipment and not on the wishful impressions of a person.[6]

Thomas Galen Hieronymus, born in 1895, worked for twenty-five years as an engineer at the Municipal Power Company in Kansas City, Missouri. Always looking beyond the daily grind at the power company, Hieronymus tinkered constantly with the frontiers of electronics and physics in his home laboratory. He had a theory that life on Earth, particularly chlorophyll plant life, depended on more than sunlight for the growth process. He postulated that there is another, unknown type of radiation from the sun that does not propagate by electromagnetism. He described it as "eloptic" energy, and in the 1930s he devised several experiments proving its existence.

His setups were orderly and scientific. In a totally dark room, several germinating pots were arrayed on a tabletop, each containing an oat seed and a small amount of dampened potting soil. On the bottom of each pot was a metal plate, connected by a copper wire to a common cold-water pipe. Atop each pot was a metal screen, and all but one of these screens, which would be the control specimen, were electrically connected to metal plates outside the lab and on trays facing the Sun. The plates were of varying sizes, and the results of growing oat plants in the dark seemed to confirm that a previously undiscovered form of solar radiation was being conducted to the plants and causing them to thrive. All plants germinated, but only the ones connected to outside radiation collectors (metal plates) survived to turn green. The specimens connected to the larger plates actually seemed scorched, while those on the smallest plates were only weakly green.

Encouraged by his experimental results, Hieronymus further postulated that all matter emits this energy, and that, on an atomic level, its wavelength depends on the atomic weight of the nuclide that is emitting. He quit his day job at the power company and devoted the

6 But science is flexible. Alessandro Volta was a professor of experimental physics at the University of Pavia, Italy, in 1800, when he invented the electric battery. The first voltmeter, used to determine that electricity was indeed being produced by the new device, was Volta's tongue. He would place the two wires, anode and cathode connections, against the tip of his taster and feel the burn. This important experiment, by definition, employed the Hieronymus effect. Some people could definitely taste the "metallic" flavor of electricity, and, over a spectrum of response sensitivity, some could not. How Volta thought to stick the wires in his mouth is not written down.

rest of his life to investigating his discovered form of radiation and developing practical instrumentation to exploit its properties.

His most famous invention, the Hieronymus Machine, was capable of sorting the elementary constituents in a sample of anything and indicating how much of which elements made up the specimen. This was doing what a mass spectrometer does, only without pulling a hard vacuum. The user would place a sample, such as an airplane rivet, onto a flat, spiral-coil of magnet wire, connected back to the instrument, and stroke his or her fingertips on a copper plate that is covered with insulating lacquer. By turning a knob slowly, the user would find places along the 60-degree arc of the knob at which the plate would become rough or sticky to the finger touch. These incidents were noted as degrees of deflection, and each could then be mapped into an atomic weight using a simple chart.[7] The rivet would then analyze as being mostly aluminum, with some magnesium and silicon traces.

This analysis was determined using a fixed, triangular glass prism to spread the radiation spectrum being emitted from the specimen into a 60-degree span. The radiation was conducted from the object being analyzed into the machine using wires, just as it had been in Hieronymus's plant-growth experiments. Inside the box, the transported radiation would escape from a small, metal electrode; be collimated by a short, narrow tunnel; and hit the prism base at a 30-degree angle, where it would be bent to an extent dependent on the wavelength of the radiation. The radiation would then be reacquired by an identical electrode connected to a swing-arm attached to the knob. The weak radiation signal would then be amplified millions of times using a three-stage audio amplifier and ultimately connected to the copper touch-plate.

T. Galen Hieronymus applied for a patent of his device, "Detection of Emanations from Materials and Measurement of the Volumes Thereof," on October 23, 1946.[8] The patent was awarded on September 27, 1949,

7 I am writing this description from T. G. Hieronymus's literal description of his device. I think that he meant to say "atomic number." Given only the atomic weight, one could derive only a vague idea of what the element is. Finding, for example, an atomic weight of 14, the specimen could be either oxygen-14 or carbon-14. An atomic number, the number of protons in an element's nucleus, corresponds only to a specific element.

8 The title of this patent, no. 2,482,773, is misleading. Nowhere in the patent does this device claim to measure the volume of anything. It is supposed to measure the element composition of materials.

and the Hieronymus Machine went into limited production. His best seller was the Eloptic Medical Analyzer, which could both diagnose and treat any medical condition in crops and farm animals.

By 1954, Hieronymus was at the top of his game. His field of science now had a name, psionics, and he had branched out into building a newer device with more signal amplification, called the "pathoclast." He was now living in Hollywood, Florida, had been given an honorary PhD in physics, and had demonstrated his device to Dr. Arthur Compton, chancellor of Washington University and discoverer of the Compton Effect. The Air Force wanted to talk to him about detecting human presence on the ground from a high-altitude airplane.[9]

At this zenith of psionics research, he got a call from John W. Campbell Jr., the influential and intellectually overwhelming editor of the magazine *Astounding Science-Fiction. Astounding* was widely read among the scientists and technologists populating the national laboratories, and Campbell, a self-proclaimed nuclear physicist, was beginning to lose interest in his favorite pseudoscience, L. Ron Hubbard's dianetics. Dianetics had started out as Hubbard's alternate answer to psychiatry but had morphed into a religion called Scientology. Religion did not interest Campbell, but psionics sounded exotic and weird, just the way he liked it. He quickly broke down any resistance Hieronymus had to revealing everything, and within days was building psionics machines in his home laboratory in Mountainside, New Jersey.

At first skeptical, Campbell with his own tests became convinced of the legitimacy of the Hieronymus machine concept. He began a new campaign to push Hieronymus and his machine in an editorial in *Astounding*, June 1956, "Psionic Machine—Type One."

This was superb advertising for psionics and the Hieronymus Machine until Campbell left the rails. During his extensive program of testing and verification of the Hieronymus Machine, Campbell noticed

---

9 Hieronymus gladly stepped up to this challenge, but instead of mounting his machine in the downward-looking bombsight window of a high-altitude plane, he instead requested *photographs* of the ground where soldiers were hidden. The Air Force complied, Hieronymus scanned the pictures with a psionics device, and he found evidence of people all over the photographs. When told that people were only in a few locations, Hieronymus explained his analysis saying that the soldiers had obviously been urinating on all the trees and had left their essence scattered hither and yon. The Air Force decided not to pursue this inquiry.

that one did not have to be touching an actual machine to exploit its principles of operation. In fact, a schematic diagram of the machine, drawn with a pencil on a piece of paper, worked just as well. He could lay a rock over the drawn depiction of the sensing coil, erase the picture of an open switch on the amplifier schematic and redraw it as a closed switch, then rub the picture of the metal pad with fingertips. When he had drawn the receiving electrode at the proper angle on the schematic, he could detect the composition of the mineral by feel. The paper became sticky. After a while, the thing would quit working, and he would have to erase the diagram of the now-dead battery and redraw it.

This discovery was written up in the magazine, and it did not go down well in the scientific community, which had been quietly discounting the claims as wishful thinking. Campbell enthusiastically interpreted this finding as proof that the Hieronymus Machine worked by means of extrasensory perception, or ESP. At that point, in the early 1960s, the Hieronymus Machine and John W. Campbell Jr. were effectively torpedoed and eventually sank out of sight. The entire concept of putting a human being's sensory perception somewhere in line in a data measurement was solidly confirmed as a dead end. Eloptic radiation does not exist, and any research scientist who would bother to test a psionics machine would find its action driven by psychology and not physics. T. Galen Hieronymus died in 1988, and his machine went with him.

Such pseudoscience was understandable coming from a radar technician, a science fiction magazine editor, and even from an electrical engineer, especially at a time when parapsychology, chiropractic, unidentified flying objects, hypnosis, and searches for Noah's Ark were seriously pursued and even funded. One can get caught up in the pursuit of new knowledge and lose track of reality, clinging only to positive evidence and neglecting any negative findings. You might expect it less from a professional scientist, but you might be wrong.

The Hieronymus Effect had its most profound incident in 1903 at the Nancy-Université, in Nancy, France. It caught Prosper-René Blondlot, professor of physics. It would go down in history as the "N-rays illusion."

In 1903, the atomic nucleus had yet to be discovered, yet what would become known as nuclear physics was crashing ahead at full speed, and discoveries were being documented on a monthly basis. Still stinging from the attention-grabbing discovery of X-rays in Germany back in 1895, Professor Blondlot and his assistants were actively striving to find another particle of radiation before they were all gone. Paul Villard, after all, had discovered the gamma ray at École Normale Supérieure rue d'Ulm in Paris just three years prior. Blondlot considered the École a good place to train high school teachers, and the major finding coming from it was something of an embarrassment to the public university system. Blondlot was determined to bring things back into balance by making a gamma-ray-eclipsing discovery with unique strangeness and novelty at his beloved Nancy University.

Blondlot was born in Nancy in 1849, the son of the professor of toxicology at the university, Nicolas Blondlot, MD. He received his doctorate in physics at the Sorbonne in 1881 and joined the faculty at Nancy the following year. He thrived in his element as a professor, winning the Gaston Plante Prize for research in 1893 and the La Caze Prize in 1899. He had successfully measured the extremely rapid response speed of a Kerr cell under electrical excitation using an ingeniously modified rotating-mirror apparatus from Léon Foucault's speed-of-light measurements.[10] After that, he measured the speed of radio waves, confirming an important prediction by James Clerk Maxwell. By 1903, Blondlot was on a roll.

---

10 The Kerr cell, invented in Scotland in 1875 by the physicist John Kerr, consists of two parallel electrode plates separated by a layer of nitrobenzene. Apply electricity across the electrodes, and the liquid develops interesting optical properties. It becomes birefringent to polarized light, refracting it off in two directions. The effect will switch on and off with incredible speed on the nanosecond scale. This property was exploited in the rapatronic camera, invented by Harold Edgerton of MIT, for use recording motion pictures of atomic bombs exploding at the tops of steel towers. These movies, only ten frames long, break an event that lasts a few milliseconds down into a slow-moving sequence, with the rapidly evolving explosion frozen in time. Watching it, you can see the fireball erupt from the bomb as the overrunning X-ray shock waves travel down the guy wires of the tower and cause them to evaporate. The tower has no time to be blown out of the way as it reduces to plasma under the spherical shock. It's a rare spectacle of two divergent theories operating in the same photograph. Quantum mechanics eat the tower, while Newtonian mechanics (inertia) make it stand still.

On February 2, he submitted a paper to the French Academy of Science, "On the Polarization of X-Rays." Blondlot was generating X-rays using a simple vacuum tube driven by a Ruhmkorff high-voltage apparatus. The Ruhmkorff transformer converted the six volts from a battery into several thousand volts. The high voltage was not exactly steady, but was produced in a ragged, pulsed alternating current using a buzzer operating off the iron core of the transformer.[11] On the positive-going section of the high-voltage waveform, or about half the time, electrons would stream off the cathode of the tube, run a couple of inches through the vacuum, and crash against the metal anode plate, canted at 45 degrees and pointing the resultant X-rays through the side of the tube and into the room.

Blondlot had a novel idea for detecting any polarization of the X-rays, or the tendency of the electromagnetic waves to vibrate in only one direction. He took a twisted pair of heavily insulated wires and wrapped the bare ends around the feed lines from the Ruhmkorff, with one on the anode lead and one on the cathode lead. He then supported the other end of the wires in front of the vacuum tube, with the bare ends pointing at each other and separated by "a very small distance," making a spark gap. He reasoned that the electromagnetic oscillations of the X-rays would interact with the spark gap, and furthermore the extent of interaction due to polarization, whatever that might be, would be dependent on the orientation of the long axis of the gap.

Blondlot found what he was looking for. When he twisted the gap, which was making a semicontinuous, sputtering high-voltage spark, the visible brightness of the spark changed, indicating that the X-ray was indeed affecting the gap and that it was orientation dependent. Moreover, using the spark-gap detector, he could observe a twist in the polarization of the X-ray using an interstitial crystal of quartz or lump sugar. His X-rays therefore behaved like visible light.

---

11 Blondlot would later employ a "rotary interrupter" to modulate the primary coil in his Ruhmkorff setup. This was a disc made of an electrically insulating material having a conducting stripe of copper foil adhered to the surface. The disc was spun at a high, constant speed by being fastened directly to the axle of an electric motor. As the disc spun, two spring-loaded electrical feelers would bear against the surface and make a periodic on/off connection through the copper strip to the battery driving the Ruhmkorff. The rotary switch setup may have improved the sputtering, inconsistent quality of the high-voltage spark.

Excited by these fascinating results, Blondlot continued experimenting, looking for other ways that X-rays might behave the same way as visible light. To preclude any cross-contamination, he shielded the cathode-ray tube with metal foil to block out any visible-light fluorescence produced by electrons hitting the glass. Using the spark-gap detector and his eyesight to judge the spark intensity in a darkened room, he found evidence of elliptical polarizations using thin sheets of mica. With this finding, he leapt to the next logical step and put a glass lens in the X-ray path. He was astonished to find a perfectly focused, inverted image of the tube's cathode, suspended in space in front of the lens. He scanned across the invisible image using his spark-gap X-ray detector, finding the sharp edges of the image at the predictable focal plane of the lens.

Coming down off the euphoria, Blondlot realized that there were only two problems with these observations. X-rays don't really brighten sparks, and X-rays do not focus using a glass lens. In fact, he believed he was looking at a new, previously undiscovered form of radiation coming out of the X-ray tube. These rays are plane polarized coming out of the tube, can be circularly polarized, and can be reflected, refracted, and diffused, but they do not interact with photographic media, nor do they make anything fluoresce.

His next, ground-trembling paper was "On a New Species of Light," submitted on March 23, 1903, in which he announced his discovery, soon to be named "N-rays" for his university. He concluded the paper saying:

> At first I had attributed to Roentgen rays the polarization which in reality belongs to the new rays, a confusion which it was impossible to avoid before having observed the refraction, and it was only after making this observation that I could with certainty conclude that I was not dealing with Roentgen rays, but with a new species of light.

So launched an accelerated effort to find the peculiar characteristics of this new type of radiation. Blondlot and his assistants, seeing a future graced with research prizes and possibly, if they dared hope, a most coveted Nobel Prize in physics, applied themselves with vigor and resolve.

Blondlot worked to identify all possible sources of N-rays. He made a chart showing which materials would conduct N-rays and which would stop them. He measured the index of refraction in different circumstances, and discovered three subspecies of N-rays. New data piled up, and on May 11, 1903, he submitted another paper, "On the Existence, in the Rays Emitted by an Auer Burner, of Radiations which Traverse Metals, Wood, etc."[12] Four days later, he submitted another paper, and six more by the end of the year. He could not have been getting much sleep. In the next three years Blondlot would submit twenty-six papers concerning N-rays and eventually publish a book. His fellow research physicists, 120 of them, mostly of Gallic origin, would collectively publish almost three hundred notes, articles, and papers on the subject.[13] It required fifty-nine pages to briefly list the discoveries concerning N-rays.

The found properties were peculiar, but so were characteristics of other contemporary discoveries. N-rays would not cause fluorescence, but they would enhance a previously established fluorescence. Aluminum, gold, platinum, silver, glass, and oak were all transparent to N-rays, but a sheet of wet cardboard would stop them cold. N-rays, to one extent or another, seemed to be emitted by just about everything, from the Sun to a Nernst lamp, with special "physiological" N-rays coming from anything alive, including human beings. N-rays were definitely not emitted by green wood or (I hope you're sitting down) tempered steel that had been *anesthetized* by first soaking it in ether or chloroform.

A tempered steel file, presumably borrowed from the school shop, that had not been put to sleep, however, turned out to be an excellent source of N-rays. Using the file as a source, it was found that N-ray flux

---

12 The "Auer burner" was invented by Carl Auer von Welsbach, an Austrian scientist, in 1890. It was a new way to use a gas flame for light, employing a mantle made of a mixture of thorium dioxide and cerium oxide. Instead of a dim, yellow flame, an Auer burner glowed brilliant white from the fluorescence of the thorium-cerium combination, and it turned out to be a strong source of N-rays. Think Coleman lantern. Carl went on to invent the cigarette lighter flint.

13 It has been said, as a slur attributed to Robert W. Wood, that "only Frenchmen could observe the phenomenon." This is an exaggeration. J. S. Hooker and Leslie Miller, both Englishmen, and F. E. Hackett, a student at the Royal University of Ireland, reported N-ray observations. Miller was the first to exploit N-rays for profit, selling a manufactured device for finding them and advertising it in *Lancet*.

enhances human eyesight. In a dimly lit laboratory, it was hard to read the clock on the far wall and tell whether it was time to put down the instruments and go home, but Blondlot found that if he held that file up to his face the N-rays would improve his vision to the point that the hands of the clock would snap into focus and the light-gathering power of the retina would increase. The same was true of a material under stress. Hold up a seasoned wood walking stick in front of your eyes, bend it almost to the breaking point, and you could almost see without the glasses.

Back in the United States, there was a mounting sense of skepticism concerning Blondlot's new form of radiation. Other forms of radiation, such as Hertz's radio waves or Roentgen's X-rays, were the result of some form of expended energy, and alpha, beta, and gamma rays were known to be generated by atomic decay, an irreversible process. If everything emitted N-rays all the time, then how long could this continue before the source was exhausted? The massive list of effects attributable to N-rays was starting to look silly.

Robert W. Wood, professor of optical physics at Johns Hopkins University in Baltimore, highly accomplished experimentalist and inventor, author of two science fiction novels, and the one who admitted to having written *How to Tell the Birds from the Flowers*, was in Britain attending a scientific conference when he decided that he had heard enough about N-freaking-rays. He had "wasted an entire morning" trying to duplicate at least one of Blondlot's experiments, and had corresponded with the French scientist making sure that he hadn't been doing it all wrong. Blondlot had been delightfully cordial and patient with him, but Wood found his body of work concerning the new radiation hard to swallow. Wood took a side trip to Nancy, France, to meet Dr. Blondlot and see for himself what was going on.

Blondlot was pleased to demonstrate all of the fundamental N-ray experiments over the course of three hours in the laboratory. First, Wood was shown a spark-gap detector with N-rays concentrated on it using a lens made of solid aluminum. The light from the spark, which is a naturally sputtering, dancing illumination source, was diffused with a piece of frosted glass. With the lights turned down, an assistant would put his hand between the source and the lens, demonstrating that the normally bright spark, enhanced by the focused beam of N-rays, would dim when his hand was in the way. Woods had to admit that he could not see this effect. To him, the spark always looked about the same, although it was randomly flickering at about 25% luminosity just due to the fact that it was a high-voltage spark. A spark tends, naturally, to heat up the air it is passing through. The hot air wants to float to the ceiling and take the spark with it, so there is a chaotic tug of war between the rising air and the need of the spark to seek the path of least resistance between the electrodes. His inability to see the difference in the brightness of the spark was attributed to a lack of sensitive eyes.

Wood suggested another way to run the experiment. "You look at the spark and tell me when my hand is in the beam." The assistants were unable to guess correctly when he had his hand in the beam and when he had withdrawn it.[14] An explanation was not offered, and they moved quickly to the next demonstration.

The lens was removed and replaced with a screen of wet cardboard having a vertical slit about 3 millimeters wide. In front of the slit was placed an aluminum prism, which would receive the 3-millimeter N-ray beam and spread it out into a spectrum. Replacing the spark gap as the detector was a piece of dry cardboard with a thin, vertical line of pre-excited, glowing phosphorescent paint drawn on it, about half a millimeter wide. The cardboard was on a track so that it could be slid past the slot with a crank-driven screw. When moved through the

---

14 This is obviously a case of the "Hieronymus Effect" taking the place of objective instrumentation, and the attributes of N-rays corresponded with the Hieronymus "eloptic" rays. The researchers at Nancy had even confirmed that the "physiological rays" to and from living things could be collected by a metal plate and conducted along a wire. Blondlot may have been deluded by his experimental results, but he was not a complete fool. He had, in fact, recorded many of his spark-gap brightness measurements on photographic plates, correctly thinking that eyes could be fooled, but not photographs. The extent that a photographic emulsion is exposed by the light from a spark over a fixed unit of time should be an unimpeachable recording. Wood saw it differently. Watching a demonstration of Blondlot's photo-recording techniques, he could see how subtle biasing of the exposure time or processing duration could throw the measurement to a consciously or even subconsciously desired outcome. There was a troubling possibility of skullduggery at work in this laboratory. It was traditional to split the monetary proceeds of a research award with the lab assistants, and if they would score a Nobel with this discovery, the reward would be substantial. The assistants, who have never been named, could have thrown out any photographic evidence that there was no N-ray effect, and kept only those that confirmed what Blondlot wanted to see.

spread of the N-ray spectrum, the demonstrators claimed that spectral lines as fine as 0.1 millimeter wide would make obvious changes in the brightness of the phosphorescent line. When Wood asked how a bundle of rays 3.0 millimeters wide, as defined by the width of the slit, could possibly make a spectrum that could be resolved down to 0.1 millimeters, he was told that this was "one of the inexplicable and astonishing properties" of N-rays.

As he cranked the phosphor detector through the N-ray spectrum, Wood had to admit that he could not see any deviation in the brightness of the line of paint. Perhaps, he suggested, it would be easier on his unpracticed eyes if they turned off the lights? The researchers agreed, but with the lights out, Wood removed the prism and put it in his pocket. Blondlot ran the detector through its entire sweep, pointing out the characteristic spectral lines as he saw them lighting up the line of phosphorescent paint, even though there was no prism in the setup to generate a spectrum. At that exact point in time, as the detector carriage reached the end of the spectrum image, the entire N-rays phenomenon turned downward and began the long slide to oblivion.

Wood was shown the effect of holding a file up to your face and being able to see in the dark. It did not work when Wood tried it, and the experiment was easy to shoot down. Wood suggested that he should hold the file in front of an assistant's face while he looked at the clock hands, while he surreptitiously substituted a block of wood for the piece of tempered steel. The assistant did not notice the chicanery and reported improved eyesight.

Wood immediately wrote up his discouraging, depressing experience at the University of Nancy and sent it from Brussels, Belgium, to the premier science journal in Britain, *Nature*, on September 22. It was published in the September 29, 1904, issue, and the word spread like the bubonic plague across Europe. In October, the same letter appeared in the French *Review Scientifique* and the German *Physicalische Zeitschrift*, and N-rays crashed slowly over the next two years.

At the end of 1904, as the N-rays controversy exploded around him, Prosper-René Blondlot was awarded the Prix Leconte, a prize given by the French Academy of Sciences recognizing important discoveries in mathematics, physics, chemistry, natural history, or medicine. It was 50,000 francs, or five times his annual salary. He continued in his post as professor of Physics for the next six years, dialing back the N-rays research but never admitting the nonexistence of his rays. He retired at age sixty-one in 1910 and lived for another twenty years in his large home at 16-18 Quai Claude le Lorrain. He wrote a new preface for the third edition of his textbook on electricity in 1927. He and the last mention of N-rays died in 1930. He was eighty-one years old. There is a street named for him in Nancy.

Robert W. Wood continued a distinguished career in optical physics, becoming a member of the Royal Society in London, winning seven awards in optics and physics, and serving as president of the American Physical Society. The ultraviolet lamp he developed for medical use is still called the "Wood's lamp," and the bright reflection from green plants in infrared photographs is called the Wood Effect. A crater on the far side of the Moon is named after him.

He died on August 11, 1955, at age eighty-seven.

N-rays are often cited as an isolated example of pseudoscience, and yet mysterious rays with similar characteristics would show up from time to time in the even more enlightened later twentieth century. N-rays were not even new in 1903. The same phenomenon had turned up in 1850, having been discovered by Baron von Reichenbach in London and revealed in his treatise "Researches on Magnetism, Electricity, Heat, Light, Crystallization, and Chemical Attraction in their Relations to the Vital Force." Before that, Franz Mesmer in Vienna had detailed the same rays in 1779 in his "Memoire on the Discovery of Animal-Magnetism." Every time they pop up, N-rays or the equivalent are eventually knocked down as scientific delusions, if not outright fraud.

Physics is a wide field of study and discovery, trying to cover all of reality. One branch of this science deals with the invisibly small "hard nut," the nucleus, at the center of the atom, the fundamental unit of condensed matter. Nuclear physics stands atop two dissimilar, conflicting theory sets working at opposite ends of the largeness spectrum—quantum mechanics for the imperceptibly small and general relativity for the astronomically big. Quantum mechanics predicts that the mass of things on the atomic scale can be diminished just by rearranging the component parts, and relativity predicts that this change will manifest itself as a surprisingly large energy release.

There is nothing intuitively obvious about this phenomenon or anything else about nuclear physics, and this inherent strangeness keeps it out on the edge, dangling over the precipice and always in danger of falling into the infinitely deep pit of the not possible. The specter of pseudoscience always hangs close.

N-rays seem impossible, at least in retrospect, but what about nuclear physics would not seem odd? The field covers everything from turning platinum into gold to the quantum entanglement of photons. So, try to stay in touch with what you consider to be reality in the following chapters, as I step you through some atomic adventures. There are inevitabilities and things that never should have happened. There are things to fear and things that should cause no alarm. There are monuments to fallibility and tiny markers for the truth. I hope to convince you that while nuclear science can be unbelievable or even dreadful, it is never a boring topic of study, conversation, or even reading.

CHAPTER 1

Cry for Me, Argentina

> "On reading the first line of Richter's papers one would think he is a genius; on reading the second line one comes to the conclusion he's nuts."
>
> —Dr. Edward Teller, 1956

If there were a Holy Grail of nuclear technology, a noble and unattainable goal for which only the bold and fearless strive, it would be power production by hydrogen fusion. Hydrogen fusion, the Grail, promises the salvation of humankind by supplying an endless source of clean, pure power. Unlike power derived from nuclear fission, fusion works at the opposite end of the scale of atomic mass, the lower end of the "curve of the binding energy," where instead of blowing apart the heavy, complex elements like uranium and plutonium, isotopes of the very lightest and simplest element, hydrogen, bind together to form heavier elements.

In fission, the mass of the fragments from the breakup does not exactly add up to the mass of the original atom, and this otherwise inexplicable "mass deficit" manifests as a burst of equivalent energy. In similar fashion, when four hydrogens bind together to form a helium, the mass of the resulting atom does not quite match the combined masses of the four original atoms, and the deficit is a burst of energy. The tremendous difference in these two modes of power production, neither of which involves burning flammable material and making carbon dioxide gas, is that the end products of fission are always unnatural, neutron-heavy nuclides. These nuclides are usually unstable, and they try various means of normalizing the neutron load in the nucleus to reach stability. This results in radioactive decay of the fission products, and the radioactivity gives fission an unattractive danger property. The waste products of fusion are only slightly more complex than the hydrogen used in the reaction, and they are not radioactive.[15] Fusion power is utterly harmless, and on top of that, the source of fuel is virtually endless. It uses hydrogen, the most common element in the known universe. On Earth there are literally vast oceans of it. It is a component of sea water.

As is the case in most high-concept nuclear technology, the heavy lifting in fusion power research was performed in secret for the development of extremely destructive devices. At this time, all of the advanced nuclear weapons in the arsenal of the United States use some form of hydrogen fusion for a fission booster, a neutron source, or as the main show. The most powerful bombs are "thermonuclear," which translates to "hydrogen fusion." There is no question that a tremendous power can be derived from hydrogen fusion, and that it leaves no poisonous residue. In the broadest view, hydrogen fusion obviously deserves Grail status, but the devil is in the details.

Lise Meitner postulated the fission process to have occurred in neutron bombardment experiments on uranium by Otto Hahn in Germany in December 1939, and given this discovery, a fission reactor was designed, built, and operating in the United States three years later. Way before that, in 1933, the Australian physicist Mark Oliphant had observed hydrogen fusion at the Cavendish Laboratory at the University

15 There are, of course, exceptions, and in the interest of full disclosure, I must mention them. The free neutron released in three known fusion reactions is technically a radioactive particle, because it decays into a proton and an electron (beta-minus decay) with a half-life of 10.23 minutes. Free neutrons can also activate other nuclides to radioactivity upon capture. Tritium (hydrogen-3) is another possible fusion product, and it undergoes a very low-energy beta-minus decay with a half-life of 12.32 years. These scant radiation sources do not hold a candle to the high-energy, million-year, mixed-mode radiation from fission products and from the immediate fission process.

of Cambridge, England. He used a 500,000-volt electrical potential to accelerate positively charged deuterium ions (deuterons) and crash them into a stationary target of tritium. Deuterium is the heavy-hydrogen nuclide, consisting of simple hydrogen (a single proton) with one neutron added to the nucleus. Tritium is the heavier-hydrogen nuclide, which is hydrogen with two neutrons added to the nucleus.

The atom created by the fusion of deuterium (D) and tritium (T) is helium-4 (He-4), the naturally occurring, inert element used to fill birthday balloons. One D-T fusion yields a respectable 17.6 million electron volts (MeV) of energy.[16]

Oliphant's experiment was a sensation, and laboratories around the world confirmed his findings with similar setups. A bonus of the helium-generating fusion reaction is a free, high-speed neutron, and an enduring application of this fusion apparatus is to use it as a neutron generator for laboratory or industrial use. By 1938, the enthusiasm was high despite a lingering, worldwide economic depression. Patents were filed in Germany and the United States for the fusion neutron generator, and at the Langley Memorial Aeronautical Laboratory in Hampton, Virginia, the first "tokomak" fusion reactor was built by Arthur "Arky" Kantrowitz and Eastman Jacobs.[17]

---

16 That energy yield, 17.6 MeV, is only 6.7×10^{-13} calories per fusion. True, you would have to fuse 1×10^{16} times to equal the 7,000 calories of energy produced from burning a gram of ethanol, but those fusions only involve 8.3×10^{-8} grams of deuterium-tritium mix. That's not much. If you "burn" a gram of deuterium-tritium using fusion, which weighs about as much as a single raisin, you are given 8.0×10^{10} calories. That's the equivalent of 84 megawatt-hours of electricity. In the hypothetical fusion transformation of that gram of material, 0.0000000037 grams of mass mysteriously vanish.

17 The Kantrowitz-Jacobs fusion device was way ahead of its time. Kantrowitz, a recent physics graduate of Columbia University, read in a magazine article that Westinghouse had bought a very powerful Van de Graaff high-voltage machine, and the buzz was that they were going to use it to build a jumbo-sized Oliphant fusion reactor. Although working on aircraft wing design, he had no trouble spinning up his boss, Jacobs, with the idea of one-upping Westinghouse with a better fusion reactor design. He proposed using a toroidal (donut-shaped) electromagnet to compress ionized hydrogen inside and cause it to fuse. This was hardly the mission of the Aero Lab, so they had to cloak it with a name, "diffusion inhibitor," in order to wrangle a budget of $5,000. The plasma was heated by a 150-watt radio transmitter, and the power-draw from the water-cooled magnet windings was enough to dim the lights in the neighborhood. Kantrowitz held the circuit breaker closed as they watched the smoke rise from the building's wiring, but they were never able to make the simple hydrogen fuse. It was a brilliant idea, and Soviet scientists would gain fame fifteen years later with their version of the tokomak. The tokomak reactor concept is still in play in the twenty-first century. Kantrowitz drifted away and founded the Avco-Everett Research Laboratory in Everett, Massachusetts, and Jacobs opened a restaurant in Malibu, California.

Fusion at the experimental level is not particularly difficult to achieve. In August 1971, *Scientific American* encouraged many a talented student to go nuclear by publishing plans for building a Van de Graaff accelerator for producing tritium and neutrons with a deuterium-deuterium (D-D) fusion. Although it is far less efficient than the D-T fusion, one doesn't have to own a tritium source to make it work.[18]

In 1935, four years before nuclear fission was accidentally discovered, Dr. Hans Bethe of Strasbourg, when it was part of Germany, was working on the theory of solar hydrogen fusion in a faculty position at Cornell University. Bethe, one of the finest theoretical physicists of the twentieth century, had been dismissed from his job as an instructor at the University of Tübingen, Germany, in 1933 because, although he was a Lutheran, his mother was Jewish. Bounced out of Germany to the University of Bristol in England, he moved to the United States a year later.

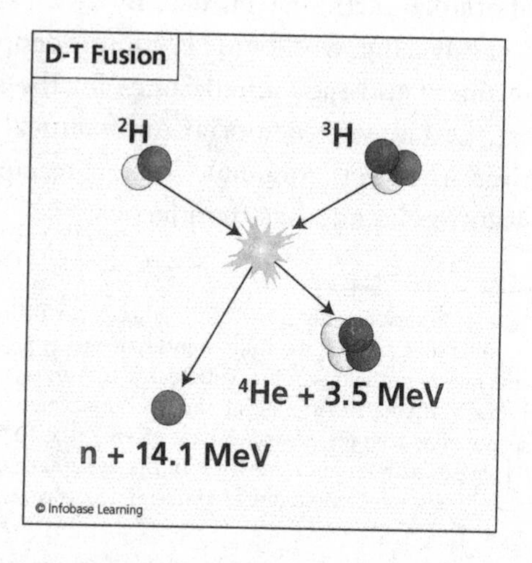

Mark Oliphant's D-T fusion reaction. The free neutron, traveling at very high speed, 14.1 MeV, is a bonus that can be exploited as a neutron generator. The helium-4 is four times heavier than the neutron and is moving more slowly, but together the two kinetic energies of the product particles are 17.6 MeV.

18 The amateur scientist was encouraged to make his own radioactive sources by bombarding various materials with neutrons generated in his home-built "machine to produce low-energy protons and deuterons." As an after-message, the article encourages the experimenter to stay out of the way of the "x-rays of substantial intensity," shield the apparatus with a double layer of solid, 18-inch-thick concrete blocks and boxes of paraffin surrounding the stationary target, wear a dosimeter and a film badge, and keep a Geiger counter turned on. Back in '71, amateur science ran wild and free.

From early optical spectra of sunlight, it had been known that the Sun seemed to be a large ball of compressed hydrogen gas with helium mixed in, but there was no plausible theory until 1920 as to why it was glowing white-hot. Arthur Eddington, a most capable English physicist, threw out the idea that four hydrogens were somehow combining to form one helium atom, and this process must involve a direct-energy conversion. He wasn't sure how.

The process of making helium-4 out of four hydrogens is not simple. George Gamow and Carl Friedrich von Weizsäcker proposed a way to make deuterium out of two hydrogens. For it to work, one of the hydrogen nuclei (a proton) has to change into a neutron, using a sort of beta-plus decay of the proton, which is something that under normal circumstances never occurs, and leftover debris from the reaction are a positron and a neutrino. It was a start, but it did not explain where the helium comes from, nor did it explain how heavier elements, which were detected by spectral analysis of larger stars, could be built up by solar fusions.[19]

The Bethe-Weizsäcker study found that in five steps, four hydrogen nuclei (protons) can eventually combine into one helium-4 nucleus. The exhaust is two gamma rays and two neutrinos, with two additional protons having "catalyst" roles in the transformation. The energy release from this process, the "PP I reaction," is 26.72 MeV, and this reaction accounts for 86% of the energy produced in the Sun. The other two reactions, PP II (14% of solar fusions) and PP III (10%), are even more complex. In addition to making tritium and deuterium as intermediate products, these reactions build up and break down beryllium-7, lithium-7, beryllium-8, and boron-8 just to make some helium-4.

---

19 George Gamow, a theoretical physicist who did pioneering work on the Big Bang theory of the creation of the universe, was born in Odessa, Russia, in 1904 and defected from the Soviet Union to a professorship at George Washington University in 1934. He was a major consultant at the Los Alamos National Laboratory in the 1950s during the H-bomb development. He died of liver failure in Boulder, Colorado, in 1968, probably as a result of processing too much ethanol through the weakened organ. Carl Friedrich von Weizsäcker, German theoretical physicist born in 1912, collaborated on the proton-proton fusion concept. They lost touch as World War II started, and Weizsäcker joined the ultimately unsuccessful German atomic bomb development project. He died at age ninety-four in Starnberg, Germany, in 2007. His German patent for a nuclear weapon, filed in the summer of 1942, was not granted.

I saw my first hydrogen fusion reactor in action in 1965. It was at the New York World's Fair in the General Electric Fusion on Earth exhibit.

The fair was held in the summers of 1964 and 1965 on one square mile of Flushing Meadows-Corona Park in the borough of Queens, New York. The theme was "Peace through Understanding" and was dedicated to "Man's Achievement on a Shrinking Globe in an Expanding Universe." The spectacle, which was boycotted by some big countries such as the Soviet Union and Canada on official sanctioning issues, was thick with American culture and technology. It was a massive Walt Disney Imagineering design, crammed with exhibit halls and pavilions sponsored by such heavy hitters as IBM, the Bell System, Kodak, General Motors, Ford, and Chrysler.

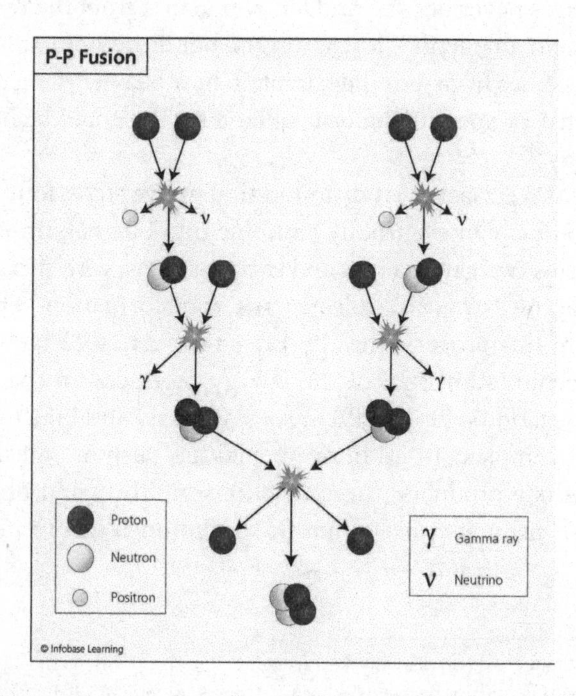

The fusion of simple hydrogen to form helium requires a complicated process that involves five steps. First, two sets of two protons combine to form two deuteriums, just as Gamow had worked out, but then another pair of protons fuse with the deuteriums to form tritiums. The two tritiums crash together and make helium-6 for only an instant, and it blows apart, leaving a helium-4 and the two secondary protons given back into the process. The total energy produced is 26.72 MeV.

General Electric had a big piece of real estate next to the Clairol hair products and Chunky Candy pavilions. It was an enormous, inverted bowl, dedicated to the use of electricity in our everyday lives. A must-see was the Carousel of Progress, which was four scenes of life in the twentieth century, inhabited by Disney audio-animatronic robots, demonstrating how far we had come in the first sixty-four years and where we would end up by 2001. The Mammoth Sky Dome Spectacular showed on a large projection screen the progress made in the development of nuclear power, while the Medallion City showed what it will be like when everything runs on electricity.

On the way out we were standing on a downwardly spiraling walkway along the inner walls of the big rotunda, and at the center was a large glass dome housing a Z-pinch fusion reactor. A bank of very large capacitors under the floor was being charged with electricity. It took thirty minutes to build up a charge, so the show was every hour and half hour. A voice came over the public address system, announcing an impending demonstration of fusion power. We were advised to turn our heads and look to the center of the floor.

The hollow center of the reactor was filled with a mixture of deuterium and tritium, and the Z-pinch machine tried to replicate conditions at the center of the Sun by suddenly switching the pent-up energy in those capacitors, all at once, across the gas mixture. The artificial lightning bolt ionized the gas, turning it instantly into a conducting plasma. The electricity conducted through this plasma created an intense magnetic field around and through the plasma, pinching it down into a blazing hot, extremely thin cylinder. Under these conditions, a few deuterium-tritium pairs might become helium, ejecting a free neutron for each fusion.

The announcer started a dramatic countdown. "Ten . . . nine . . . eight . . . seven . . ." The crowd hushed, and I could now barely perceive a hum, as the transformers and rectifier arrays crammed the last measure of electrons into the capacitor bank, running hot, vibrating, and gradually loosening a few rivets. ". . . three . . . two . . . one."

BANG! It sounded like a filing cabinet had been tipped over and hit the floor a few inches from the back of my head. The glass hemisphere flashed nuclear-blue as the more sensitive members of the crowd flinched and exclaimed. There were a few high-pitched screams.

On the wall was a huge digital display showing the total number of neutrons counted since the reactor had first banged over a year ago, with neutron counts implying fusions. I cannot remember the number, but I think it was in the millions. I watched as the numbers spun and a new total was displayed. It went up by more than one hundred fusions.

The reactor had generated at least 4.5 watt-hours of energy so far. Compared with the tens of thousands of watt-hours used every day to charge the capacitors to throw a lightning bolt across the core, that was not very much. The announcer voice thanked us for watching the demonstration and assured us that more work was necessary before General Electric would be powering our home with fusion. It would take as long as the next thirty years to build commercial fusion power stations.[20]

Thirty years came and went, and there was a disappointing lack of progress toward lighting up America with fusion power. Hundreds of experimental reactors were built worldwide, and many dollars were spent. Billions and billions, like the number of stars and galaxy clusters in a Carl Sagan lecture. What was the problem? Why, with all the science and engineering effort concentrated into it and all the hideous success with fusion weapons, were we still in 1995 thirty years away from fusion power?

While achieving hydrogen fusion is not difficult, doing it on a continuous basis at a rate that will generate net power, in which more power is recoverable from the reactions than goes into creating the fusion environment, seems beyond our current technical ability. The Sun, from which we have derived our reactor model, is 109 times wider than our planet, having a volume 1,300 times that of Earth, and it weighs 2.19×10^{27} tons. This enormous mass produces

20 There is some question as to how many of the detected neutrons coming out of a Z-pinch device are actually the result of fusions. The Z-pinch effect was thoroughly studied by British scientists from 1957 through 1958 at the ZETA (Zero Energy Thermonuclear Assembly) experiment in Hangar 7 at Harwell. (Half the hangar, which was quite large, was used to house the capacitor bank.) After initial announcements of success in creating a "sun in a bottle," further tests indicated that most of the detected neutrons were not due to fusion, but were neutrons bounced out of the plasma stream by energetic protons. In the short duration of the lightning bolt, there simply was not enough time to build up a temperature high enough to produce hydrogen fusions.

a pressure of 9,400 pounds per square inch and a temperature of 15,700,000° Kelvin. Under these conditions, the hydrogen in the Sun near its center is reduced to plasma, with the electrons stripped off the atomic nuclei. The hydrogen plasma is 150 times heavier per unit volume than water. Temperature is a measure of the speed at which the atoms in matter are traveling as they randomly collide with one another. At this temperature in the Sun, the protons are traveling at 1,400,000 miles per hour.

The only reason that hydrogen does not fuse with abandon is that the proton (the nucleus) is electrically charged, and things charged with like polarity (positive, in this case) avoid each other. They repel, and the force of repulsion increases rapidly as the distance closes. For fusion to occur, two nuclei must get close enough together for the strong, attractive nuclear force, which binds together the protons and neutrons in an atomic nucleus, to overcome the electrostatic force. For two hydrogens, this distance is 1 to 3×10^{-15} meters, which means they are practically touching each other. A head-on collision at 1,400,000 miles per hour can achieve this condition, and fusion occurs. In the Sun, it occurs 9.2×10^{37} times per second, giving a net power output of 3.85×10^{26} watts.

However, proton-proton fusion has an extremely low probability of happening, even in the best of conditions. Fusion at the center of the Sun is so unlikely, the power density is only 7.85 watts per cubic foot. To put this in perspective, that level of power density describes the metabolism of a reptile. A large box turtle, awake, generates about as much heat as a like volume of tightly compressed plasma at the center of the Sun. The only reason the Sun gets hot enough to glow is that it is so big. An active compost heap the size of the Sun would generate as much power, and it would warm the Earth just as surely.

Generating electricity by fusion in a large concrete building on Earth will therefore not be emulating the Sun. A fusion power reactor will have to establish conditions of pressure and temperature that are orders of magnitude greater than exist in the Sun, and a simple proton-proton fusion scheme, which has the least probability of fusing, has no hope of success. The fusion reaction with the highest probability (5.0 barns), is the one that Oliphant achieved in 1932, the deuterium-tritium reaction.[21] The next best, at 1.2 barns, is the proton-boron-11 reaction, and the deuterium-deuterium fusion is down on the list, at a maximum 0.11 barns.[22] A reliable deuterium-tritium fusion will occur at temperatures over one billion degrees Kelvin, which is sixty-four times hotter than the center of the Sun.

At the end of World War II, when civilian nuclear power seemed an inevitability, these cold facts concerning hydrogen fusion were well-known. The use of a single fusion pulse, initiated by the pressure and temperature of an adjacent atomic bomb explosion, was in crash-program status as Dr. Edward Teller's "super bomb" for military purposes, and no serious development of controlled fusion for power generation was considered worth the effort. It was a far-off concept, perhaps to be examined later.

Everything changed suddenly in 1951, when controlled, continuous fusion for power production became an obsession that continues to span the globe. On March 24th of that year, a stimulus blasted across the technical world from way out in left field, from an unexpected place and an unknown perpetrator. The kick that started fusion research was from Argentina, and the expert who surprised the world was an obscure third-tier scientist, part of the human fallout left over after the smoke cleared in the defeat of Nazi Germany, Dr. Ronald Richter.

Ronald W. Richter was born on February 21, 1909, in Eger, Bohemia, which was then in Austria, later to be annexed into Czechoslovakia.

---

21 The reader is reminded that the probability of a nuclear reaction is expressed as the effective cross-sectional area of the involved nucleon. The bigger something is, the easier it is to hit with a projectile. The unit is a cross section of 1×10^{-24} square centimeters, which, in subatomic terms, is "as big as a barn."

22 Having been given the cross section for the deuterium-tritium fusion reaction, you may expect me to cite a cross section for the proton-proton fusion occurring in the Sun and stars. At this writing, that cross section has yet to be measured. In fact, a proton-proton fusion has not been accomplished in a laboratory setting. A definitive paper that explains this cross section, "Solar Fusion Cross Sections" in *Reviews of Modern Physics*, Oct. 1998, has thirty-six authors and is fifty-seven pages long. Suffice it to say that this cross section is "terribly small," to quote the principal author, Eric G. Adelberger. If you must have a number, then use 4.5×10^{-24} barns. With a probability that low, one wonders how anything ever gets done in the universe, and the "sun in a bottle" concept of fusion is truly impossible. The proton-proton fusion does not scale down, and an Earthbound fusion power reactor will depend on some other reaction mode, such as tritium-deuterium.

He was an only child, and as such he enjoyed the undivided attention of his parents and as much prosperity as one could attain in Europe straddling World War I. At the age of twelve, he entered high school in Eger and graduated in 1927. He was fascinated by all aspects of physical science, and his parents, who managed a coal mine, that summer gave him a laboratory setup on the mine property where he could indulge in research topics of his choosing.

He was accepted into Prague University in 1928 to study physics, chemistry, and, at his parent's urging, geophysics. His undergraduate interests were wide to the point of attention-scattering, ranging from astrophysics, quantum mechanics, and nuclear physics to gravimeter systems and the Earth's magnetic field. As a grad student he was seen as a loner, rarely participating in anything, but impressive in an aloof sort of way. Dr. Kurt Sitte, a fellow student at Prague and a year behind him, remembered him as furiously jotting down notes in class, using weird signs and symbols, seeming more sophisticated than his peers. His class attendance was erratic, and his performance in advanced classes was nothing notable.

Professor Reinhold Fürth of the Department of Experimental Physics was his thesis adviser, remembering him as "a moderately gifted scientist with an excessively active imagination and an incapacity for self-criticism." Richter was bouncing up and down at having read a magazine article about "delta-rays" coming up out of the middle of the Earth, and he was bent on basing his dissertation on detecting them. Fürth and his recently appointed research assistant, Sitte, managed to talk him down off that cloud, and he wound up doing a more sedate and reasonable "Investigations on Photo-Voltaic Cells with Soft X-Rays." Still, he managed to inflate the investigation with as much drama and hyperbole as possible. Further work on this topic was continued by another student, Hans Felsinger.[23] Richter graduated with a Doctor of Natural Science degree on March 2, 1935.

---

23 Heavy rumors have it that Richter never published his thesis, and his work was found to be erroneous and riddled with "spurious evidence" by Felsinger. I find these allegations hard to believe, but I must admit that Richter himself could not cite his own paper, but he could point to Felsinger's article in *Annalen der Physik* in 1937. If he managed to graduate with no publication and a seriously flawed series of experiments, it was the fault of Reinhold Fürth and the University of Prague and not his.

Richter was fortunate to have graduated in 1935, even though the world was in the middle of a deep economic depression. The new government of Germany was purposefully divesting itself of all Jewish influences, and it was thus stripping out a population of technical specialists, from theoretical physicists to watchmakers. There was a resulting vacuum of scientific and engineering talent that would have to be filled. Germany, even at this early date, was forming plans to take over the world, and the resulting conflict was predicted to be unusually technical and innovative. Employers would take anybody with a technical degree, even space cadets from satellite countries, such as Ronald Richter. He had no trouble job-bouncing through World War II.

Upon graduation, he started a private research laboratory in Eger with funding from a coal-industry consortium arranged by his parents. Germany was already planning to manufacture gasoline using coal as source material, and Richter's pitch to the sponsor involved plans to research catalyst-controlled coal hydrogenation and a new coal-cracking process. Given the freedom of being solely in charge and with his mind amped up with the postdoctoral buzz, his ideas and research were original and breaking some new ground.[24]

Roaming at will, he plunged into a fascination with high-power electric arcs as a source of concentrated, high-temperature plasma available in no other process. Drifting out of coal research, he saw the electric arc as something that could be used to make exotic, high-temperature alloys and compounds. The problem with arcs that gained all the attention was a lack of control at the flashover boundary. As power is applied to two carbon electrodes, facing each other and separated by an air gap, tapping the carbons together starts a bright electric arc that can be expanded by moving the electrodes farther apart. At a sufficiently high voltage, electrons jump the gap, causing the gas between them to become a superhot, superconducting

24 One of Richter's projects in 1935 was a scanning microscope using electrons, protons, or deuterons looking for fine surface details of a specimen held in a hard vacuum. I have found no confirmation of this claim, but, if true, then Richter's scanning electron microscope (SEM) work beat Manfred von Ardenne's SEM prototype by one year. Richter referred to his setup as an "image-converter activity contrast microscope." The first commercial SEM was built by the Cambridge Scientific Instrument Company thirty years later, in 1965.

plasma. More voltage and more amperage make the gap hotter and produce a larger volume of plasma. Turn up too much power and the well-behaved electron stream breaks down into a sudden mini-explosion and blows out the arc.[25] For that reason, arcs had to be run below their highest possible temperature, and much work was going into feedback controls that would detect an oncoming flashover and adjust the voltage down until the crisis potential dampened. In the age of vacuum-tube electronics, this was not easy.

Richter saw it differently. To him, the arc breakdown was not a problem; it was a feature. He recognized the breakdown as a plasma shock wave, detectable as a very loud, floor-shaking bang. It meant that the sudden disturbance in the plasma between the electrodes was moving faster than the speed of sound, and this meant that a region of unusually high pressure and accompanying temperature must exist in the middle of the apparatus, at least for a split second. He purposefully over-drove the arcs, so as to consistently cause the shock wave, and he designed a focusing reflector to further concentrate the effect at the center of the disturbance. It was difficult to measure exactly what pressure/temperature situation he was creating, as the effect was literally over in a flash.

He, along with every other physicist in the academic world, was well aware of Mark Oliphant's fusion experiment using deuterons hitting tritium.[26] Richter's density-enhanced plasma blob would, in

---

25 At least two popular applications for the carbon arc remained in place through the 1960s: searchlights and motion picture projectors. These two uses for the arc have now disappeared, taken over by xenon lamps. In World War II, carbon arc searchlights were used all over Europe to illumine the bottoms of night bombers and allow the visual aiming of antiaircraft artillery. The carbon arc was first demonstrated in 1802 by Sir Humphrey Davy.

26 Before we leave this subject, Oliphant's experiments were remarkable in that there was no commercial source of deuterium in 1933. He was given a few drops of it by American physical chemist Gilbert N. Lewis, who had expended great effort to separate it from tap water. Oliphant discovered tritium, a by-product of his particle accelerator experiments, and then used countable atoms of it in his fusion setup. Tritium, an artificial nuclide not occurring in nature, was first manufactured in minute quantities a year later at the Cavendish Lab by Earnest Rutherford and Paul Harteck. Harteck and Johannes Jensen invented the ultracentrifuge isotope separator in 1943, working in the German atomic bomb program. Thousands of these devices are now used by the Iranian government under suspicions of a nefarious agendum.

theory, do the same thing to heavy hydrogen nuclides that Oliphant's linear particle accelerator would do. It would throw them at each other at high speed, overcoming the repulsive e-mag force and causing a percentage of them to fuse. High temperature means high speed, and it was the same effect as achieved in a particle accelerator, only from a completely different setup. Richter reasoned that he could measure the physical properties of his plasma shock wave by using it to cause fusion. It would not, of course, be a net energy-producing fusion, but it would produce a burst of secondary gamma rays, and those he could detect and count with a Geiger-Müller instrument located external to the plasma. The bigger the gamma-ray burst, the bigger was his shock-wave effect. Tritium was hard to come by, but he bought a vial of Norsk Hydro heavy water and spritzed an aerosol of it into the electrode gap.[27] A deuterium-deuterium fusion has a lesser cross section than Oliphant's deuterium-tritium reaction, but it is still a fusion. It worked as planned, and it was an excellent idea.[28] Monitoring the gamma-ray bursts, Richter was able to optimize his shock-wave reflectors and experiment with other ways to enhance the arc effect.

Having concentrated on his plasma-shock experiments, Richter quickly lost his sponsor's attention, and his paycheck collapsed. In 1937, he started another research laboratory connected to the Berlin-Suhler-Waffen und Fahrzeugwerke (weapon and vehicle factory) at the Gustloff-Werke in Suhl, Thurlinga, Germany. He partnered with Dr. Hugo Apfelbeck and Otto Eberhardt, working in a top-secret underground facility while living in a small house at Mühltorstrasse 11 in Suhl. He was supposed to be working on hydrogen storage systems, but most of his time was consumed investigating ways to measure the

---

27 The year before, in 1934, the Norsk Hydro-Elektrisk plant in Telemark, Norway, began purifying and selling 99.6 percent pure deuterium oxide taken from water in the Rjukan Waterfall. The plant, built in front of a nitrate fertilizer factory, had a production capacity of 12 metric tons per year. The product came in flame-sealed glass vials. Each vial contained 5 grams of heavy water.

28 I should say, it was an excellent idea if this really happened. Unfortunately, no dated notebooks from these experiments seem to exist, and we are relying on Richter's written account given to the U.S. Air Force in 1957 (AFR 190-16, declassified April 26, 1999). If true, it's a grand story, and it is a perfectly logical setup for Richter's postwar work in Argentina.

temperature in his plasma shock waves.[29] It was a happy if less than productive time.

His comrade and best friend, Eberhardt, was killed in an auto crash on January 31, 1939, and seven months later, on September 1, Germany invaded Poland, starting the Second World War. Richter bounced to the Junkers Aircraft Factory in Dessau, Germany, and was put to drawing pay on a project to find airplane vibration problems at supersonic speed. He was delighted to find that the instrumentation used in the supersonic wind tunnel at Junkers was just what he needed to further measure his arc shock waves.

A year later, Richter was passed off to Dr. Busemann and Professor Dirksen at the Aircraft Research Facility Hermann Göring, Department LC1, near Braunschweig. With the war running hot and shrill, he found himself constantly under surveillance and prodding by the Gestapo, the secret police that did not trust a single loyal German, much less a Czechoslovakian working on military secrets. They confiscated his passport, making it impossible for him to leave the country. Sometime in the autumn of 1942, while he was traveling by train to another aircraft factory, the Gestapo arrested him on charges of spying for the British, which probably would have paid better than what he was actually doing.[30]

Richter could talk a very good game, and he was able to wrangle an apology out of the authorities in Berlin by using impressive nuclear physics jargon and suggesting that his extreme talents be put to good use in the atomic bomb program, which apparently suffered from many information leak-points. (Technically, Richter should not have been aware of the top secret project's existence.) There in the Gestapo office, he completed a phone call with R. Abraham Esau, a physics professor

---

29 Remember this sentence when we get to chapter 3. There is a coincidental cross-linkage here, and a possible missed opportunity by Ronald Richter.

30 It is interesting to note that at this juncture, August or September 1942, Richter finally reported success in his fusion scheme to measure plasma shock-wave performance. This project had been following him from job to job, and the fact that he was working on it in Braunschweig may indicate less than 100 percent performance on assigned tasks. In his unpublished memoir, he claimed to have invented a three-stage lithium-6/neutron reaction cycle to be used in a hydrogen bomb, a neutron bomb, and a gamma-flash bomb at around this time. No mention of these inventions has been discovered in wartime documentation.

at the University of Berlin with an impressive fencing scar, who happened to be head of the physics section of the Reich Research Council, and pried open a position in the ongoing nuclear fission effort.

He was assigned to Baron Manfred von Ardenne. Von Ardenne was a self-educated physicist and electronics engineer. He was blessed with a massive inheritance and his own private, lavishly outfitted research laboratory in the basement of his home at Berlin Lichterfelde Ost. He was credited as the father of German television, which covered the 1936 Olympic Games, and in 1943 he was working on a magnetic uranium isotope separator under contract with the German post office. Backed up against a wall was his 60-ton cyclotron, named "Elle," with its guts out on the floor; his custom-built scanning electron microscope bolted to another wall; and a desk groaning under the weight of a mass spectrometer. He put his new hire, Richter, in charge of his Van de Graaff high-voltage machine and hoped that he wouldn't break anything.

The personalities of von Ardenne and Richter meshed together like two trains meeting head-on in a dark tunnel, and von Ardenne had the bigger locomotive. Richter was terminated after a few weeks, and the Deutsche Versuchsanstalt für Luftfahrt (German Research Institute for Air Transportation) at Berlin Aldershof was encouraged to take him. He was assigned measurements of subsonic vibrations using airplane models in the wind tunnel.

By this time, Richter claimed to have improved "the detectability of shock-wave-induced nuclear reactions by developing nuclear reaction schemes based on the chain-reacting consumption of lithium and boron isotopes." The chief of the research department of Army Ordnance, a Colonel Geist, listened briefly to Richter's exciting findings and his pitch for further investigations, but he lacked sympathy. "We have a war to win here! Get back to work at Aldershof, or the Gestapo will make an ashtray out of your skull."[31]

---

31 Or something to that effect. The use of lithium and boron in a fusion reactor is not as loopy as it may sound. The Z-Pulsed Power Facility at Sandia National Laboratory in Albuquerque, New Mexico, uses an electromagnetic pulse vaguely similar to Richter's setup to generate X-rays. The purpose of this device is to simulate the X-ray environment generated in a thermonuclear explosion, and it does so by fusing small amounts of deuterium. Seeing a possible application of the "Z Machine" as a pulsed-power production reactor, Sandia is considering using lithium and boron fusion to enhance the effect.

Within weeks Richter had signed a consultant contract to work for the AEG research institute in Berlin, developing lightweight storage batteries. As an alternate activity, he managed to hide out in the AEG transformer plant at Berlin-Oberschöneweide and work on his arc plasma experiments in relative privacy, using lithium borrowed from the battery project. Allied bombing raids by this time, 1944, were making life in Berlin extremely unpleasant, and AEG had bigger problems than Ronald Richter.[32]

By May 1945, the European war was over, and the ancient land of the Germanic tribes was painted with chaos and dread.

Richter, who was not a member of the Nazi party and was not even a German, could not get arrested, much less land a paying job. He was not on the war criminals list, nor was he on the list of desirable scientists for which the Americans and Soviets were competing. Repeated attempts to gain a US visa were not productive. In 1946, he was in France looking for work in nuclear physics, and according to him, he loudly refused a position in the French Atomic Energy Commission because the president, Nobel Prize winner Jean Frédéric Joliot-Curie, was a notorious Communist. The Americans seemed unimpressed by this gesture.

Desperate, he went to Holland and then to England where, in 1947, by pure luck his path intersected with Professor Kurt Waldemar Tank in the lobby of a London hotel.

---

32 This account of Richter's employment history during World War II is based on his detailed, written application to the U.S. Air Force for a research grant in 1956 (declassified in 1999). Richter probably believed that USAF Air Intelligence had ways to check up on everything he claimed, so this information is probably as correct as can be found. His unpublished memoir, written decades after the war, paints a different picture. According to this document, Richter was heavily involved in the German atomic bomb development program, and he was the only scientist who argued that graphite should have been used as a moderator in the nuclear reactor experiments, instead of the hard-to-produce heavy water. He claims that in 1944 he was called to Adolf Hitler's headquarters in Berlin to state his case for an immediate change in the reactor design, for which he had personally argued with everyone from Walther Bothe (the scientist who declared graphite unusable) to Karl Brandt (Hitler's personal physician). After the war, the Germans blamed a lack of heavy water moderator for their having no success in achieving a working fission reactor, while the Americans enjoyed spectacular success with using graphite instead of heavy water. There is no other evidence to support Richter's claims of involvement in or knowledge of the German atomic bomb research.

Kurt Tank was born in 1898 in Bromberg, which was in Germany at the time, into a military family. After a splendid showing in the Imperial German Cavalry during World War I, he graduated from the Technical University of Berlin in 1923 and sought his fortune in the crippled postwar German aircraft industry. He wound up a chief designer for Focke-Wulf Flugzeugbau AG, where he was responsible for the highly successful Fw 190 fighter plane and the Ta 152 interceptor. At the end of World War II, as the world collapsed in debris around Focke-Wulf, he was working on the radical design of a jet-powered, swept-wing fighter plane, the Ta 183 Huckebein.[33] Tank loved designing out on the wet, naked edge of technology, and his Ta 183 was definitely out there. He slipped quietly out of Germany after the war with a suitcase filled with microfilmed copies of his unfinished Ta 183 plans, looking for a place to land.

The two behind-the-front fighters regaled each other with their war stories, and Richter topped his off by disclosing his latest brainstorm: a jet engine with a pulsed auxiliary source of thrust, his plasma-shock fusion reactor running on lithium and boron. Tank was captivated. The main thing wrong with jet propulsion, without which he would design no further aircraft, was the excessive fuel consumption. Richter's idea employed the lightest elements in the universe in very small quantities. The extra thrust would not lift a plane off the ground, but the efficiency of his proposed auxiliary energy release was millions of times better than oxidizing hydrocarbon fuel. (Richter may not have mentioned the megawatts of electrical power necessary to produce the brief plasma condition.)

Tank was on his way out of Europe—out of the bomb-cratered, army-occupied landscape and to the Land of the Free, to the last place on Earth where a man could walk down the street in his snappy SS uniform without engendering harsh rebukes or sniper fire. Kurt Tank was going to Argentina, the new home of 800,000 expatriate Germans, and he promised Richter that he would find a way to get him out of Europe to join him.

Argentina, mirroring the United States, had in the last quarter of the nineteenth century encouraged a mass immigration of Europeans to fill

---

33 *Huckebein*, Kurt Tank's advanced jet fighter, was named for a Nazi-era cartoon character: a troublemaking raven. Think Woody Woodpecker in German.

out the empty western regions and bring an ingrained sense of civilization to the New World. By 1908, it was the seventh richest country in the world, with a per capita income right behind Switzerland and ahead of Denmark, Canada, and Norway. The average income in Argentina was 1.8 times higher than that in Japan. World Wars I and II, in which Argentina did not participate in a military way, only enhanced the economy, and fortunes were made selling beef and grain to both sides of the conflicts. In the early 1940s, Argentina sold .45-caliber pistols to Great Britain (the Ballester-Molina HAFDASA) while hosting a Western Hemisphere spy ring for Hitler's Third Reich. Money piled up, Buenos Aires danced the tango, and Argentina was at peace with the world.

Argentina managed to grow a vibrant economy despite its leadership. Most presidents were elected by military coup. Foreign policy was directed by a cadre of Vatican-connected medievalists striving to create a Hispanic Catholic Nation, counterbalancing the materialistic Northern Hemisphere and erasing the unholy consequences of the French Revolution. The country was definitely in bed with Nazi Germany right up until January 26, 1944, when it looked like they were going to lose the war and pressure from the United States to join the Allies intensified.

On February 24, 1946, Colonel Juan Domingo Perón was elected president of Argentina. Perón, the former vice president and dictator-in-charge of the Fascist-posturing GOU (United Officers Group, a conspiracy of military colonels), had been arrested on October 13, 1944, for the crime of being popular. This and having the lovely, wildly popular Eva "Evita" Duarte Perón on his arm all but guaranteed his election.[34]

Perón, assuming office on June 6, was a forward-looking, progressive leader. He immediately paid down all of Argentina's foreign debt, and soon just about everybody in the country was working and paying modest income taxes. Perón's goal was to catapult Argentina not just into the twentieth century but into the future. He did not want to catch up with the United States in its world-beating technology, science,

---

34 Eva was the best thing that ever happened to Perón. A former actress of questionable reputation, she moved into Juan's house in the Polermo Chico neighborhood after kicking his fourteen-year-old futon-mate, known only by her pet name, "Piranha," to the curb. They were married in a civil ceremony on October 18, 1945, followed by a flashy, very public church wedding on December 9.

and manufacturing; he wanted to exceed it. He immediately started a jet-fighter-plane project in Córdoba, with French war criminal/aero-engineer Émile Dewoitine in charge.

Around Christmas in 1947, Perón's trusted operative, Gallardo Valdez, smuggled Kurt Tank, his suitcase full of microfilms, and a busload of valuable technicians, out of Germany right before the refugee window in Denmark was closed. Tank was given a new name, Mr. Pedro Matthies, a large budget, and Émile's job as director of the jet-plane project.

Early in 1948, Perón caught the atomic fever. It was a topic of conversation worldwide, and he, prodded by Enrique Gaviola, director of the Córdoba National Observatory, decided that Argentina should be in the game. Gaviola agreed, but he was not sure that they had the right infrastructure or, more important, the load of scientists that would be necessary. A country like Great Britain was currently trying to match speed with the Americans, but they had about ninety thousand scientists. Including himself, Gaviola could count twenty scientists, right here in Argentina. A public relations effort was given immediate priority to attract scientists to the new, progressive Argentina. Werner Heisenberg, George Gamow, and Enrico Fermi were offered free vacations to Argentina, a land where scientific studies would be free of secrecy and military applications, where atomic power would know nothing but peace, but nothing seemed to be jelling. Getting desperate, Perón called in Kurt Tank from his work in Córdoba for counsel.

Tank's advice was simple. Get Dr. Ronald Richter down here, if you really want to shake up the Americans. He has some fantastic ideas, and he knows what he is talking about.

Right after the end of hostilities in Europe, with all the chaos and confusion, it was easier to smuggle in German refugees, but things were not so simple anymore. Perón plugged Richter into his "ratline" extraction service. In May 1948, Richter received word from Tank that he was wanted in Argentina for nuclear work. He had picked up some work at the French Petroleum Institute in Paris, but it was not hard to break the contract and disappear.[35] By August, he had received instructions for the transfer.

---

35 Richter had been striving furiously for a transfer to the United States as a nuclear physics asset, exchanging letters with a Colonel Peters in the USAF, but he was having trouble starting a fire under the Americans. The offer from Argentina was urgent and solid, and Richter had to abandon his shot at America for the time being.

In a group of former Focke-Wulf technicians, Richter was told to contact Dr. Gerhard Bohne in Munich, who directed him to a Croatian tavern in the city where he met a mysterious fellow in a black leather coat named Lavic. Lavic loaded the group into a caravan of American army jeeps, driven by occupation soldiers who were augmenting their incomes, and shot them through a soft spot in the Austrian border, where they were booked into a hotel. So far, so good. The group members were then issued Red Cross passports for displaced Croatians with an Italian residence permit folded inside. All they had to do was glue their photographs into the inside page and sign the Croatian name under the photograph. The group then traveled by train to Rome through Milan and Genoa, eventually meeting Ivo Omrcanin, a former Ustasha official who spoke flawless German and knew Krunoslav Draganović, a Croation war criminal/priest with solid Vatican connections and headquarters in the pontifical College of San Girolamo degli Illirici.[36]

The group spent the night practicing their name pronunciations in the Centrocella convent for Croatian nuns and loaded into a DC-4 airliner owned by Juan Perón's FAMA airlines the next morning. It was a long flight, with stops in Madrid, Casablanca, Dakar, Natal, and Rio de Janeiro on the way to Buenos Aires. The plane rolled to a final stop on August 17, 1948, and the weary travelers were met at the airport by Nazi spy/glad-hander August Siebrecht, welcoming them to Argentina.[37]

One week later, on August 24, 1948, Dr. Ronald Richter was received by the Excelentísimo Señor Presidente de la Nación, General of the Army, Juan Domingo Perón. From the first moment, it was a match made in heaven, or at least on another planet. The two men had identical

36 The Ustasha was the Nazi-controlled puppet government in war-occupied Croatia, famous for efficiently cleansing the country of Serbs, Gypsies, and some thirty thousand Jews. There remains much mystery and controversy as to what happened to the Ustasha treasury, believed to have been $80 million in gold confiscated from murdered Serbs, Gypsies, and Jews, transferred to the Vatican for safekeeping after the war. Vatican authorities deny any connection.

37 I have found no indication of how Richter's wife, Ilse, joined him in Argentina. She may have had no travel restrictions and simply waited until he sent her travel money from Argentina and flew down on a commercial flight. Richter could not speak a word of Spanish, the official language of Argentina, but Ilse spoke several languages, and she often acted as his interpreter.

personality traits, temperaments, body language, and complementary dreams of conquest. Richter's dream was to be locked in a building alone where he could pursue arc-fusion to his heart's content, and Perón's vision was to lock a German in a building until he could place Argentina at the front of the nuclear power quest. It was two minds melted together and floating above the humdrum noise of reality. "What I have in mind," said Richter, "is the creation of a tiny sun. The immense energy of the Sun results from the thermonuclear reactions fueled by hydrogen, the most abundant element in nature." Perón felt as if he could understand everything Richter said, and it sounded magnificent. "All we have to do is make two or three discoveries," Richter added.

Perón was sold. He gave Richter the corner of a building at the aircraft factory in Córdoba, an open budget, and a salary of $1,250 (US) per month.[38] Richter had quite a mountain to climb. He was confident that he could make deuterium fuse at the focus point in the plasma shock wave caused by an overdriven arc, but to make power, it had to be a continuous burn and not just a momentary flash. For this to happen, the reaction had to "go exponential," meaning that the heat energy produced by fusion in the confines of the plasma blob was enough to cause further fusion, removing the need for the megawatts of power necessary to overdrive the electric arc. To make usable power, the reaction had to sustain itself at extremely high temperature and pressure using feedback from the fusions. He was certain that he was on the threshold of an exponential reaction. He needed to do some adjustments, and he was already planning to use movable cadmium rods to control the neutron output and therefore bring the exponentially increasing power under control.

His second discovery would be a way to contain a blob of fusing deuterium plasma heated to a gazillion degrees. If the blob were really generating power, then it would instantly melt through any material it touched. There is simply not a material that can remain solid when subjected to such a temperature. Either that, or if the blob had yet to attain high power, its temperature would be quenched out by conduction when it hit something solid. Obviously, the blob of plasma would have to be suspended in midair.[39]

38 In 1948, that was a movie-star salary.

39 Due to extreme secrecy, Richter's scheme for isolating the plasma from nearby solids

Richter found trouble working in Tank's shop in Córdoba. He lacked privacy and security, and he did not want anyone looking over his shoulder. He insisted that two trusted assistants be imported from Germany, Dr. Wolfgang Ehrenberg and an old friend, Heinz Jaffke. Still, Richter seemed strangely fearful of espionage and technical interference.[40]

Richter's lab caught fire one night in early 1949, and all points of tension released in one big tantrum. "Sabotage!" he cried. Although the official report attributed the small conflagration to casually configured electrical circuits, Richter's rage was not containable. Perón, alarmed at the destructive disturbance in his atomic power project, assigned Colonel Enrique Gonzalez, director of the Migrations Bureau, to take care of it using any and all measures.

Gonzalez flew Richter all over Argentina, looking for a place that would suit him. After scouting over the desert landscapes, which did not appeal to the demanding and autocratic semi-German, as they were running out of places to look, Richter gazed down at the freezing waters of Lake Nahuel Huapi in a majestic landscape near the Chilean border.[41] It was spotted with small islands, and the prospect of working in the isolation of a deserted isle appealed to him. Slightly inland of a heavily wooded island he saw the village of Bariloche, which from the air looked like a set used in *The Sound of Music*.

---

is not clearly known, but a letter he wrote to the editor of *Scientific American* (unpublished) contains a clue. In his brief period of activity in Córdoba, he investigated ball lightning, a rare atmospheric electrical phenomenon that causes a glowing sphere to float in the air for several seconds, then disappear, often with an explosion. There are some excellent theories, but the nature of ball lightning remains unknown at this time. Richter could have spent the rest of his life trying to figure out ball lightning, and considering its use in a fusion reactor is an impossibly long stretch. A ten-minute film of one of Richter's overdriven arcs, erupting violently between two cross-pole electrodes, separated by about ten centimeters of air and reigniting several times, was made in the Córdoba laboratory. Unfortunately, the film is considered lost.

40 Richter and Ehrenberg had killed time together in von Ardenne's lab back in 1943. They tried bombarding mercury fulminate, the highly explosive chemical used in cartridge primers, with protons rapidly exiting von Ardenne's Van de Graaff particle accelerator. One asks, "Why?"

41 I have found out a great deal about Ronald Richter and his fusion experiments, but there is simply not room in this chapter to reveal all. As a demonstration of the depth of my research, I propose the following Trivial Pursuit question: German scientist Dr. Ronald Richter built a nuclear fusion laboratory and reactor in Argentina in 1950. What was the name of Richter's beloved Siamese cat? Answer: Epsilon.

He was enchanted. Here, on the secluded place half a mile from the shoreline, Huemul Island, he would establish his atomic laboratory and build the world's first fusion power reactor, and his headquarters would be in Bariloche.[42]

Exercising presidential authority, Richter took the largest house in town, belonging to Major Carlos Monti, head of the Bariloche military garrison, as his residence. From the porch, Richter and his wife, Ilse, had a magnificent view of the Andes Mountains, a hundred miles away, over the sky-blue Lake Nahuel Huapi. Spreading himself very thin as program theoretician, experimentalist, engineer, radiation safety officer, architect, and administrator, he commanded a 24-hour-a-day building project on Huemul Island, using three hundred soldiers in the Second Company of Engineers plus one hundred masons, carpenters, and electricians. Captain Pasolli was in charge, assisted by Second Lieutenant Fernando Manuel Prieto. Bulldozers arrived on July 21, 1949, and soon came barges, heavily loaded with bricks, lumber, cement, concrete mixers, and excavation tools of all types. Dirt flew.

First, soldier barracks were built so that the workers could sleep on the island. A canteen, a kitchen, a warehouse, and four laboratories were constructed, beginning with a small two-bedroom house with an attached equipment shed. Secret missions were sent to smuggle in exotic materials and supplies, such as lithium, from the United States and England. Money flowed with no constraints. In one week in November, $17,500 was spent on miscellaneous purchases.

On April 8, 1950, Perón and Evita paid a surprise visit to the construction site, first being entertained by Ronald and Ilse, acting as interpreter, at their lovely home in Bariloche. The soldiers on the island were dazzled by the rare occasion, and the two dignitaries were shown everything. Pedrocca, a particularly enthusiastic bulldozer driver, was given a watch, and Captain Pasolli received a brace of pistols for his fine

---

42 The place-name *Huemul* is said to mean "antelope" in a few references, but in researching this topic I found it not to be true. When laborers were clearing the right-of-way for a trail to the first plateau on the island in 1949, they found a wooden box containing human remains and marked "Chief Huenul 1902." Huenul, later changed to Guenul, was a Chilean fruit-grower who had lived on the island with his family since the 1880s. His canoe, also marked with his name, was found later at the bottom of the lake, and it now rests in the Parques Nationales. Huemul is a misspelling of the name Huenul.

work. Evita ordered that the soldiers' pay be increased. Richter was given his ID card, indicating Argentine citizenship and waiving the two-year residency requirement. As they peered down into the newly excavated site of the fusion power reactor, the two visitors were moved to tears.

Richter had tried some crazy things in his illustrious career, but to build the power plant before any theory of how to maintain a continuous fusion had been nailed down was pure madness. Unlike any other large-scale technical project in the history of mankind, he had no questioners. There were no technical monitors, no design evaluators, no peer reviewers, and no external specialists. He was it. He specified a hollow concrete cylinder, 36 feet high by 36 feet in diameter and 12 feet thick, sitting on a 60-foot by 60-foot concrete slab, with brick walls 3 feet thick and 50 feet high surrounding it, to be built using absolutely no iron. Fourteen thousand cubic feet of concrete were poured in one 72-hour marathon.

In June 1949, once the massive structure had cooled down from the exothermic hardening of the concrete, Richter came to look. He stared into the vacuum of space for a while, then pointed to the 2-inch venting tubes sticking out of the concrete. "Those are iron," he opined. "Tear the whole thing down and rebuild it, this time using eight-inch fiber-concrete tubes instead." He further suggested that next time, it should be built underground, excavating the bedrock underneath it. The way this thing was built, the chimney that had to be on top of the reactor would interfere with the roof girders.

A stunned hush fell over Huemul Island. This was a turning point. The large work crew, laboring long and hard for an exciting, exotic project that would kick the world's ass, had been taking orders from this exacting, precise, autocratic, enigmatically eccentric, infantile stereotype of what they expected a German to be for months. Unfortunately, he was out of his mind. Enthusiasm sagged. His military code name became "Colo" ("loco" spelled inside out). Work continued at pace, and the reactor was pulled down under presidential order.

⚛

On May 31, 1950, Perón issued decree no. 10936, establishing the Argentine Atomic Energy Commission (CNEA). In late July, Richter's daughter, Monica, was born, and his cat, enjoying the good life, now weighed ten pounds. There were exciting developments on Huemul Island, and a breakthrough was expected any time now. The army had a watchtower on the island, machine guns, searchlights, and two gunboats to circle the island in oppositely orbiting paths. At a party, Richter, perhaps hitting the peppermint schnapps a little too hard, blurted out that all the men in Laboratory 2 would turn up sterile, and if one of his exponential power experiments went a little wrong, the island would become a blob of molten glass.

⚛

Richter's goal was to fuse lithium-7 and ordinary hydrogen together, making two helium-4 nuclei and 17.28 MeV of energy per fusion. He hoped to further devise a way to achieve a self-sustaining fusion process, in which the energy produced by a momentary start-up would create enough temperature and pressure to keep the fusion process going as long as it was fed with lithium and hydrogen.[43]

What exactly was Richter doing on the island? Richter never published any of his research, and his Huemul activity was excessively secret, but there are sources from which a complete description of his primary experiment can be pieced together. His plan was to produce a lithium-hydrogen fusion burst using a directed shock wave from an overdriven electrical arc. This initial fusion would release enough energy to raise the temperature in the "plasma-zone" between the arcing electrodes to the point where lithium-hydrogen fusion would continue to grow on its own, without external power being applied. To approach this condition, Richter's apparatus would have to reach a temperature of 150,000,000° Kelvin.

It is not necessary to reach this temperature just to make some lithium-hydrogen fusions. To make a condition under which there is

43 What we know about Richter's work comes from a long paper written by Wolfgang Ehrenberg and published eight years later, "Problems and Possibilities in Nuclear Fusion." Dr. José Antonio Balseiro's report to President Perón on September 15, 1952, concerning Richter's experiments, completes the picture. The only scientific result to be published in connection with the Huemul project was a paper concerning the use of fractional distillation for heavy-water enrichment, written by Ehrenberg and Jaffke. It was published in 1953 in an obscure German magazine, *Z. angew. Physik 5*. Richter's later reminiscences dwell on political impediments and instances of sabotage and offer very little to describe his experimental setup.

a 1 percent chance of fusion, a temperature of only 40,000,000° Kelvin is necessary. Richter's electric arc produced a temperature no higher than 4,000° Kelvin in its hottest region, but he believed that he had found a way around this energy mismatch. He could force a low-energy resonance in the cross-section curve of the lithium-hydrogen reaction using Larmor precession of the lithium-7 atom.

Every interaction on the atomic scale has a likelihood, depending on the kinetic energy or temperature of the particles involved, and this probability is shown on a graph of probability (in barns) vs. energy (in KeV or degrees Kelvin). For the lithium-hydrogen fusion, this curve is generally zero at the low end of the energy scale, increasing steadily as the energy increases. Finding a "resonance" on this curve means that there is a sudden, upward jump, or a large blip, that makes the probability of interaction very favorable at the energy associated with the resonance.[44]

Richter reasoned that he could use the magnetic resonance property of atoms to force the spinning lithium nuclei to wobble vigorously, an action referred to as the "Larmor precession," thus simulating the kinetic energy necessary to crash the protons (hydrogen) and lithium-7 together and fuse, skirting around the inconvenient need for 150,000,000 degrees of temperature. To achieve this effect, a static magnetic field of specific strength had to be established through the electric arc, with a continuous short-wave radio signal of specific frequency bearing down on it.[45]

---

44 Resonances are common in nuclear cross sections. One of the first found is on the very low-energy scale of the uranium-235 fission cross-section curve, indicating that there is a huge probability of fission at room temperature. Otherwise, the curve gradually increases up into the MeV range (thousands of KeV). This resonance makes light-water fission reactors possible.

45 If you have ever been subjected to an MRI scan in a hospital, then the hydrogen atoms in your body have experienced Larmor precession, the existence of which was predicted by the Irish physicist Joseph Larmor. The presence of the wobbling is detectable by perturbations in the radio-frequency radiation used to cause it, and the frequency required to make the atoms precess is dependent on the value of the magnetic field. The MRI, magnetic resonance imaging, machine used for medical diagnostics scans through your body in a straight line, detecting the concentration of hydrogen at places along the line, then rotates the angle of the magnetic field and takes another scan. After having taken many scans through different angles, an image of the cross section of your body can be constructed using a tomographic algorithm. The magnetic field intensity varies with the depth of the scan, so the hydrogen detection can be mapped into a depth measurement by noting the radio frequency at which the hydrogen concentration is detected.

The experimental reactor consisted of two carbon electrodes separated by about 15 centimeters of air. One electrode was hollow, and as an electrical charge jumped the gap, a synchronized puff of lithium-hydride was blown through the electrode and into the arc by pressurized hydrogen gas. A brief fusion event was accomplished by charging a large bank of capacitors until the voltage was sufficiently high to ionize the air in the electrode gap, at which time there was a bright flash and a floor-trembling detonation. It took about half an hour to charge the capacitor bank, using a kenotron rectifier and a big, high-voltage transformer.[46]

The electrode gap was set between the poles of a very large, water-cooled electromagnet, which provided the static field necessary for the Larmor precession. The source of the radio-frequency radiation to excite the lithium atoms into wobbling was bolted into a nearby rack of equipment. A control panel provided adjustments for the various electrical parts, including the magnet and the arc-current. The gap and magnet were surrounded by a heavy shielding of lead and concrete.[47]

Instrumentation of the experiments consisted of several devices. The magnetic field surrounding the gap was monitored using two crossed coils of wire, with each connected to a Tektronix 511 oscilloscope.[48] Two photocells, one with a red filter, were connected though

---

46 Richter's kenotron was almost surely a General Electric KR-3. It was a long, glass thermionic diode tube with a bulge in the center and a Bakelite base at both ends, commonly used to make DC current in X-ray machines. At the time, it cost a budget-tweaking $248.

47 Richter claimed that the shielding of the gap was to protect the research staff from X-rays. There was no way, however, for an electrode gap operating at atmospheric pressure to produce X-rays, regardless of the voltage used to ionize the air. The only way to accelerate electrically charged particles to a speed that will produce X-rays is to do it in a hard vacuum. Oddly, the circular, 12-foot-thick wall surrounding the reactor in the building that he had torn down was not a radiation shield. It was to protect personnel and sensitive equipment from destruction due to ultrasonic sound developed by the spark gap.

48 The 65-pound Tektronix 511 DC-coupled oscilloscope cost $795.00 FOB Portland, Oregon. That is the equivalent of about $10,000 in 2014 money. Richter ordered an even dozen of them. He cried sabotage when they were delivered to the Buenos Aires Physics Institute by mistake.

DC amplifiers to the vertical and horizontal axes of a pen plotter. One Geiger counter was connected to a rate meter, showing gamma-rays-per-second. Another Geiger counter was connected to an analog scaler, integrating the total number of gamma rays, and this signal was connected to the vertical axis of a Speedomax pen plotter. Another red-filtered photocell signal was amplified and connected to the horizontal axis of the Speedomax. The red-filtered photocells would detect the flash of light from the fusion event and move both horizontal axes on the plotters. The unfiltered photocell charted the total strength of the photon flash during the fusion. The scaler connected to the Geiger counter recorded, indirectly, the total number of fusions.[49] An optical spectroscope was used to estimate the temperature produced in the arc by noting the momentary spread of spectrum lines due to Doppler broadening during an arc-flash. It is not clear what the oscilloscopes were supposed to show about the fusion, but they certainly looked cool.

In addition to the Larmor precession, Richter was counting on his "plasma shock," a blast of energetic ultrasonic sound that would shake the building, to make fusion happen. This sound made by his electrical discharge was the well-known "singing arc," discovered by the British physicist William Du Bois Duddell back in 1898. An arc oscillates wildly, due to a natural instability, and it can make a howling noise or even ultrasonic sounds. Richter's lack of knowledge of such electrical esoterica was his weak point, so he was delighted to find Heriberto Hellman, a German electrical engineer formerly employed by AEG, living right there in Buenos Aires.

With Hellman he could discuss magnets, extremely high electrical currents, and ways to suspend his plasma-blob in midair. On December 9, 1950, as the project gained speed, he ordered Hellman to build 3 magnet coils, 3 feet in diameter and capable of conducting 150 amperes, a couple of large high-voltage induction coils, a power generator, some big carbon electrodes, and to find 1,500 feet of coaxial

---

49 There were no gamma rays produced by the fusions. There were only alpha particles (heliums) generated, and an alpha particle in air has a range of about one centimeter, preventing it from reaching any kind of remote radiation detector. Richter was counting on *bremstralung*, "braking radiation," caused when the alphas slammed into the air, to register on his gamma-sensitive Geiger counters.

cable. The total cost was $312,500. He dropped a request for a 10-million-watt power switch, to Hellman's relief.

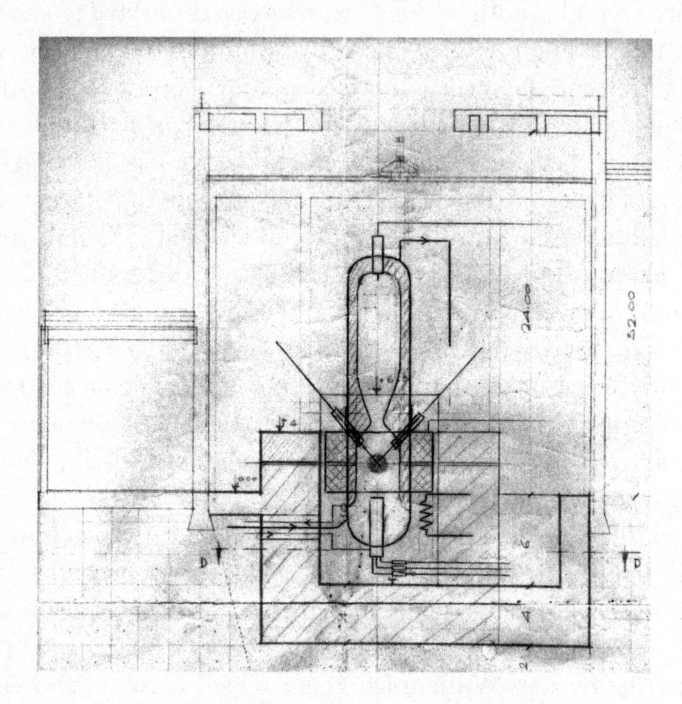

Richter's diagram of his large fusion reactor. The reactor pressure vessel was 18 meters high, with an outer diameter of 4 meters and a wall thickness of 60 centimeters. A magnet coil was wound around the bottom end of the vertical reactor, with two high-voltage "cross-pole" electrodes in the center of the magnetic field. Gases to be fused were injected at the bottom. How he planned to extract the energy and make use of it is unknown.

On February 16, 1951, experimental results apparently indicated that the heat from start-off fusions had caused further fusions and increased the temperature in the electric arc. This was the breakthrough. Richter's assistant, Jaffke, triggered an arc detonation in the small reactor, which was in a concrete shield 10 feet long and 6 feet in diameter. The concussion made everything—the oscilloscopes, the pen chart recorders, and the spectroscope—jump off the floor. Watching a ground glass at the focus of the spectroscope, Jaffke saw the sharp spectrum line of a light source blur into a round halo. He interpreted this as a sudden increase in the speeds of the light-scattering plasma particles in the arc, throwing off the spectral line by the Doppler effect, and indicating a temperature that the arc was incapable of producing without excess fusions.[50]

Richter was elated. They ran the experiment multiple times and used photographic plates to record the spectroscope image. Perón was advised of the success, and he scheduled a press conference for March 24, 1951, to make a triumphant announcement to the world.[51]

It was a small press conference. There were twenty people including Perón and Richter, and only Argentine journalists were invited. The president began by apologizing for scheduling this at 10:00 Saturday morning, suspecting that many journalists worked late at night. He introduced Dr. Ronald Richter, rustled his papers, and started his prepared statement, saying, "On February 16, 1951, controlled thermonuclear reactions were achieved at a technical scale, at the Atomic Energy Pilot Plant on Huemul Island, near San Carlos de Bariloche." He went on to specify that Argentina, in its collective wisdom, had decided to jump ahead of certain western powers and concentrate on the peaceful use of hydrogen fusion rather than make bombs or anything with atomic fission.

The journalists did not quite know what they were hearing. Terms such as "thermonuclear" were familiar only to a few people in the world, and nobody in the audience knew anything about atoms or hydrogen. Richter got up to answer questions, using an interpreter, but the journalists had been caught completely off guard, and intelligent-sounding questions were difficult to phrase. Jumping at a chance when

---

50 The "halo" effect did not indicate high temperature. It indicated that the spectroscope was shaken so hard by the arc explosion, it caused the beam of light to orbit around until the floor stopped vibrating. The fact that Richter and Jaffke interpreted this as spectral broadening indicates a high level of delusion. They were seeing what they desperately wanted to see.

51 At this same time, Perón was taking care of a problem that Richter had produced. The new garrison commander, Colonel Fox, decided to make an inspection tour of Huemul Island, apparently unaware of Richter's banning of any visitors. Richter met him as the boat docked, held a pistol on him, and backed him off the end of the pier and into the cold Lake Nahuel Huapi. The only way El Presidente could keep the army from executing Richter was to assign him presidential powers, which was probably unconstitutional.

Richter mentioned that his reactor produced a "controlled atomic explosion," someone asked if it could be heard.

"Well," said Richter, "that would depend on whether there is a storm at the time."

"Has it been heard at Bariloche?"

"No."

The written speech was sent to all news media, and given the communication systems of 1951, it circled the globe with stunning speed. Every nuclear scientist within arm's reach was dragged away from whatever he was doing and asked to comment on Argentina's claim to have achieved atomic power by fusion. Nobody had ever heard of Ronald Richter, and the statement was maddeningly vague about what exactly he had done, but there was a deep-running suspicion that the announcement was a hoax. One German working alone in Argentina had built a fusion reactor in nine months? No way. Fermi, Heisenberg, and Gamow all delivered doubt and disbelief.

A few were not so sure. Sir John Cockcroft, head of the Harwell Atomic Energy Research Establishment in England, thought it not impossible. French physicist Funet-Caplin claimed that he had used these exact methods to achieve fusion back in June and September 1950. Richter had obviously stolen his ideas.

In Argentina, there was a measured cry of protest among the small community of scientists and nuclear experts, none of whom had been included in this project. Richter had promised full cooperation with the Argentine universities and inclusion of students, giving them valuable hands-on experience with the birth of a new atomic technology, but none of this had happened. This one project had drained out every bit of scientific research money that might have been available in the government budget, and even the enormous demand for concrete had halted the construction of everything from bridge piers to foundations for housing. The scientists found Richter's secrecy noxious and unnecessary.

For the next year and a half, Richter resumed his eccentric ways, keeping a tight lid on the work at Huemul and spending money like it was coming out the end of a fire hose. Outside the laboratory buildings, to which no outsider had entrance, there was no visible progress toward a power station. He wanted to move the entire enterprise a few miles to the other side of Bariloche, to the Indio Muerta (Dead Indian) desert. He was also visiting the American Embassy in Buenos Aires on an erratic basis, and Perón, being just as suspicious as Richter, was growing concerned. Under pressure from all sides, he agreed to a High Commission visit to the Huemul Island facilities and to take back an objective evaluation of the project.

The investigating committee consisted of physicist Dr. José Antonio Balseiro, who was recalled from Manchester University, England, where he was doing his postdoctoral research; two engineers, Mario Báncora and Otto Gamba; an army officer, Captain Beninson; and a Catholic priest, Fr. Juan Bussolini, from the San Miguel Observatory. The group, these five investigators and twenty staunch Perónist congressmen, arrived on the island on Friday, September 5, 1952, and were assured full access to the atomic power work. They witnessed a demonstration of the fusion reaction, the photo darkroom, and Richter's desk.[52] On Monday, the team was treated to a demonstration of deuterium-deuterium fusion using the small experimental reactor. A noncommissioned officer named Eguireum, who had assisted Richter since the previous April, poured heavy water into the electrode gap with a coffee spoon while the arc detonated. After watching another demo using the lithium-hydride, the group visited Wolfgang Ehrenberg's lab, where he was working on heavy-water separation. Having seen enough, the committee returned to Buenos Aires to record their observations.

The reports from Balseiro and Báncora, addressed directly to President Perón, were devastating, finding the famous atomic power plant project a mixture of dubious science, vague and unfounded theories, incompetent experimentation, and a dose of pure fantasy.

Richter's great discovery, that Larmor precession could substitute for high temperature, was obviously not used in any demonstration. The radio-wave source, necessary to make the atoms wobble, was in another room, and it was not connected to electrical power. His

---

52 At the first BANG of the reactor, the twenty congressmen dived for cover. Seeing precious lithium poured out on the floor under the electrode gap, a technician, Bertolo, hastened to suck it up with a vacuum cleaner. The highly corrosive lithium compound quickly dissolved the insides of the vacuum, the motor shorted out, and it exploded. This time, the team of five investigators jumped for shelter.

radiation-detection equipment was mounted behind shielding, and there was no way it could detect any activity in the reactor. The radiation counters did jump during a detonation, but it was because of the big jolt in the building's power system and not gamma rays. The counters would register phantom radiation even if the electrode gap was pulsed dry, without any fusion fuel in it.

When asked about the claim of intense ultrasonic sound increasing the arc-temperature, Richter launched into a complaint about his budget, saying that he could not afford equipment to measure the ultrasonic pulse. In that case, how did he know that there was so much ultrasonic sound coming out of the thing?

"I knew, because it caused fatigue and neuralgia in us as we worked near the reactor," Richter replied.

Photographic plates used to record the spectrograph focal plane during the experiments were developed as the committee watched. None showed any spectral line broadening. Activation foils in place during the deuterium-deuterium fusion, used to record neutrons released in these fusions, were subjected to radiation counting afterward and showed zero activity. The heavy-water separation activity had resulted in a quantity of water said to contain deuterium. "How much?" they were asked.

"We don't know. We never bought a mass spectrometer with which to measure it." If the entire operation over the past years had been a fraud or a scam leveled at the Argentine government, then it was one executed with astonishing clumsiness. After another in-depth evaluation of the project, delivered in September 1952, the project was quietly shut down. There were grand plans to make Huemul Island into an extension of the university system, researching cosmic rays and other nuclear topics. Eventually, all the hundreds of scientific instruments and equipment were scavenged for other projects, and the buildings were given to the Argentine army. They were used as targets for artillery practice. In four years, Richter had burned the equivalent of $300 million in today's money.

Richter never got his visa to come to the United States. He was jailed in Buenos Aires for three nights for contempt of Congress, but he never served time for having embarrassed Juan Perón. Finding no work in Argentina but unable to leave, he started selling his household possessions for food. He regretted saying *auf Wiedersehen* to his custom-ordered, practically new Cadillac convertible. Wherever in the world he tried to drum up work to expand any of his grand ideas, he was shot down over his performance on Huemul Island. He last popped up in the 1970s, reported to have been seen in Libya and Iran, hawking exotic atomic weapons designs. He died in his home back in Buenos Aires on September 25, 1991, largely forgotten even in the nuclear physics world.

Perón was kicked out of office on September 21, 1955, by a military coup, and spent time in exile in Spain, keeping the chemically preserved corpse of Evita, who died of cancer in 1952, in his dining room. He returned to rule Argentina in 1973.

⚛

Kurt Tank led the team that produced the Argentine fighter aircraft the Pulqui II, based on his Ta 183 design, making Argentina the fourth country to build its own unique jet warplane. Unfortunately, the Soviets had a copy of his prints from Focke-Wulf, and they built hundreds of MiG-15s, bearing an odd resemblance to the Pulqui II. After the Korean War reached cease-fire stage, you could buy a surplus MiG-15 for a fraction of what it cost to build a Pulqui, and the market collapsed. Tank found work as the director of MIT, the Madras Institute of Technology, in India.

Dr. Lyman Spitzer, astrophysicist at Princeton University, returned from a ski trip after absorbing the fusion-power announcement from Argentina and wrote a proposal for the United States Atomic Energy Commission. The crazy news from Huemul Island was enough to start him thinking that Princeton should be where power from controlled hydrogen fusion is developed, and within a few months he had a million dollars, a technical staff, and a building for his secret creation, the Model A stellarator magnetic pinch reactor.

Meanwhile, in Great Britain, John Cockcroft had a similar idea for exactly the same reason. He started the Z-pinch reactor experiment at Harwell, named the ZETA reactor, and the Brits poured their hearts and souls into it.

⚛

Neither reactor design was able to do significantly more fusing than Richter's setup, even with decades of trying, redesign, retrying, and hundreds of millions of dollars of government investment. The last stellarator reactor design was canceled in 2008 for lack of monetary support. Many alternate designs have taken its place, as laboratories all over the world have tried to do what Richter did, only better. In the last half of the twentieth century, billions of dollars were spent without once making a net power output from fusion for a sustained period of time. The quest still lives.

Did Richter ever make fusion? Yes, he did. He and everyone examining his methods missed the fact that when he shot a strong electric field across the electrode gap, he was establishing a powerful magnetic field around it, collapsing the plasma into a thin line and causing hot-fusion for a very short time.[53] It was the same thing I had witnessed at the General Electric Pavilion in the World's Fair, and it was the same effect that caused the British government to prematurely announce that they had made a sun-in-a-bottle on January 24, 1958. Richter's ideas were not much crazier than a long list of fusion reactor designs that followed.

Did Richter's reactor have a chance of eventually generating usable power? No.

---

53 I cannot back this up with any written material, but in the mid-seventies I absorbed rumors from Savannah River that the AEC was working on a pulsed neutron generator for nuclear weapon triggers that would replace the deuterium-tritium accelerators in use at that time. The nagging problem of the D-T reaction was that the T, the tritium, had a twelve-year half-life, and that affected the shelf life of the bombs. They were working on a small, hollow globe made of gold with a two-electrode spark gap at the center. It was filled with deuterium, and when set off with a "Richter-pulse," it would make a shower of neutrons by fusing a small percentage of the resulting deuterium plasma. Perhaps it was true.

CHAPTER 2

AFP-67 in the Dawson Forest

"A stands for atom; it is so small
No one has ever seen it at all.
B stands for bombs; the bombs are much bigger.
So, brother, do not be too fast on the trigger.
F stands for fission; that is what things do
When they get wobbly and big and must split in two.
And just to confound the atomic confusion
What fission has done may be undone by fusion."

—from Dr. Teller's alphabet rhyme
for children, 1957

RONALD RICHTER HAD SOMETHING THAT a nuclear physicist sorely needs: an ability to sell a large, speculative experiment sequence to an organization that has enough money to afford the risks, which is usually the government of a prosperous country. In this rare case, Richter was also able to skip over all the traps and barriers that protect government from scientists and go straight to the top, pitching his high concept directly to the chief executive. The result was a quick, decisive lunge by the president, ultimately resulting in a painful percentage of the national budget being drained away with nothing to show for it.[54]

Why did the president of Argentina, Juan Perón, jump at this questionable scheme to make electrical power with fusion? There was no particular energy crisis in South America at the time, and Argentina was awash in natural resources. Perón's motivation was to impress his enemies, rivals, citizens of Argentina, and the technically arrogant nations in the Northern Hemisphere by finding a vacant spot in the scientific landscape and diving into it. Any energy that the thing could make was of little consequence, as long as he could show the world that Argentina was leading the way. The effort was supposed to crown his legacy and put Brazil in its place.

One would think that this wild occurrence could only happen once in modern history, but one would be mistaken. It has happened right here in the United States, twice in my lifetime. In the last instance, the nuclear physics–based program was the Strategic Defense Initiative (SDI), or "Star Wars" as it was popularly called, announced to the public on March 23, 1983, by President Ronald Reagan. Reagan's foreign-born scientist, his equivalent to Dr. Ronald Richter, was Dr. Edward Teller, a Hungarian transplant known for having convinced the nuclear establishment to develop the hydrogen bomb back in the 1950s.

Teller's polished skills and peculiarities were similar to Richter's. Like Richter, Teller could talk a magnificent game, and both men would glow with confidence and expertise while pitching a project at exactly the right pace and technical level, making deep eye contact and adjusting to match the listener's rapture level. Both men were

---

54 Richter was also a nutcase. In support of this assertion, I submit his written statement to the U.S. Air Force in 1956, in which he claimed that in his hydrogen fusion experiments in Argentina he encountered an inexplicable energy pulse. He attributed this measurement to an exciting encounter with "zero-point energy." The quantum vacuum zero-point energy, as predicted by Albert Einstein and Otto Stern in 1913, is the reason why helium-3 doesn't freeze at absolute zero temperature and is the lowest possible energy that anything can have. The concept has, so far, been a good plot device for science fiction writings and a dreaming point for theorists, but only Richter has been crazy enough to claim in writing to have seen it in an unrelated experiment.

dreamers, flowing with ideas beyond human capability, and neither had the vaguest concept of engineering or the work necessary to bring a dream to reality. Teller was the more expansive, pushing projects ranging from digging a quickie ocean-connecting canal through Nicaragua using H-bombs to building "clean" nuclear weapons, spreading no fallout. Richter was differently ambitious, but both men had a German PhD, and in the postwar world this somehow elevated their status as nuclear experts.

Teller had talked his government into establishing the Lawrence Livermore National Laboratory in California back in the 1950s with the sole purpose of developing better thermonuclear devices, and as lab director he oversaw the successful rollout of the warheads used on Polaris missiles to be fired from submarines. In 1976, his favorite project involved fracturing underground oil-shale deposits in Colorado using atomic bomb blasts to stimulate oil and gas production. This was "fracking," a long time before it was cool, and now it is done routinely, only using hydraulically forced fluids instead of repurposed nuclear warheads.

Ronald Reagan was elected and assumed the presidency on January 20, 1981. He had known Teller since 1966, when he was given an extended tour of thermonuclear weapons research at Lawrence Livermore. This tour had affected him, and he came to office with a desire to defuse the worsening Cold War with the Soviet Union using purely defensive means. Up until this point, the only way to prevent a nuclear war with the Soviets was to display an ability to destroy them at the same time they destroy us. As a defense policy, this seemed madness, and Reagan wanted to develop an ability to shoot down any bomb-carrying missile fired at the United States with the push of a button. If the threat of a surprise attack by intercontinental ballistic missiles could be neutralized, then there would be no reason to keep a stockpile of humanity-destroying weapons cocked and ready to launch.

A committee of Reagan's wealthy friends and scientists, the "High Frontier," formed before the election to map out missile defense strategies, and Edward Teller was a prime motivator. Antimissile defense had been tried before, of course, and had been on the minds of strategists ever since the first German V-2 missile was fired at England in the last months of World War II, but nothing practical had come of thirty years of development.

It turns out that a missile, coming down at you at several times the speed of sound, is hard to hit. There are portions of a missile trajectory in which radar can find and track its trajectory, and it would seem possible to then direct an antimissile to intersect its path and destroy it with an exploding warhead. The high speed of the attacking missile and the unavoidable delays in painting it with a radar beam, calculating a position, and moving the controls on an interceptor made this an impossible task under anything but the most favorable conditions, and the chances of getting close enough to a warhead to disable it were practically nil. At the time of Reagan's election, the best thing the committee had was BAMBI, Ballistic Missile Boost Intercept, a plan to hit a rocket as it was taking off (before it had a chance to build up speed) using hundreds of little rockets set to ram it. Given the inaccuracy of little-rocket guidance, each would be equipped with a 60-foot-wide wire net with menacing steel pellets attached. Needless to say, such a system had never been tested, even though it had been kicked around since the Eisenhower administration.

There was nothing sexy or even mildly attractive about the BAMBI plan, and everyone on the committee wanted something very high-tech, like a laser beam. Just aim a laser beam at the missiles as they came into optical range and melt them out of the air. If you can cut off a steel bridge beam with a laser, then why can't you slice a missile in half with it? It's a clean kill, and it leaves no weapon debris forever orbiting the earth. It sounds good at first, but a ground-based visible-light laser has no chance of burning the paint off an enemy missile as it reenters the atmosphere. By the time it reaches the offending vehicle, the power density of the beam has been modified by the atmosphere to the point where it is not effective. All it takes is a white, fluffy cloud to completely defuse such a laser weapon. Or, make the missile shiny and the laser beam will reflect off it without causing harm.

Teller's Lawrence Livermore National Laboratory had been hoping to develop an appropriate laser weapon since the late 1960s. To cut metal a few inches away, a high-powered visible-light laser was sufficient, but making a laser work outside the visible-light spectrum and far beyond it, in the hard X-ray region, was the ticket for knocking out missiles. An X-ray is a higher frequency, smaller-wave-length electromagnetic radiation, and while it was still photon-based like visible light, it is as much as ten thousand times more powerful per photon. So powerful, the Teller-Ulam hydrogen bomb concept uses an X-ray wave front to collapse a deuterium-tritium mixture to the point of sustained fusion, which is no mean trick. Ordinary air would diffuse and block it, so an X-ray beam laser could not be used as a ground-based system, but in the near-vacuum of the upper atmosphere or low orbit, where an aggressor's missile would have to fly, it could be melted by a well-aimed X-ray laser lofted by a rocket. Teller pushed the concept, and despite a loud minority of naysayers expressing doubts, the High Frontier group locked onto it.

Teller and High Frontier had their first formal meeting at the White House on Monday, September 14, 1981, in the office of Edwin Meese, Reagan's Counselor to the President. Many ideas were discussed, but Teller's exotic, extraordinarily ambitious X-ray laser concept gathered much attention. As he talked, he stressed a need for "assured survival" of the United States rather than "assured destruction" of the foe. This phrase resonated with the president's philosophy, but others in the group saw many flaws in Teller's argument for the advanced laser weapon. An X-ray laser had never actually been turned on and aimed at a metallic object.

Teller's first meeting with President Reagan was 2:00 P.M. on Friday, January 8, 1982, in the Roosevelt Room, with the High Frontier group sitting around a large mahogany conference table.[55] His spirited presence dominated the meeting. Reagan left feeling a new resolve, and the ball toward an antimissile shield for the United States started rolling.

Teller, with his reputation and ideas preceding him, gained a private audience with Reagan on September 14, 1982. The well-seasoned physicist was seventy-four years old. Still a ball of fire, he limped with his prosthetic foot, spoke with a distinct Hungarian drawl, and flashed

---

55 This meeting was classified SECRET, and was not recorded on the president's official schedule. Some participants do not remember Teller being in the meeting, but Karl R. Bendetsen, captain of the antimissile group, assures that he was. Bendetsen, who was very successful in private life and a member of the Bohemian Club of San Francisco, was the army colonel who directed the internment of 110,000 people of Japanese origin during World War II.

his untamed gray eyebrows.[56] His clock was ticking, and he was anxious to accomplish something that brought peace to the world while he was still in control. Reagan had turned seventy-one, and his goals were similar. Teller asked for $200 million to Lawrence Livermore to start work on the X-ray laser, sweetening the request by absurdly suggesting that they would have a working system in four years, by 1986, and full deployment by 1989. Richter would have been proud of him.

Teller got the budget he asked for, and a lot more. Eventually, $25 billion would be flushed away in the badly focused Star Wars program, chasing fanciful antimissile systems that now seem almost comical. Teller's constant reassurances of brilliant X-ray laser success to the president, based on null or very questionable experimental results and increasingly ridiculous promises of performance, kept the game going until Reagan ran out his second term and Star Wars dropped dead far from the finish line.

The biggest disappointment was Teller's favorite project, the X-ray laser. It had not seemed a bad idea. LASER is an acronym, standing for Light Amplification by Stimulated Emission of Radiation. It is meant to amplify electromagnetic radiation, such as light or X-rays, by releasing a massive burst of specific stored energy upon stimulation by a similar energy. The premier example is the ruby laser. Energy from a very bright light source is absorbed into the crystal lattice of a man-made ruby, filling it to the breaking point, until one more photon hitting the lattice will cause all the pent-up stress to break free at once, in an avalanche of red light. The light bounces back and forth between mirrors located at the far ends of the rod-shaped ruby, until the released red light is bright enough to break through one of the mirror surfaces (purposefully made to be the lesser of the two reflectors) and shoot forward as a tight column of intense energy. The effect is like turning

56 In 1928, Teller was enrolled at the University of Munich, Germany, to study physics under Arnold Sommerfeld, and on July 14 he was on summer break and was heading to the mountains to go hiking with some friends. On a three-car tram heading for the train station, he was loaded down with a heavy pack, and he was daydreaming. He realized as the tram was pulling out that he hadn't noticed that it had stopped at the station, and he lunged for the door. Falling on his face on the pavement as the tram gathered speed, he rolled over to look back, seeing a loose hiking boot lying by the steel track. His foot was in it.

one photon into billions of photons, as if it were "amplified" by the laser device into a destructive beam.

The X-ray laser uses the same basic principle but is a different configuration. For one thing, there is no X-ray mirror. X-rays will penetrate solids, and they will not reflect as visible light does. The multiple reflections in a ruby laser organize the traveling photons into a straight line, and the X-ray mechanism does not have this feature. The X-ray laser rod must be extremely long and thin to get the collimating effect of the ruby. An X-ray laser rod is approximately one meter long with half the diameter of a blond hair, about 50 microns. This makes an extraordinarily thin column of energy, and to make it mean enough to destroy a missile, hundreds or even thousands of the tiny, delicate rods are bundled together.[57]

The lasing medium is powdered selenium, which must be suspended in space with the metallic particles separated from one another by a consistent distance. This difficult task is accomplished by mixing the selenium in a high-tech foam made from, believe it or not, seaweed, making the weapon biodegradable.[58] The extreme length-to-diameter ratio of each laser rod takes the place of multiple end-mirror reflections for the task of forming the X-rays into a tight, collimated beam that can travel hundreds or thousands of miles without spreading out and losing the energy density necessary to kill a missile.

This radical, free-thinking design for the X-ray laser was not Teller's. It belonged to 24-year-old Peter L. Hagelstein, who had just earned his PhD in electrical engineering at MIT. Hagelstein, a reclusive, shy young man, worked himself to exhaustion developing the idea, doing a

---

57 A feature of the X-ray laser that is rarely mentioned is that it shoots out two destructive X-ray beams, one out the front of the rod-bundle and one out the back. The front beam takes out the enemy bomb-vehicle, but the back beam wipes out the gimbal aiming-mount that has pointed it in the right direction. This is of no concern, as you will find out presently.

58 The foam is SEAgel (Safe Emulsion Agar gel) made of agar derived from kelp and red algae. After mixing with the selenium powder, the gel is freeze-dried to remove the water content, and the result is a foam that is technically lighter than air. (This is true only in the vacuum of space or a test chamber, in which the air has been removed from the micro-voids in the gel.) This material is perfect for spacing the selenium particles, because it almost isn't there. It poses a minimum impediment to develop an intense X-ray signal in the laser rod.

lot of computer simulation on the laboratory's supercomputer cluster.[59] His laser would be code-named EXCALIBUR. Later, in one of Teller's more diffuse fantasies, it would be elevated to SUPER EXCALIBUR, for marketing purposes.

There was a dramatic flaw in the X-ray laser concept: the source of unfocused, high-energy X-rays that it would concentrate into a deadly beam was a 1.3-megaton hydrogen bomb. X-rays are the first radiation out of the box as a nuclear weapon fires, just because electrons, which must be accelerated to make electromagnetic radiation, are lightweight and easy to move, and they are available in the structure of all atoms making up the bomb. The blast of high-energy X-rays, emerging before the bomb has a chance to come apart, is the key to the laser sequence. This fact was heavily undermentioned to President Reagan in his briefings, and Reagan would confidently say in speeches that the Strategic Defense Initiative was nonnuclear, trying to erase qualms about using atomic bombs in space to wipe out enemy rockets. In fact, an X-ray laser weapon could only be used once, as it would vanish into a brightly glowing plasma cloud milliseconds after it was fired. Aiming the laser rod at the speeding target hundreds of miles away, which would not be a trivial task, would have to be spot-on. There would not be a second chance. Each EXCALIBUR vehicle would aim as many as ten rods at different missiles before the H-bomb went off.[60]

The big problem was that the United States had signed an agreement with the Soviet Union that neither side would ever explode a nuclear weapon in outer space, and this put a big crimp on an X-ray laser testing program, if not the eventual deployment of the weapon. The

---

59 In 1986, seeing that the X-ray laser program had descended to crisis mode, Hagelstein resigned from the O Group, left the Lawrence Livermore National Laboratory, and joined the faculty at MIT in the Electrical Engineering and Computer Science Department. Teller was not pleased. He was, in fact, enraged. Falling back on his previous experiences with traitors, he accused Hagelstein of letting Hitler and the Nazis get the upper hand, and finished by leveling the most cutting damnation he could think of: He accused Hagelstein of being a pharmacist.

60 As questions concerning the effectiveness of the EXCALIBUR weapon and its deployment strategies grew, Teller inflated the number of laser rods per vehicle, first to 100, then 1,000, and as high as 100,000 independently aimable devices surrounding one H-bomb X-ray source. The geometry of such a cluster of gimbal-mounted lasers, each getting a full X-ray broadside, boggles the mind.

only way to legally test any nuclear explosion was way underground in Nevada, and it was time-consuming and expensive to do so. One explosion could cost $100 million. A very deep shaft had to be excavated and then instrumented with one-shot equipment to measure the power and focus of the X-ray beams, and cross-shafts for the beams had to be hermetically sealed and evacuated to simulate a far-off target in outer space. A bomb hole could only be used once. Make another measurement, and you dig another hole. It would take a lot of underground shots to test this unknown technology, make it work, and then optimize it. In 1986, scientists at the test site were asked, "How long will it take to finish the X-ray laser tests?"

"Given the time it takes to set up one test? One, maybe two hundred years."

The X-ray laser work went dark, but under the new president, George Herbert Walker Bush, taking office on January 20, 1989, the Strategic Defense Initiative still burned brightly. The X-ray laser was replaced with an upgrade of the BAMBI system, named Brilliant Pebbles, which sounded better than Smart Slingshot Projectiles. Hundreds of little rockets would be fired off, shotgun style, each with its own rocket-exhaust seeker system and explosive warhead. Hopefully, they would choose not to seek each other.

Dr. Teller moved on to push an asteroid-killer spacecraft equipped with an extremely large thermonuclear warhead. He died on September 9, 2003, at the age of ninety-five. An asteroid, 5005 Teller, is named after him.

President Bush's redefined SDI program did not last long. As the smoke of what appeared to be an enormous boondoggle cleared away, questions remained. How had a representative government, geared up and down with checks, balances, General Accounting Office, science advisers, a House and a Senate in opposition to everything, and a vast population always needing a new bridge or a cure for cancer, managed to keep this questionable science project going for so long on so much money? Were we really expected to have an installed shield against missile-borne nuclear warheads, or was there another strategic purpose?

The deep mission of the Strategic Defense Initiative may have been to destabilize the Soviet economy and bring it down. By 1986, the Soviets themselves were convinced that this was our goal, and they were very nervous. News of the outrageously high-tech X-ray laser development progress, which was always classified secret, constantly leaked into the Soviet Union by the usual means. The supposed airplane news magazine *Aviation Week & Space Technology* carried information from "informed sources" detailing exciting news from the SDI program with specifications and technical descriptions of the hardware and the intentions. In addition to known embedded Soviet agents being allowed to do their thing, *Aviation Week*, long known in the secret-project community as *Aviation Leak*, was used as a controlled information pipeline to the enemy, giving them a confusing mixture of truth and deception. They never knew for certain what was fact and what was fiction, but altogether it caused constant concern. Even reports of X-ray laser test failures and setbacks in *Aviation Week* could not be trusted. Maybe just the opposite was actually happening? These crazy Americans were upsetting the delicate balances of the Cold War, and ruinously expensive counter-developments would be necessary.

Leaked news of an X-ray laser focusing "breakthrough," increasing its effectiveness by six orders of magnitude (an erroneous report), hit the Soviets hard. Our underground tests were obviously yielding stunning results. Leader Mikhail S. Gorbachev announced on July 29, 1985, that the Soviet Union would suspend the underground testing of nuclear weapons, and he politely asked the United States to join the moratorium in the name of peace and harmony. We declined.

The Soviet government had to throw money at its own X-ray laser project, without a lot of Teller-driven enthusiasm, but mainly they decided to upgrade their largest and most effective military asset, the Soviet ground army, with a dazzling array of new, expensive, high-tech tools and weapons.[61] It was a big waste that the shaky

---

61 In the 1990s, I was a member of a team that analyzed, evaluated, and tested many of these foreign devices and systems. This work was classified SECRET and TOP SECRET. It's a pity, but I have not obtained permission to reveal what we were working on. I will go as far as to say that I was made cautious to learn that all but one member of the team that had developed one of these devices had died working on it. Yikes. What had killed them? An uncontrolled pulse of radioactivity? An explosion? The remaining scientist was interrogated, and the answer was startling. In celebration of a successful test, they drank the coolant and died of liver failure. The extremely poisonous liquid (ethylene phosphate) unfortunately smelled exactly like vodka with a hint of berry flavor. The survivor was a recovering alcoholic who did not join the toast, and he lived to tell the tale.

Soviet economy could not stand, and by 1991 the Evil Empire had collapsed. The Union of Soviet Socialist Republics was a grand experiment that had a fair trial lasting seventy years, but various pressures exerted by the United States and its fundamentally flawed economic system had finally brought it down, and Star Wars may have been the last straw.

As a strategy for upsetting the Soviet Union, this action was not even new. It had been tried before in a program that spanned another decade, the 1950s, and into it was lost the 2015-equivalent of the same $25 billion without producing anything usable. It was a badly organized, high-tech weapon development effort that went under many names, but the one that stuck to it was the Aircraft Nuclear Propulsion project, the ANP.

The concept of flying an airplane on nuclear power was bandied about shortly after fission was discovered in 1939, before the first chain-reacting atomic pile was successfully tested. As the American effort to develop an atomic bomb attained highest priority status in 1942, all engineering was concentrated on plutonium production reactors and gravity bomb design, but you could not keep scientists and engineers from thinking about the exotic atomic applications that were sure to come. The idea that produced the most off-topic discussions was not the civilian reactors producing electrical power, but was the concept of an atomic-powered airplane. Even Ronald Richter in his wartime fantasies skipped civilian power and went straight to the aviation angle.

Warplanes were extremely important for the Second World War, and both sides made as much use of flying machines as possible. The Allies, the United States and Great Britain, used them to drop bombs on Germany, and Germany used them to shoot down the Allied bombers. Germany produced dozens of radical airplane designs for this purpose, building everything from jet fighters to liquid-fueled rocket planes, but all the airplanes used in the war had a common flaw: They burned up fuel very quickly. The faster an airplane flew, the more rapidly its fuel was consumed. The fastest manned aircraft in the war, the Messerschmitt Me 163 Komet rocket-powered interceptor, could go 596 miles per hour, but the engine only ran seven and a half minutes before the fuel was gone. You could not very well fly it to England and destroy the allied airbases with it.

The limited range of American heavy bombers, such as the four-engine B-29 used to drop the atomic bombs on Japan, meant that tens of thousands of soldiers died capturing otherwise worthless islands in the Pacific Ocean, trying to put a runway close enough to the home island of Japan to fly there with a full bomb-load and make it back to the base. It was agonizing for the nuclear technologists in Los Alamos to hear the daily news of the conquest of tiny Pacific islands and the lives it cost as the development of the bomb to end the war proceeded as quickly as possible. A way to give airplanes unlimited range using fission, doing away with the need to get close to a bombing target, was on many brilliant minds. It was certain that a gram of uranium-235 could produce millions of times more energy than a gram of high-octane aviation fuel.

The disadvantages of fission power in an airplane quickly followed the energy-density advantage. For one thing, airplanes have to fly, and as such they have to be lightweight, made of thin aluminum and chunks of magnesium hollowed out to prevent heaviness. Fission reactors, on the other hand, have to be built of concrete made super-heavy by substituting chunks of steel for gravel, topped off with thick layers of lead. These heavy shielding components do not contribute to the energy efficiency. They are added to prevent the machine from sterilizing everything within a quarter of a mile, eliminating all living things with the fraction of fission radiation that is wasted in the energy conversion process. No matter how you rearrange the design, an air-plane with a fission-powered engine would be more hazardous to the people flying it than to those standing on the bomb-aiming point. On the ground, you could at least see it coming and jump into a hole. Crash one atom-powered bomber at the end of the runway on takeoff, an accident that could occur on a Pacific island airbase several times in an hour using conventional heavy bombers, and the area would have to be evacuated for a few hundred years while the scattered and embedded fission products decayed away.

With the end of the Second World War came another war with a different enemy and entirely different strategies. In this conflict, the long-running Cold War, the two sides did not rain explosives down on each other, but they felt a need to be constantly poised and ready to do so. The conventional thinking at the time was that heavy bombers were the correct mayhem-delivery system, and they would have to be in the air at all times, lurking just outside the international borders and ready to strike. The limitations of the size of a bomber's fuel tanks was the critical point, and for this reason, the run toward atomic-powered flight graduated from wishful thinking to a serious development effort.

The fun started in 1946, when a contract was drawn up between the U.S. Air Force and Fairchild Engine & Aircraft Co. for NEPA, the Nuclear Energy for the Propulsion of Aircraft program, to study the feasibility of the nuclear-powered bomber concept. Fairchild, cannily anticipating such a project, was already in place at the Oak Ridge, Tennessee, headquarters for the recently established Atomic Energy Commission (AEC). Under the new, twisted rules, any system having nuclear components, be it a bomb or a powered vehicle, could be paid for and borrowed by a military arm, but it belonged to and was under the control of the civilian AEC. Not satisfied by the two-year Fairchild study, which seemed a little bit too enthusiastic, the AEC commissioned another report in 1948, this time to be conducted by the distinguished Lexington Committee, this time headed by the Massachusetts Institute of Technology.

Depending on how you read the Lexington report, the nuclear-powered strategic bomber would either be the savior of western civilization or its downfall. The idea did have some attractive features, such as an ability to keep a bomb in the air indefinitely, but the necessary lack of radiation shielding and crash-site decontamination could prove to be problems.

The Air Force read that as a yes, proceeded, and in 1949 engine experimentation and development began at Oak Ridge National Laboratory. It was a heady time in nuclear engineering. New types of reactors were being designed without precedents, and this one to power a jet engine was a particular challenge. Instead of running at hundreds of degrees to make steam, this one had to run at thousands of degrees to take the place of flaming jet fuel. No previously developed nuclear fuel, even uranium oxide or carbide ceramics, could run at this level of temperature without melting. The creative solution was to run the aircraft power plant on liquefied fuel, and the molten salt reactor concept was born.

The uranium fuel, instead of being formed into rigid metallic rods, was dissolved in sodium fluoride salt, which would melt in the reactor core and act as both the fissioning material and the heat-transfer medium. The liquefied mixture was circulated through the critical geometry of the reactor vessel and into a separate heat exchanger. White-hot heat exchangers were supposed to take the place of the burners in the center of an axial-flow jet engine, propelling the aircraft forward with compressed, superheated air. It took five years to design and build this radical machine, and by the time it was tested in 1954, General Electric, under an AEC contract signed in May 1951, had developed a full-blown, direct-cycle nuclear jet engine at the National Reactor Testing Station in Idaho. It ducted the compressed air from the front sections of two GE J47 jet engines, ducted it through the core of a very hot reactor running with chromium-uranium fuel, and fed it back through the exhaust turbines of the engines. After a brief test schedule to prove the now-obsolete circulating-fuel concept, the molten-salt Aircraft Reactor Experiment back at Oak Ridge was canceled and buried in secret classification.[62]

By 1955, Convair, an aircraft manufacturer in San Diego, California, had converted a damaged B-36 Peacemaker to carry around an operating nuclear reactor. The Peacemaker, with a 230-foot wingspan, was the biggest prop-driven airplane ever made, and it was the only bomber large enough to carry the 42,000-pound MK-17 thermonuclear weapon. It could, in fact, carry two of them. The Air Force plan was to modify the B-36 airframe design for nuclear power. GE provided the reactor to simulate its nuclear jet, but it

62 The Aircraft Reactor Experiment (ARE), with its lavish instrumentation displayed on an entire wall full of pen-chart recorders, cost millions of dollars to build, but it only ran for nine days. First criticality was achieved at low power at 4:45 in the afternoon of November 3, 1954. It ran perfectly at full power (2.5 megawatts), demonstrating that the molten-salt idea was golden. Colonel Clyde P. Gassen, chief of the Nuclear Powered Aircraft branch of the Wright Air Development Center in Dayton, Ohio, ceremoniously palmed the red SCRAM button at 7:04 P.M. on November 12, shutting down the reactor and the project. The molten-salt fuel concept has resurfaced as a possible design for the advanced Generation IV commercial power reactors.

was to run at a modest three megawatts, located in the airplane's cavernous bomb bay. This was the first stab at finding out what radiation would do to the necessary materials in an airplane and to the operating crew of human beings. The plane was redesignated the NB-36H "Crusader."

An enormous block of borated rubber and lead, weighing 12.5 tons, acted as a shadow shield around the flight crew, which included a reactor operator. Radiation was free to escape from the top, sides, bottom, and front of the plane, but as long as the crew was shadowed by the rubber lump, they were perfectly safe. It was eerily quiet for a plane with ten screaming engines, but there were only three small windows, made of foot-thick lead glass. For pilots used to flying a B-36 under a big, bowl-shaped greenhouse, flying this odd contraption was an interesting experience.

Crusader ran forty-seven missions over Texas and New Mexico (always followed by a B-29 packed full of marines with parachutes in case something went wrong), running the GE reactor and dumping the unused three megawatts of heat into the atmosphere. Nobody was killed, but the experiment raised more questions than it answered about flying a reactor in an airplane. A lot more systems and materials testing would be needed.[63]

Parallel to the ANP program, another branch of the Air Force was working on intercontinental ballistic missiles to deliver nuclear weapons quickly and without any pilots, and the strategic bomber concept was beginning to look antique. The nuclear B-36 idea was scrapped, and a fresh sheet of vellum was tacked to the drafting table. The newly hatched project would be Weapons System 125A (WS-125A), consisting of a uniquely designed nuclear bomber using

---

63 Meanwhile, the navy impaled itself on its own, independent nuclear airplane development program, code-named the Princess Project. They already had the airframe picked out. It was a four-stories-tall behemoth of a seaplane, the Saunders-Roe SR.45 Princess, built in East Cowes, Isle of Wight, UK. It was over 55 feet high and weighed 190,000 pounds before the nuclear reactors were installed. The engines were to be Pratt & Whitney closed-loop, liquid-metal-cooled reactors, a design dropped by the Air Force for weight and complexity issues. The navy was spared by the Department of Defense, which shut the project down before an SR.45 had been purchased. There was no apparent purpose for such an aircraft other than to show the Air Force how it is done.

the GE X-39 nuclear engines under development in Idaho. Its mission would be to cruise outside the Soviet Union at subsonic speed and then dash inland in a supersonic rush to make craters at predetermined spots.[64]

A lot of preliminary work finding unknown radiation effects would be necessary. For this purpose there was the Nuclear Aircraft Research Facility (NARF) at Convair in Fort Worth, Texas, consisting of the reactor taken from the NB-36H airplane and mounted in a concrete-lined hole in the ground. A shield cover could be rolled back to let fission radiation bathe a test subject and simulate the environment in and around the nuclear bomber. The reactor ran at only three megawatts, and it could only radiate small objects. There was a 10-megawatt test reactor at the Wright Air Development Center in Dayton, Ohio, but it was enclosed and did not replicate the open-air flying experience.

A new facility was built about 50 miles north of the Lockheed Aircraft Corporation plant in Marietta, Georgia. It was designed to be the ultimate testing station to find the effects of radiation from an unshielded jet-engine reactor on everything from rats to radios, and it would define what imprints there were to be expected on the ground below a nuclear bomber, in the air in which it flew, and in the cockpit. A neglected plot of ground, 10,000 acres, far from dense population, was chosen as the site the Georgia Nuclear Aircraft Laboratory (GNAL), known as Air Force Plant 67 or AFP No. 67.

The land was just south of Dawsonville, a town known for bootleg-whiskey runners/early NASCAR drivers. In the late 1930s, Roscoe Tucker of Dawson County had begun buying up small farms in the Dawson Forest, and by 1950, he had a rectangular tract consisting of some cleared fields, woodlots, the confluence of two rivers and a creek, and a dozen carelessly concealed, taxation-avoiding cottage

---

64 At this point in 1954, the Aircraft Nuclear Propulsion effort was technically dead. The X39 engine was two GE J47 jets using a common nuclear reactor core as the heat source. It weighed a lot more than two conventional, jet-fuel-burning J47s, but the lack of thousands of pounds of gasoline to run them made up for it. An X39 was almost as powerful as a pair of normal J47s, but it took four J47s plus six Pratt & Whitney "Wasp Major" R-4360-53 engines (2.83 megawatts each) just to get a B-36 off the ground and eventually accelerate it to 230 miles per hour. The goal of a supersonic nuclear bomber was outside reality, but bear in mind the hidden agendum of spoofing the Soviets, and it begins to make sense.

industries producing ethanol.[65] There were some dirt paths and a rickety steel "postal" bridge over the Etowah River, built to support a one-horse mail carriage. The tract had been burned, logged, lived in, plowed up, punctured all over with shafts and tunnels, washed away by hydraulic mining, and haunted by the spirit of a woman dressed entirely in black, searching for her lost son. It was a land of a thousand secrets, full of hidden waterfalls, evidence of Paleolithic inhabitants, busted stills, europium-151, and Cherokee tears. In August 1956, Lockheed cut Tucker a check for his land and then generously donated it to the U.S. Air Force.

This piece of land had a history. In the early nineteenth century, North Georgia was occupied mainly by Cherokee Native Americans having acquired some lifestyle aspects of the ever-encroaching white settlers. In the Dawson Forest, they lived in houses, planted corn and cotton, and enjoyed the quiet, agricultural life.

All was peaceful until August 1, 1829, when a newspaper in Milledgeville, the *Georgia Journal*, ran a short notice under the headline GOLD. A gentleman in Habersham County had run across two previously discrete mining parties pulling gold out of the ground along the southeast slope of the Appalachian Mountain range. It had been noticed that gold was present on the downslope of the mountains in North and South Carolina, but this was proof that the deposit extended along a northeast-to-southwest line clean into Georgia, and the pickings were unusually good. The Cherokees, who happened to live on top of the gold, had known about it for a long time without overwhelming interest. They had demonstrated how to mine it to the Hernando de Soto expedition from Spain back in 1540, but nothing came of it. Now, the news of gold mining hit a particular nerve at just

---

65 In 2008, I participated in an attempt by Georgia Public Broadcasting to film a documentary about the Georgia Nuclear Aircraft Laboratory. We tromped all over the 10,000-acre site, filming me standing in front of ruins and explaining what they had once been. The producer had done her homework and had found a former moonshiner who knew a lot of unwritten history of the place. I was surprised to learn from him that illegal micro-distilleries had been active on the property as late as 1977! I had been assured by old-timers who had worked at the laboratory that all the moonshiners had been cleared out by the Corps of Engineers as they bulldozed everything to the ground and cleared the land for construction. They had not. The morning after we finished filming, the producer was fired in a budget crisis after nearly thirty years working for GPB, and her film project was never completed.

the right time, and all hell broke loose. The first fully developed gold rush in the history of the New World commenced.

In a short time, thousands of gold miners were sifting the sand in every creek on the south slope, and the boomtowns Auraria and Dahlonega, a day's walk north of the Dawson Forest, exploded into cities. By 1830, 300 ounces of gold were being mined per day on Cherokee land. The Cherokees, caught up in the excitement, laid down their plows and proceeded to dig up the Dawson Forest. The "Dahlonega Belt" happened to lay right across it, and it was blessed with the Amicalola River, the Etowah River, and Shoal Creek providing the water necessary to process gold-bearing dirt.

The State of Georgia saw an opportunity to both organize the mining operations and increase the state's treasury. The Dawson Forest was divided into 40-acre plots and sold to speculators by lottery. It was a fine idea, except that Indians happened to be living (and mining) on the land, and of that problem came the infamous Trail of Tears. Native Cherokees were persuaded by bayonet point and bribes of pure Georgia gold to relocate to a reservation laid out in Oklahoma, and it was a hard, tragic walk to get there.[66]

By 1840, easy pickings were gone, and gold mining in Georgia became more a serious, profit-loss business and less an adventure. In 1849, gold fever broke out on the West Coast. Most of the Georgia prospectors stopped digging and rushed to exotic California as quickly as antebellum transportation could carry them.

The gold mining industry continued to turn a profit in the Dawson Forest for another half a century. Up until about 1903, pits, mines, and ditches were dug and hills were dissolved using high-pressure water, altering the landscape and keeping the deer away.[67] After the

66 There were no heavily wooded mountains and no gold to be mined in Oklahoma. Cherokees from Georgia who had become skilled in the gold-mongering art left the reservation and trekked to California with the rest of Dahlonega after gold was discovered in the Sierra Nevada Mountains in 1849. The gold-mining town of Cherokee, California, was named for them. Their skills would drive many mining operations, and the Cherokees moved on to Colorado and eventually to the Klondike, chasing the yellow metal and impressing the amateurs.

67 When we were working on the GNAL documentary in Dawson Forest, we stumbled across an old gold mine. At first, all I could see was a heap of dirt, heavily invested in

miners wandered away to other pursuits, an eerie calm descended. The vegetation began to erase the scars, and wildlife cautiously returned. Quietly, the alternative industry of batch-processed distilled spirits took hold of the forgotten, secretive Cherokee land. That was about the state of affairs when Roscoe Tucker sold the whole caboodle to Lockheed.

Construction of the Georgia Nuclear Aircraft Laboratory proceeded at a war-footing pace, and it was, of course, classified secret, as was everything that the government did that had "nuclear" in the title. As a rule, people living in the area lacked a subscription to *Aviation Leak* and therefore had no idea what was going on. The irritating lack of information was filled in with rumor and wild speculation. The most imaginative story was that they were building a center to analyze, repair, and test-fly alien spacecraft that were found crashed near an air base back in the southwestern desert. Concrete trucks and bulldozers rolled in day and night. Strange lights at night and sounds filtered out of the woods. Panic stirred one night when the sky over the center of the secret lab glowed an ominous crimson. A column of red rose high in the sky, and it was visible in Dawson, Pickens, and Forsyth counties. Families gathered to pray, and churches enjoyed unusual attendance with people anticipating and fearing the End Times.[68]

---

pine trees and brambles. We were a few hundred feet off the road, and the woods were dense and undisturbed. Behind the heap, which turned out to be a pile of mine tailings, was a neatly cut entrance into a tunnel, completely hidden unless you were right on top of it. Unfortunately, we had left the lights back in the truck, and all we had was a flashlight. Venturing into the tunnel, it looked almost as if it had been dug yesterday. You could still see tool marks on the walls and veins of sparkly quartz. The only inhabitant was a mouse, who gave us a "And who are you?" look. Our local guide piped up and said that it was an exploratory tunnel dug by the Kin Mori Mining Company back in the day. We could stand up in it, and it had a hard-packed, flat floor.

68 Some people knew just enough about what was going on to speculate that an enormous nuclear reactor was in the middle of the property, and worry that it had gone out of control and would kill everyone soon with radiation. Reality is much tamer. Many acres of pine and oak trees had been bulldozed and gathered into a mountainous pile on a hill at the center of the lot. It was set afire and allowed to burn down for several days, sending up a column of smoke that would be illuminated by the fire from below at night. It looked as if the entire sky was aflame. To just burn it was easier and faster than trying to get security clearances for an army of loggers and pulpwood truck drivers.

Construction was on a vast and complex scale. First on the schedule was an electrical substation and a barbed-wire-topped, chain-link fence enclosing the rectangular perimeter of the site. A similar barrier, the "Lethal Fence," was built surrounding the materials test reactor in a circle 3,600 feet in radius.[69] With the naked reactor running at full power and simulating a nuclear aircraft engine, the radiation would be potentially deadly to any living thing above ground inside that circle, and the exclusion area for anyone without a dosimeter to monitor radiation exposure was 8,100 feet from the reactor, or right up to the perimeter fence.

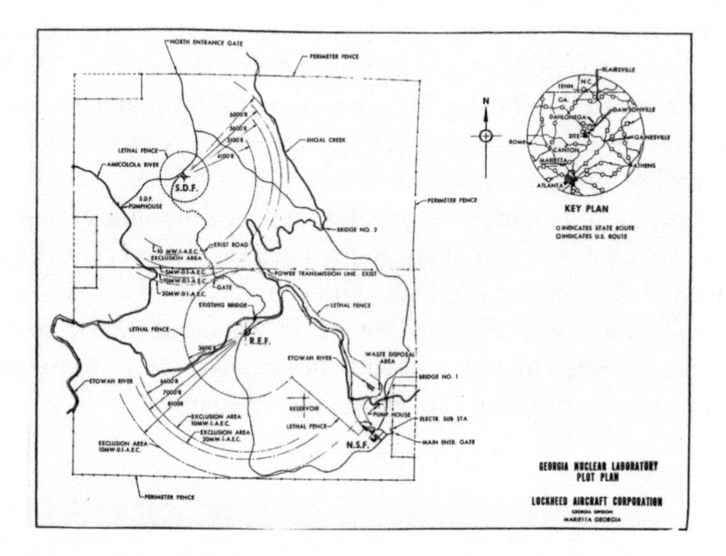

A map of the Georgia Nuclear Aircraft Laboratory. N.S.F. (Nuclear Support Facility) is the administration and laboratories building; R.E.F (Radiation Effects Facility) is the main, 10-megawatt test reactor; and S.D.F. (Shield Development Facility) is the reactor hung in the air. It was purchased but was never installed.

---

69 Using Google Earth, go to latitude/longitude coordinates 34°21'56.64"N/84°10'05.64" with an eye altitude of 10,698 feet. Look closely and you can see a perfect circle etched into the heavy forest cover. That is the right-of-way for the Lethal Fence, still there. At the center of the circle there is a bare spot on the ground. Zoom in and look east. You can see a rectangular bright spot. That is the above-ground portal to the buried reactor control room. It's a concrete cube, with the doorway blocked off to keep curious people from falling down the elevator shaft to the basement floor, 30 feet down. Just beyond it, to the east, is the foundation for the REF reactor building, completely overgrown with trees. A borated-concrete emergency shadow-shield, to be used if a worker were caught in the open when the reactor started up, has been appropriated by graffiti artists.

A 2.5-mile-long, narrow-gauge railroad was installed, with concrete and steel trestles over the rivers. A complete water purification plant with a pump house taking water from the Etowah was built, and a large, cylindrical water tower was erected on the highest ground. This would provide enough gravity-driven cooling water to keep the reactor from melting in the event of a major plumbing accident for several weeks even if electrical pumping power failed.

The administration building and radiation labs were housed in a large, L-shaped, one-story brick building. It was intended to be a two-story building, but at the last minute in February 1957, the Air Force saw danger of the ANP program being cut from the federal budget. They reduced the building budget from $28.7 million to $13.6 million, and an entire wing and the second floor were left off, although the slab had already been poured.[70] The entrance gate was reduced to a one-man guard shack. The building contained the meteorology lab including weather radar, instrumentation shop, chemistry lab, radiation counting room, health physics headquarters, the facility manager, and an Air Force rep's office. A hundred feet west of the building was a concrete slab supporting a 320-foot meteorology tower festooned with instruments, and across the street was the lighted parking lot. On a busy day in 1957, there were 112 cars and a bus parked at the GNAL.

The main road from the entrance ran southeast to northwest. Just past the administration building was the firehouse and central receiving, and northwest of it was the warehouse. A stone's toss north of the warehouse loading platform was the water works, having seven rectangular purifying basins in a heavy concrete structure[71]

---

70 The slab under the administration building, or Nuclear Support Laboratory, is still there, and it is used as a parking lot for visitors with large horse trailers. See it on Google Earth at 34°21'00.34"N, 84°08'32.43"W, and eye altitude 1,479 feet.

71 The concrete warehouse floor and loading dock are still there, visible on Google Earth at 34°21'04.61"N, 84°08'31.06"W, at eye altitude 1,086 feet. The water works are still there but are hidden by foliage. At the northeast limits of the cleared area was an odd structure. It looked like Stonehenge, with a central spot and monoliths in a circle around it. I've never found the site remains or mention of it in the collected documents.

An aerial view of the GNAL. The administration building is in the foreground, and in the lower-left corner is the meteorological tower. In the top-left quadrant is the Radiation Effects Laboratory, consisting of the hot cells and the machine shops. At the top is the water works.

At the end of the entrance road was, and still is, a massive concrete blockhouse, three stories high, containing the high-radiation hot cells and a shielded high bay that was capable of housing three railroad cars of irradiated objects. Entire wing sections, landing gear assemblies, or pieces of fuselage could be irradiated to simulate exposure from the nuclear engines at the naked reactor and moved into this facility by train. The tracks approached from the reactor area from the north and split at a switch, heading into the blockhouse doors or into the huge machine shop connected to the west side of the building.[72]

72 The rectangular blockhouse of the Radiation Effects Laboratory (REL) is visible from the satellite on Google Earth at 34°21'03.11"N, 84°08'37.79'W. Zoom in to eye altitude 1,298 feet. You can see a hole punched in the roof by "grave robbers," reminiscent of the Arabic adventurers who first penetrated the Great Pyramid of Giza. The very large blower room is still attached on the south side, and the loading dock is almost intact. You can walk on the floor of the machine shop, but it is hidden from view by foliage. Penetrations on the east face of the buildings have been blocked off with steel plugs. These were for the viewing ports and the remote manipulators.

Using the machine shop, aircraft assemblies could be assembled and mounted in place on a flatcar. There were sufficient tools and fixtures to build airplane assemblies from scratch, if desired. From there they would be pulled out using a small diesel locomotive, hauled north over a trestle, and looped down to the reactor for radiation exposure. The cars, each 20 feet long and 8 feet wide, were made of pure aluminum, a material chosen because it would not activate into something dangerously radioactive under the intense neutron blast from the irradiation process.

Looking into the railroad-car gallery were seven hot-cell stations—each consisting of a 2-by-3-foot lead-glass window at standing eye level, 4 feet thick—and two heavy-duty Model D Central Research Laboratories manipulators. By operating these devices outside the high-radiation environment of the hot cell and looking through the glass, a person could handle objects on the other side of the thick concrete wall in perfect safety. Each manipulator was a mechanical hand, wrist, and swinging arm that would precisely mimic an operator's every motion as he or she held the external control sticks. Test objects, activated to high radiation levels by reactor exposure, could be examined, dismantled, measured, photographed, and evaluated still sitting on a flatcar rolled into the "hot work area."[73] A bridge manipulator, a muscular robotic arm (General Mills Model E2) capable of delicately lifting a jet engine, ran on rails down the length of the area, and above that was a 15-ton traveling crane. A human being could not be in the area with a radioactive object, so remotely controlled zoom-and-pan television cameras were mounted on both the crane and the bridge manipulator. An operator, trained for two years in using the manipulators, could turn a screwdriver, wield a cutting torch, or pour a milliliter of liquid into a test tube using the mechanical hands, under 2,000,000 curies of gamma radiation.

The Radiation Effects Facility (REF) materials test reactor was a water-cooled, 10-megawatt unit, resting 30 feet underground in a vertical, concrete shaft filled with shielding water. Designed by General

73 The windows were not four-feet-thick blocks of solid glass, but were two thick panes of lead glass with a manganese-salt solution between them. Everything through the glass had a yellow tinge. Working on an object in the hot cell, behind the glass, was like taking apart a cuckoo clock with five-foot-long tools looking through a very large aquarium.

Electric Vallecitos, California, the reactor used standard-type "MTR" fuel, made of fully enriched uranium-235 alloyed with aluminum. Electric pumps in an underground room circulated water coolant through the reactor core and through a redwood-slat cooling tower about a hundred feet east of the reactor building. Farther up the hill from the tower was a cylindrical water tank to make up any coolant loss from accidental core leakage, and it was gravity-fed by the main water tank at an even higher level. A similar tank was located to the west, downhill from the makeup tank, designed to hold coolant that was possibly radioactive from a fuel-failure accident and keep it out of the environment. It was blocked by a concrete shadow-shield on the east side, facing the reactor building.[74]

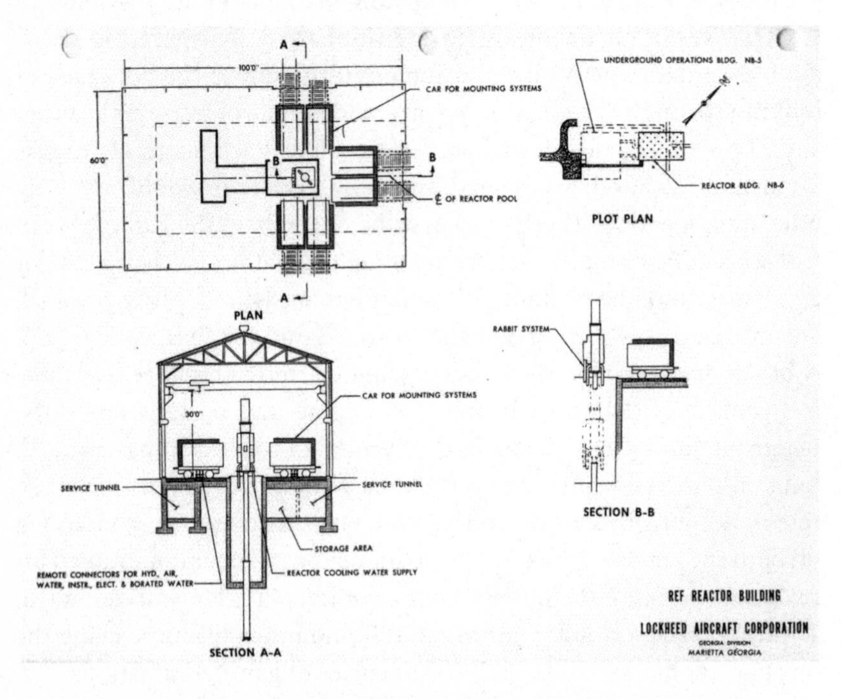

Plan and elevation views of the 10-megawatt Radiation Effects Facility reactor.

---

74 The holding-tank shadow-shield, the cooling tower foundation, the makeup and main water tank foundations, and the control-room air conditioning condenser base are all still there on the grounds, but none are visible from the air. It's best to find them on foot in the winter, when the foliage is thin and the snakes are sleeping. Radioactively contaminated water in the holding tank was drained into a rock-lined "seepage pond" and just allowed to evaporate and soak into the ground. That was an accepted way to get rid of contaminated water back in the day, but you certainly could not do that now.

The REF reactor was purposefully designed to have minimal radiation shielding, simulating the shield-free situation in a nuclear airplane, and this made it unique among fixed-base nuclear reactors. During a materials test run, the reactor was raised out of the shaft and six feet above ground by a remote-controlled hydraulic ram, and it would blast everything within reach with a withering dose of radiation. The radiation pattern could be aimed using electrically controlled water tanks surrounding the core, and it was usually pointed northeast, away from the control-room parking lot but directly into rail cars on the far end of the building.[75]

The approaching rail line split in front of the test facility into separate lines, so that as many as six cars of material on separate tracks could be tested at once, with the cars backed into bumpers a few feet away from the reactor. The intention was to roll a car into the test position, retreat with the engine, soak the test objects with radiation, then actuate the two hydraulic bumpers remotely, giving the car a sincere push and sending it down a gentle incline, 1,000 feet, coming to rest on the tracks in the woods. There, it could cool off for a while before being re-coupled to the locomotive and hauled back to the hot-cell building. That never worked. Instead, the locomotive driver would bump-couple to the car after irradiation using the front coupler while cowering behind the mass of the diesel engine, pull the car in reverse across the river and down to a "cooling-off" area, and drop it there for a few days while the short-half-life nuclides decayed down. He would then re-couple and continue south, crossing the Etowah on the second trestle, and push the car into the hot-bay for examination of the tested object.

The Critical Experiment Facility (CEF) or "staging reactor," was a duplicate design, although it was limited to 80 watts of power, and it

---

75 At one point an unusual radiation pattern was required, and the only way to configure the reactor was to have someone work on it directly and up close. Unfortunately, the reactor had already been run at full power for days at a time, and the fuel was blazing hot with fission-product radiation. No human being would be able to stand near it, raised out of the shaft, ever again. This eventuality had not been considered in the facility design. The man who was most familiar with the aiming mechanism and who knew how to reconfigure it was Mel Dewar, and he volunteered to go in on a flatcar behind a pile of lead bricks and quickly apply a screw driver to the mechanism. I have not run across his dosimeter record, but I would guess that he absorbed his allowed dose for multiple years in the minutes he spent working on the aiming problem.

was located within a few yards of the access road from the administration building. Both reactors were the same core design using MTR fuel. Experiments in fuel and moderator configurations could be conducted using the staging reactor at the bottom of a pool of water without the danger of radiation from accumulated fission products or neutron activation of surrounding structures, because this reactor ran at near zero power.[76] The reactors were housed in corrugated metal buildings, made completely of dazzling stainless steel, so that the metal could not rust and dust off, spreading radio-contamination into the soil. The stainless steel was transparent to neutrons, and it gave negligible outside protection from the fission gamma rays.

Radiation damage to neighboring communities and anyone outside the perimeter fence due to full-power testing was avoided by the physical distance between the test-reactor and civilization, but a possible problem was neutron activation of the existing argon-40 in the air around the reactor to radioactive argon-41, a beta- and gamma-ray emitter with a 1.83-hour half-life.

Escaped neutron flux was also a unique consideration. Neutrons don't fly at the speed of light in straight lines, like gamma rays. They are just another gas, like the argon, and flow lazily in exactly the same way. They have a half-life of 10.23 minutes and decay into ionized hydrogen gas, emitting a respectable 0.782 MeV beta particle. Neutrons are a unique threat in that they can be captured by just about any element and produce a neutron-heavy, radioactive nuclide. The building surrounding the reactor was probably enough to contain potentially dangerous fission-product gases, such as iodine-131. Under the wrong wind conditions with the reactor running at full power, radioactive argon and neutrons could flow down the Etowah River valley and into inhabited areas downstream. For this reason, a number of meteorology towers were erected around the main reactor, each collecting wind-speed, temperature, and wind-direction information for mapping

---

76 The CEF building foundation is still there, complete with its 15-foot-deep reactor pool. From satellite view, it looks like a patch of plant growth in a bare spot, just off the main road that passes by what was the front gate of the GNAL, and winds down to the REF reactor site, at 34°20'50.30"N, 84°08'50.34"W, with an eye altitude of 1,116 feet. The pool is filled in with debris and dirt, and acts as a water-retaining planter in which a lot of plant life is thriving.

the air currents. The reactor could not be started unless atmospheric conditions were just right, and smoke generators were used to see where gas from the reactor was going and how fast. These unusually detailed meteorological data were also used to call in weather predictions to the local radio stations, earning public relations points in a distrusting community.

A potential safety problem that no fence could prevent was small planes flying over at an altitude lower than 5,000 feet with the reactor running at full power. Although it was an air-restricted zone, private airplanes were capable of anything, and it was important not to kill civilians. A microphone was installed atop a tall, steel tower guy-wired to the ground in an open field next to the reactor building. One of several unique automatic SCRAM functions for this reactor was the sound of a Piper Cub engine picked up on the microphone.

Buried 30 feet southwest of the reactor building was the operations building. Being outside in the vicinity of the reactor while it was running would be fatal, but all was safe in the concrete building with the top-floor roof buried under 5 feet of earth. To enter the building, you had to park in a small lot off the road to the left and down the hill. Inside the arched concrete entrance you would select a scooter, then drive down the 660-foot, lighted tunnel.[77] At the entrance to the building was the scooter-storage room where you could park your vehicle. On the main floor were offices, the reactor-control area, computer rooms, an instrument repair shop, and dozens of racks of data-recording equipment. Downstairs was the machine shop, air-handling equipment, and a records room. Constantly working sump pumps kept it from flooding. A service tunnel led electrical cables back to the reactor via a pump room under the stainless-steel building.

---

77 Entering the reactor control complex, which was a most secret place, you didn't have to open or close any doors. There was no code to be entered into a locking mechanism or even a door to keep in the air conditioning or the heat. It was an open, unrestricted tunnel. This was a standard way to build an underground control room, and I've been in the similar facility in Idaho. The tunnel was so long, there was no weather passing though the portal, and the security check was up the road at the gate through the Lethal Fence. The Park Service now tries to keep the entrance blocked with a covering of dirt, but many adventurers have been able to dig through it and make their ways to the buried control building. You have to use an inflatable raft to see the upper floor, as it is always under water. Just about everything is stripped out, but there are still equipment racks and electrical parts bolted to walls.

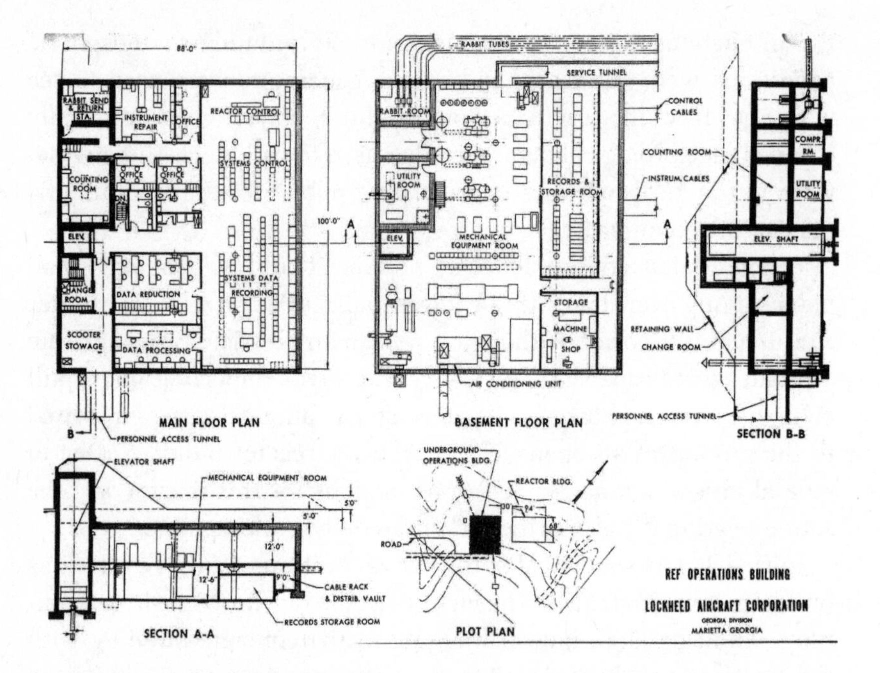

Plans and elevations of the buried control room for the REF reactor.

Although everything in the reactor building was remotely controlled, with TV cameras for surveillance, eventually someone would have to make the trip to the reactor for maintenance, such as for working on the balky hydraulic systems. To be anywhere near the reactor, which you could only do with it stowed and shielded by water in the 30-foot pit, you had to suit up from head to foot in a dust-proof garment, which you would don in the changing room on the main floor, then take the hydraulic elevator or the stairs up to the outside portal. Open the heavy steel door and you were within easy walking distance of the reactor.

By December 14, 1958, the major construction was completed, reactor operators were trained, and, as a final step, the fuel was loaded into the REF reactor, one assembly at a time using the overhead crane. The detailed, expensive work over the past two years paid off as the reactor was brought successfully to criticality, and all systems ran smooth and normal. The first aircraft system to be tested was an AN/ARC-34 ultra-high-frequency radio transceiver to be used for plane-to-plane communications in the nuclear bombers.[78]

In June 1959, the Radiation Effects Reactor was first run at full power for an extended time and managed to kill every living thing within 1,000 feet, with the possible exception of weeds. Pine trees seemed the most sensitive, while crabgrass seemed immune. Oak trees became confused as to when leaves should be shed and new buds developed. In one extended power run, the landscape around the reactor received a radiation dose equivalent to the fallout from large-scale nuclear warfare. The landscape dosage was up to 100,000 rads. The lab personnel noticed a phenomenon, which they called "instant taxidermy," in which if a small animal, such as a toad or a bird, happened to be near the reactor building during a power run, the unfortunate creature would perish along with all the bacteria on, in, and around it, leading to interesting preservation. One big question for the nuclear aircraft program concerned the effect that a fly-over would have on crops and farm animals, and much animal work was planned. Many rats died for their country at the GNAL. The largest organism to be hosed down with radiation was a mule. Imagine, if you will, a dead, thoroughly irradiated mule that will not degrade normally over time.[79]

There was another extended, full-power run in August 1960. By that time, the ANP program was in deep kimchi, as the rapid development of

78 The ARC-34, "Little Eva," had been around since 1952, and it was not exactly state-of-the-art equipment, as had been promised for everything in the advanced aircraft. It did, however, use vacuum tubes, which did not seem affected by high radiation. Semiconductor transistors and diodes suffered miserable deaths when subjected to the kind of radiation that would be present in the nuclear bomber. The Soviets were aware of such findings when they orbited the first Earth satellite, Sputnik, in 1957, into the unknown radiation environment of outer space. The radio transmitter in Sputnik ran on dry-cell batteries and vacuum tubes.

79 References mentioning any accident at the GNAL have not been discovered. There is a tale once told by a retired nurse who worked at a clinic in a nearby town. A sick young man was brought to the clinic by some very concerned personnel from the GNAL. He had symptoms suggesting acute radiation poisoning, and he died at the clinic in a few days. The implication is that a curious civilian defeated the Lethal Fence and approached the reactor building during a high-power run, which would have produced the observed symptoms and death. The clinic and its records are long gone.

guided missiles was obsoleting the need for manned bombers of dubious utility. The Air Force was digesting the data from the lab in Georgia and other test sites, and was beginning to realize the hazards involved with flying this nuclear-powered gizmo. Rubber airplane tires, as it turns out, turn liquid under radiation bombardment, hydraulic fluid turns into chewing gum, and solid-state electronic circuitry stops working. It was seriously suggested that older pilots be used, beyond the age for fathering children, to minimize genetic mutations due to the radiation exposure in the cockpit. The WS-125A program was canceled, and the latest findings caused the AEC to forbid the Air Force from flying the nuclear bomber over the United States. On April 1, 1960, the GNAL contract with Lockheed was modified. The lab was put on standby, and its $3 million annual budget was cut to $750,000.

Hoping to save the program, the Air Force formulated the CAMAL plan, whose acronym was torturously derived from "continuously airborne missile launcher and low-level system." The newly elected President Kennedy, under the steely-eyed fiscal advice of his Secretary of Defense, Robert S. McNamara, officially killed the ANP research and development effort on March 28, 1961.

Lockheed Georgia was now the owner of the whitest white elephant in existence, having two nuclear reactors sitting idle. You can't just turn off the lights in the control rooms, put padlocks on the reactor buildings, and mothball the lab, hoping for the dawning of a better day. By federal law, you have to maintain the reactors in perfect working status with hermetic security and trained operators, health physics, and a mechanical maintenance staff. Otherwise, you have to pull everything down to the ground and manage any residual radiation above normal background. Either option is a financial sinkhole.[80]

80 There were more than two reactors. The GNAL was designed to have a shielding development reactor (SDR) on the extreme north side of the lot. It was to be suspended in the air by four steel towers, simulating a nuclear bomber flying over the ground shortly after takeoff, and it would bathe anything underneath it with a simulated blast from the nuclear jet engines. When the GAO got wind of this plan, which was sucking money out of the budget, the construction was halted immediately. There was a duplicate SDR already built at Oak Ridge, and there was certainly no need for another one. Lockheed had bought the reactor and had it delivered. A road and a bridge to the remote location were in the works. The reactor and its fuel load, which cost $952,000, was stored in the CEF pool and used for spare parts. Another reactor, a remotely controlled Godiva II plutonium sphere mounted on a flatcar, was to be used to simulate the radiation spectrum of an atomic bomb explosion for aircraft systems testing. It was not delivered. The $400,000 spent on the road and bridge were a write-off.

Not to be discouraged, Lockheed Georgia changed the name to Georgia Nuclear Laboratory (GNL) and cast a wide net, scrounging for new nuclear horizons. In 1963, Dr. Werner von Braun from the George C. Marshall Space Flight Center at the Redstone Arsenal in Alabama visited the laboratory site, personally examining the setup for nuclear rocket component testing.[81] A contract was signed for using the reactor to test semiconductors operating under radiation bombardment, and results were reported in September 1964.

In 1965, the radiation-induced polymerization of wood products was an exciting subject in nuclear circles, and the GNL jumped in with its extensive capabilities to blast big structures with radiation. Wood could be infused with a monomer plastic and then exposed to penetrating gamma rays. The result was a supermaterial having better hardness, strength, and general durability than any other wood product. Irradiated wood seemed indestructible, and the REF reactor radiation could be filtered to give the optimum gamma energy and dosage to an entire railcar stacked with wood. Lockheed Nuclear Products, a branch of Lockheed Georgia, named it "lockwood," and the first big shipment went to the new AEC headquarters building in Germantown, Maryland, to be used in the floors. By September 1965, GNL was processing 6,000 board-feet of lockwood per month, and it was being used in products from cigarette-burn-proof tabletops to bows and arrows.[82]

In 1967, GNL had a big hot-cell building that was underutilized, and Lockheed turned it into a 250,000-curie cobalt-60 gamma-ray irradiator chamber. A slotted well, running parallel with the sides of the flatcar hall, was built in the floor and populated with a rectangular steel rack covered with coin-shaped pellets of cobalt-60. Cobalt-60 is an excellent source of radiation, decaying with a half-life of 5.271 years and emitting a 1.333 MeV gamma ray. It is used for everything from medical equipment sterilization to shrink-tubing manufacture,

---

81 Bob Boyd, meteorologist, health physicist, and reactor operator at GNAL, was assigned to show Dr. von Braun the facilities. Boyd took him straight to the REF, and led him to the center of the floor, pointing to the 10-megawatt reactor resting submerged in the water. Von Braun took a glance down into the deep pit and commented, "Seventeen thousand gallons." He hit it right on the nose.

82 A furniture salesman was impressed by the properties of this incredible product until he learned how it was made. "Does it grow hair?" he asked.

and Lockheed hoped to exploit the fact that it had the perfect setup for large-scale application. Things to be irradiated were positioned in front of the slit in the floor using the overhead crane system. With the source buried in the floor, it was perfectly safe. With everyone evacuated and behind the four-foot-thick radiation barrier, the rack was lifted out of the floor and turned 90 degrees to face the irradiation subject. The 10-megawatt reactor, still operating under AEC license R-86, may have been used to activate inert cobalt-59 into cobalt-60.[83]

By 1970, the enthusiasm for nuclear industries was on the downhill slide, and Lockheed decided it was time to shut down the GNL. The extensive meteorological lab had already been dismantled and moved to the Lockheed Nuclear Products (LNP) location at 3272 Peachtree Road in Atlanta. For a while, the new site specialized in weather forecasting and air pollution studies. Now, the radar tower base is covered over by Brides by Demitrios in Buckhead, and LNP is long gone.

⚛

On April 12, 1971, Lockheed applied to the AEC for authority to dispose of the internals of the 10-megawatt reactor. The railroad tracks were pulled up, the trestles were dynamited, and, as much as possible, the buildings were bulldozed down. A big cobalt-60 encapsulation project for Westinghouse Astronuclear Laboratory was killed and transferred to the Georgia Tech Research Reactor in Atlanta, along with 2,000,000 curies of cobalt-60.[84]

---

83 Another GNL employee, Henry Cotten, mentions that he may have prevented a serious dilemma by noticing that the deployment cable on the cobalt rack was damaged and on the verge of letting go. If the thing had jammed in the up position, there would have been no way to send in a person to fix it.

84 There are stories, only stories, of a fire in the hot cells at GNL with cobalt-60 in use. The vaporized cobalt-60 was sucked into the air-blower system and piped to the machine shop, covering everything with the highly radioactive dust. It may have been cheaper to default on the big NASA contract with Westinghouse and blow the site down than to try to clean it up. Air ducting was through concrete tunnels underground, and when the demolishable parts of the building were brought down, the cobalt-60 dust was ground into the dirt. This could explain why cobalt-60 radiation was still measurable in and around the hot-cell blockhouse forty years later. Lead bricks scrounged at the blockhouse demolition site in 1971 were brought to Georgia Tech for use in the reactor building, but they set off criticality alarms at the back door with 250 rads/hour gamma and beta radiation. It raised questions. The Lockheed-Georgia Corporation strongly insists that there was no fire.

In 1972, Lockheed sold the property to the City of Atlanta, which wanted to build a second municipal airport on the site. This proved impractical, due to the amount of grading that would be necessary to make a landing strip, and its use as a new water reservoir has been bandied about. For now, the Dawson Forest is a wildlife preserve.

The ground where the big reactor was is unique in that it shows traces of europium-152. Europium-151, present in scant quantities in the ground, was activated into europium-152 by the action of the unshielded nuclear reactor. It gamma-decays with a 13.52-year half-life. Not many places have Eu-152 in the ground.

Did the nuclear bomber program, with its costly, frantic developments, secret yet lavishly disclosed in *Aviation Week*, accomplish what may have been its hidden agendum? As a Cold War ploy, it may have had an effect on the Soviet Union, but at the time, it wasn't quite clear what was happening. The December 1, 1958, edition of *Aviation Week* carried a big article titled "Soviets Flight Testing Nuclear Bomber; Atomic powerplants producing 70,000 lb. thrust are combined with turbojets for initial operations." A full-page artist's concept showed a sleek, scary-looking nuclear bomber, designated "54," making a 45-degree climb off the page.

The detailed report was compiled from many "foreign observers" who reported seeing it flying around the Moscow area and warming up on the ground for at least two months. It was an impressive 195 feet long, with short, radically swept wings implying a cruising speed of twice the speed of sound. Two outboard nuclear-powered jets were the main thrust, but two conventional jet engines were included for the purpose of takeoff thrust and accelerating to hypersonic speeds. The Soviet engineers had accomplished a "major breakthrough in shielding techniques," so ruinously heavy radiation protection for the crew was not necessary. For the past eight years, Soviet technical literature had been leaking information about their tests of nuclear rocket engines and their brilliant progress.

Who was spoofing the hardest? The descriptions of the Soviet nuclear jet were strangely incredible. They had managed to package the nuclear reactor into the same space used for the fuel burners in a normal jet, and our engineers thought it impossible. You could build a tiny reactor, but it wouldn't have an opening down the center of it big enough for the air that the jet was using to generate thrust. The American version had a separate, off-axis reactor to heat the air, and it was twice the diameter of the jet engine. The Soviets complained that the problem was that the compressor shaft overheated, because it was in the middle of the tube-shaped nuclear reactor. Engineers rolled their eyes.

A diagram showing the reactor installed in the belly of a Tupolev bomber to test the way radiation spreads in the airframe. A thick radiation shield is shown in back of the crew cabin, and pipes lead to the two inboard engines. These enhancements were on the drawing board, but not necessarily in the airplane.

The truth remained murky until 1992, when the Soviet Union ceased operation and secret files began spilling. The Soviet engineers had a nuclear bomber program that exactly paralleled ours. They were struggling to build a workable engine using an off-axis reactor, very similar to the HTRT-2 developments at the National Reactor Testing Station in Idaho. They mounted a test reactor in an existing bomber airframe, just as we did with the NB-36H Crusader.[85] Theirs was the Tupolev Tu-95LAL. Our nuclear engine was the GE X39. Theirs was the Kurnetsov NK-14A. Neither engine was ever installed in an airplane, and both programs were canceled in the same year for the same reasons.

So passed the Cold War and all of its interesting intrigues, but the atomic adventures were just warming up, and the quest for fusion power never waned.

---

85 The only available picture showing the reactor that was mounted in the Tu-95LAL shows it underneath the airplane fuselage, sitting on the tarmac at an odd angle. It looks as if it broke free of the lifting apparatus and was dropped accidentally. It appears very large, implying that the "breakthrough" shielding of beryllium oxide and paraffin are around the reactor and not around the crew.

CHAPTER 3

Inside Cold Fusion

"Everything you know is wrong."

—The Firesign Theatre, October 1974

It is interesting to hear the thirdhand stories from deer-stalkers hunting in the Dawson Forest. There are accounts of deer, obviously mutated by ionizing radiation playing havoc with their genetic material, having six legs or snow-white pelts. There are pygmy deer the size of squirrels, giant deer that can knock over trees, and even a Cyclops deer with one eye in the middle of its forehead. There are reports of oak leaves the size of an elephant's ear and pinecones that you don't want to stand under. There is not a spot on Earth with a better chance of having strange flora and fauna than the Dawson Forest, but not one artifact or photograph has yet to support these claims. This and the fact that giant eight-foot-long ants have not taken over New Mexico frustrate nuclear dramatists to no end. There is simply no physical evidence, and these accounts are most likely the fabrications of active imaginations.

The adventure that I will spin in this chapter is no more believable than accounts of rats the size of feral pigs turning over trash cans in Dawson County, but I swear it really happened. I was right there in the middle of it, and I can prove it.

On Friday, March 24, 1989, life was good. I was a senior research scientist at the Georgia Tech Research Institute (GTRI), with an office in the basement of the Electronics Research Building on Ferst Street.[86] Across the street loomed the seven-story Joe Howie Physics Building, and right behind us on Atlantic Drive was the gleaming white containment structure for the high-flux nuclear reactor, the Frank H. Neely Building. My office had once been the "Iridite Room," filled with open-topped barrels of chemicals used to put a golden color on things made of aluminum and prevent corrosion. With explosive growth in the early 1970s, the machine shop and small labs on the basement floor had been converted into private offices for senior personnel. I had a repainted steel desk, a combination-locked filing cabinet, a chalkboard, one piece of chalk, a dial telephone with lighted buttons, and bookshelves to the ceiling groaning under the weight of printed material. The walls, painted a dingy light green, were cinder blocks, and my PRI Deluxe scintillation counter in the corner kept track of the gamma rays streaming out of them. My window looked across the parking lot and into the reactor's redwood cooling tower.

Next door to me was Al Becker, head of research security and filled near the bursting point with secrets left over from World War II. He had served in "Wild Bill" Donovan's OSS (Office of Strategic Services) during the war, and he was good for an oblique tale of wartime intrigue if things were slow. Across the hall was a cypher-locked lab full of much newer secrets. Things were seldom slow. If I wanted to talk to somebody, I went upstairs.

There was a personal computer in everyone's office, and we had access to the proto-Internet, but digital communication services were

---

86 GTRI at that time was divided into six or seven laboratories housed in five buildings on campus, a clump of buildings and a long-field testing range at Lockheed Georgia, north of town, and satellite offices in strategic places, such as the Redstone Arsenal in Huntsville, Alabama. Our laboratory, the Electronics and Computer Systems Laboratory (ECSL), had about one hundred people inhabiting the three-story Electronics Research Building. The budget for GTRI was about $100 million in 1989.

different back then and somewhat crude. There was no universal e-mail. We had a system installed by IBM called "profs." It allowed us to send and receive messages on campus, but there was no profs connection to the outside, and the World Wide Web was lame compared to what it has become. The nearest thing to an Internet was the "Newsgroups." The Newsgroups was a form of large, universal BBS (Bulletin Board System). Anyone could sign on to a specific subgroup, such as "alt.fusion," and post a message or read postings left by other interested parties.[87] The personal computer connection to it was by modem through a mainframe in the Richard Rich Computer Center. It was simple, eight-bit ASCII text, with no graphics. I had a cellular telephone, but it was bolted to the driveshaft tunnel in my car, between the front seats.

This morning I was sitting in Darrell Acree's office with my feet up on his desk, sipping a can of Dr Pepper from the drink machine in the break room. Darrell had BA and MA degrees in mathematics and an MS in electrical engineering from the University of Kentucky. I considered him a valuable resource in subjects ranging from quantum electrodynamics to stoichiometry, and I could count on him for an enlightening discussion to begin my workday. We were in the eerie, dangerous stillness that exists between projects, and we would have to quickly scrounge up something to support our salaries.

Darrell's office was on the main floor, but it was small and windowless, and it was at the nervous edge of the human traffic circle that flowed through the building. On one wall was an unframed print of van Gogh's *The Sower*, on another was a plotter rendition of an MH-53J helicopter, and taped to his filing cabinet was a portrait of Jed Clampett from *The Beverly Hillbillies*.

The first topic of discussion this morning was an odd phone call I had gotten at 7:20 the previous night from Mark Pellegrini. Darrell

---

87 There were hundreds or perhaps hundreds of thousands of subgroups. There was no directory, nobody took the time to count them, and more subgroups were being added as the seconds ticked by. That morning on the "profs" campus mail system I found and deleted a memo from a project director reminding us of special security requirements at Warner Robbins Air Force Base, a press release announcing the results of someone's project over in the electromagnetics lab, and a copy of the last progress meeting of project A-8265.

had apparently received a similar phone call from Mark about something he saw on the MacNeil/Lehrer news report on the local public broadcasting station, channel 30.

Mark had a BS and an MS in "double-E" (electrical engineering) from Georgia Tech, and like the rest of us, he had worked at GTRI since graduation on myriad projects that at times seemed far afield of his formal education. In this racket, one had to be flexible. We were, for lack of a more accurately descriptive term, "techno-whores." We would do anything, as long as it didn't break the skin. Working at GTRI meant that we were paid by the State of Georgia, but the State was paid by external sources needing specialized research and development for which we were the lowest bidder. Most of the work was for the Department of Defense, and the traditional strongpoints of GTRI had always been radar measurements, advanced radar design, and foreign technology exploitation. We did a lot of secret work, and there were pockets of extremely focused radar expertise in the many GTRI buildings. Darrell's office at 8:30 Friday was not one of them.

Mark's calls concerned an apparent breakthrough in fusion-power research. Two electrochemists at the University of Utah in Salt Lake City had staged a scientifically strange news conference at noon on Thursday, breaking to the world that they had achieved a net power output from deuterium-deuterium fusion using an electrolysis cell the size of a water fountain cup. They had repeatedly referred to the process as "cold fusion," to separate it from everybody else's fusion reactor.

The common mode of fusion-power attempts to that point had been trying to outdo the Sun with millions of degrees of heat and nucleus-crushing pressures. In 1989, the United States government was going to put another $350 million into a continuing series of experiments trying to coax a net power out of one of the very large, complex, and power-hungry "hot fusion" machines scattered from Princeton to Lawrence Livermore. Forecasts were that it would take "another thirty years" of funded work to see positive results. The United States was competing with countries from Great Britain to Japan to see who could pour the most money into hot fusion. Worldwide, probably $10 billion had been spent trying to get more power out of a fusion reactor than was put into trying to achieve fusion. To claim to have jumped over forty years of heavy research and found the Holy Grail of nuclear science in a glass of water seemed strange, to say the least, and a bit reminiscent of Ronald Richter's adventure down in Argentina.

It was also very irregular and Richter-like to make a scientific announcement in a news conference. The usual way is to publish a paper describing the breakthrough experiment in a refereed journal. A panel of anonymous referees would read the paper before publication, making sure that the claims were scientifically reasonable and adequately explained. There had to be enough detail that another scientist could use the paper as a guide and reproduce the experiment as described, confirming the claims.

Mark's description of this miracle was too sketchy to make out what was going on. He did say that they were making neutrons, and that is a signature of deuterium-deuterium fusion. There is no chemical reaction that will release neutrons from the nucleus. If a neutron escapes, then there has been a profound disturbance on the nuclear level involving a million times more energy than the most energetic chemistry can achieve. I confidently dismissed it as two chemists having rediscovered one of the many ways to make neutrons using compact setups.

"That's done all the time in A-bomb triggers. It's not new," I opined. "You can make neutrons with small systems, but there's no way it's an energy source." At that time, the only system that could achieve a net power from fusion reactions was a thermonuclear weapon.

Darrell probed me for an opinion about the weird cold-fusion announcement from Utah, but honestly, I needed some details to even tell what they were doing, much less criticize it. We were debating the topic when Rick Steenblik bounced into Darrell's office, looking even more enthusiastic than usual.

"Well," he began, "What do you think?"

Darrell glanced at me, thinking, "Let's slow this boy down before he starts spinning on his major axis." He leaned way back in his chair, making the springs squeal. "Think about what, Mr. Rick?"

"Cold fusion! Haven't you heard?" Rick's kinetic enthusiasm could make me feel sluggish in comparison, but it could also stir the thought process. Sometimes knowing Rick was like being urged forward, being tugged at, but sometimes it was like being dragged by a car with my hand caught in the door.

"Oh, that. Yeah. Jim here thinks it's spark fusion. There's nothing to get excited about."

"No! No! It's not spark fusion.[88] Didn't you see the piece on MacNeil/Lehrer? It's electrolysis. Deuterium into palladium. I have a tape of it, if you'd like to see it."

I sat straight up. "You've got a tape? Where'd you get a tape?"

"The MacNeil/Lehrer hour is on twice. It comes on channel 30 at seven o'clock and on channel 8 at nine. I saw it first at seven, and then I taped it at nine."

"There's a VHS recorder upstairs in the Quiet Room. I'll get a key to the room, you get the tape, and we'll meet you there."

I had met Rick Steenblik several years ago, when he was working in the Technology Applications Laboratory (TAL) on the other side of the campus. He was a research engineer with a BS in mechanical engineering from Georgia Tech. He had grown up, to some extent, in Hialeah, Florida. I went to see his power plant simulator routine running on a modest Commodore 64 microcomputer. It was impressive, but I was blown away by his original design of a stereographics method. He pulled out a pair of handmade glasses having two prisms of unequal refraction indices bonded to each of the lenses. He brought up an ordinary-looking graphic on his Commodore screen, showing solid geometric figures, crudely rendered in color. Using the glasses, the figures jumped off the screen. The method did not separate left and right images to achieve stereo vision. It translated the color of the image into depth using the convergence angle of the eyes. Bluish objects were far away, and reddish things looked very close.[89] It was ingenious. By the time he had worked a few years at GTRI, first as a

---

88 The day before, I had regaled Rick and Darrell with a brief lecture concerning compact Z-pinch fusion reactors used as neutron generators, and the same day I had received an unsolicited catalog from the Sigma Chemical Company, which sells deuterium oxide (heavy water), lithium deuteride, and several other deuterated chemicals. Darrell would later interpret these happenings as prophetic.

89 For a complete explanation, see US Patent number US4597634A, "Stereoscopic process and apparatus," at www.google.com/patents. Richard A. Steenblik holds thirty-nine US patents and uncounted foreign patents. His "Unison" product is protected by twelve unusually voluminous patents, containing more than 1,200 claims, or about 100 claims per patent. On average, a US patent contains 18 claims. In comparison, the author's name is on a measly two patents.

student assistant, he already had patents pending. His first invention was a spiral solar-power sunlight collector.[90]

The working model of his spiral was now on the wall in his windowless top-floor office, crowded with three desks, a bicycle, and a large aquarium. Rick had left GTRI in 1984, hoping to find an environment that was less on the constant verge of a financial crisis, and he took a design engineer job at Auto Ventshade Company. There he was free to think and create to his heart's content, as long as he created Ventshade products for automobiles. After a couple of years, he came back to GTRI, and he was earning a master of science in physics in the building across the street while working full-time. His specialty was optics.

The Quiet Room was thoroughly sound-insulated on the walls, the ceiling, and the floor. It had been configured years ago as a place to interrogate mynah birds in private. (Don't ask.) Now, it was used for secret briefings or "read-ins" and as a storage space for unused office furniture. Chairs were piled up on the far wall. We took three chairs down and set them in front of the television monitor. Darrell closed the door. Rick switched on the VHS recorder, fed it the tape, and touched the Play button.

The crack of an atomic bomb detonation came over the speaker as on the screen a fireball turned into a mushrooming cloud of dust. Rick jumped forward and turned down the sound. The voice of Charlayne Hunter-Gault, a journalism school graduate of the University of Georgia, came on over the rumbling sound of the bomb's shock wave. "There are two types of nuclear energy," she began. "Fusion and fission. The most familiar is nuclear fission. . . ." The video turned to shots of the rising mushroom cloud over Hiroshima, Japan; the

---

90 The sunlight-focusing collector reminded me of a Fresnel mirror, but it wasn't. It was ingeniously unique. It was a spiral cut from mirrored Plexiglas, tensioned to make the outer edges of the spiraling plastic lean in and cause incoming sunlight to reflect and concentrate at a point in front. He carefully laid his newly completed working model on the backseat of his car, so as not to disturb the focusing, and he went to perform some other task. When he returned, he found the car filled with opaque smoke. The sunlight through the window glass, focused neatly on the headliner, had significantly reduced the car's resale value. The windows were up, limiting the available oxygen inside and occluding the sunlight with the smoke, or the car would have burned to the ground. He and Georgia Tech benefited greatly from his stereo-vision patent, named "ChromaDepth."

Three Mile Island meltdown; and the smoking wreckage of an RBMK graphite reactor at Chernobyl. The brief commentary pointed to the contrast between the destructive, dangerous, and expensive action of nuclear fission and the benign, gentle consequences of nuclear fusion using hydrogen isotopes. A final bit of film after the Chernobyl shots showed an aerial view of Elugelab Island being obliterated in 1952 by an 11-megaton tritium-deuterium fusion event. The casual viewer may have missed the joke.[91]

The scene then shifted to the studio, showing Hunter-Gault sitting in front of a picture of a Rutherford lithium atom, made obsolete in the 1920s with the advance of quantum mechanics. The caption, "Fused with energy," was above the picture. She continued, "The advantages of fusion energy are many. It is less radioactive. Therefore the waste disposal problem is reduced, if not completely eliminated, and the power released by fusion would be inexpensive and abundant, capable of supplying the energy needs of the entire world. Here with us now to explain their work and its implication are the two researchers."

The scene cut to a close-up of Dr. B. Stanley Pons, an electrochemist and former restaurant manager who enjoyed cooking and had unusually large lenses in his glasses. He was forty-six years old; was born in Valdese, North Carolina; and earned his PhD at the University of Southampton in England.

"Stanley Pons is a professor of chemistry at the University of Utah. His collaborator . . ."

The scene switched to Dr. Martin Fleischmann, retired professor of electrochemistry and occupier of the Faraday Chair at the University of Southampton. He also liked cooking, and the two men enjoyed the sport of snow skiing. He was sixty-two years old; was born in Karlovy Vary, Czechoslovakia; and got his PhD at Imperial College in London. He spoke clearly with a hint of a European accent, his comb-over was looking thin, and his eyeglass frames were the type we wore back in the early 1960s. The plastic was dark across the top but clear on the

---

91 We had started the tape 13 seconds into the news segment, and we had missed the opening sentence, putting us right there at the detonation of the Trinity test of the atomic bomb in 1945. See the first seven minutes of the interview on YouTube at https://www.youtube.com/watch?v=00IFpIBpa9Y. The video quality of Steenblik's VHS tape was much better than this YouTube rendition. We watched it twice that morning.

bottom halves of the lens frames. They both had the pale, serious, "put up wet" look of active research scientists.

". . . is Martin Fleischmann, a professor of electrochemistry at the University of Southampton in England. Gentlemen, to you both, congratulations. Your discovery is being hailed as a breakthrough. Professor Pons, how accurate is that?"

"Well," Pons began, modestly. "It's, uh, sort of a breakthrough in the field of nuclear fusion. Uh, we have a cell which is comprised of a block of metal, which is immersed in deuterium oxide, which is heavy water, and the amount of heavy water present on the earth is enormous. It is virtually an inexhaustible source of fuel, if it can be used for a fuel." Pons had sort of leapt over any explanation of how this block of metal was causing two deuterium nuclei to fuse together, something to which nature was opposed, but we remained patient, straining to hear critical details.

The interviewer continued with what was for us a useless question. "What makes it heavy water?"

"Unlike normal water where each of the hydrogen atoms in the water have a single proton, the hydrogen atoms in heavy water have both a proton and a neutron. So it's a heavy isotope, if you like, of water."

"And, you were saying?"

"In this particular cell, we use an electrical current to change the water into deuterium gas."

The camera switched to a close view of the cold-fusion cell, lying on its side on a tabletop. We leaned forward in our seats and squinted. On brief glance, it looked like a test tube inside a larger test tube, with a white Teflon stopper on top and a Teflon ring inside to support the end of the inner tube. It was hard to tell, but we got the impression that the anode of the electrolysis cell was at the center, long, thin, and vertical, and the cathode was a wire wrapped in a helix, using the inner glass tube as a support. Out the top, wires and vents emerged.

Pons continued. "These are then forced into the lattice, the metal lattice, by the current, and are highly compressed in that lattice. They are compressed to the point and are retained close enough to each other and for a long time, that fusion occurs."

The interviewer nodded in apparent understanding. "All right. I want to get into the details of that experiment . . ." (Finally!) ". . . in the simplest terms possible, but first let me ask you . . . Mr. Fleischmann, this is also being hailed as the ideal energy source. Is that the case?"

Looking distinguished and professorial, Fleischmann answered. "Yes. There would be many advantages in using it as an energy source, because as was referred to in the run-in to this program, the reaction would be clean, the fuel supply would be, and, this Professor Pons has said, the fuel supply would be plentiful, and it could, in this embodiment, be carried out in a very simple manner."

". . . the fuel supply being inexhaustible," added the interviewer. "Does that relate to this 'heavy water'?"

"Yes," continued Fleischmann, "it's the content of heavy water in the sea which would be the fuel in this instance."

"Does it indeed have the potential of transforming the world's energy source?" Where were these details, albeit simple, that we were getting to?

"If the engineering problems can be solved, certainly. Yes."

"Is that a big *if*?"

"Wha . . . ? I beg your pardon?"

"Is that a big *if*?"

"Well, in any scientific investigation there are always the problems of a science, and there are the technical problems. But, we do not see such massive technical problems as there might be in some of the other approaches that have been tried so far."

"You have described this process as 'ridiculously simple.' Something that could be done in a freshman chemistry class. Mr. Pons, you did this in the kitchen, right?"

Pons gave a nervous chuckle. "I think the kitchen thing has been blown up a little today . . ."

"Well, it's pretty sexy. I think maybe that's why . . ."

"Yeah. Well, the . . . uh . . . it's the simplest of electrochemical cells. It contains two electrodes. It contains a large palladium electrode that serves as the device for containing the deuterium. It contains another electrode, an anode, which is wrapped around that but is electrically isolated from it except by the solution between them, and you simply pass a current between the two electrodes."

"How did you know you were creating nuclear fusion?"

"First by the enormous amount of heat that was generated. There is no known chemical process or other process that we're aware of that could explain such huge amounts of energy, and subsequent to that we have detected particles that are associated with nuclear fusion reactions over and above normal background. Those particles would be neutrons."

"How long did this go on?"

"We have sustained cells for several hundred hours over the last few years. This latest experiment we've been running one or two hundred hours at very high energy outputs."

"Mr. Fleischmann, was there anything else that you could see or feel during the course of this investigation that gave you more information that you were generating fusion?"

"No. I think that Professor Pons has given really the correct description. The main indication that we had nuclear fusion was the extremely large release of heat energy from the cell."

"Scientists the world over and governments have spent billions of dollars with very sophisticated equipment trying to generate the incredible amount of heat that . . . that would simulate the heat in the Sun to create this reaction. What led you to think that you could do it at room temperature?"

"The conditions in this cell are completely different to the conditions which are now investigated in the conventional approaches to nuclear fusion. I think one can best explain it in simple qualitative terms by saying that if you pass an electric current into the cathode under the conditions which we have used, then if you try to achieve the same conditions in the cathode by compressing of the gas, you would need a billion-billion-billion atmospheres. That is, a billion-billion-billion times the pressure at the surface of the earth, and it seems it is this enormous compression of the species in the lattice which made us think that it might be feasible to create conditions for fusion in such a simple reactor."

The billion-billion-billion, or ten to the twenty-seventh power, was a big number to throw at us, and it made the intended impact. Fleischmann went on to say that when captured in the crystal lattice of palladium, any isotope of hydrogen had an effective density greater than it had in the form of a gas. It was, in fact, greater than hydrogen compressed into a liquid, higher than liquid hydrogen compressed into hydrogen ice, and still higher than hydrogen ice compressed into its metallic form. That was a stunning statement, and we could see how it could mean fusion. At that kind of pressure, you could fuse tapioca pudding. These electrochemists had twiddled with hydrogen fusion at the opposite end of the spectrum from where everyone else had been trying to achieve more than a momentary event. They had looked where nobody had thought to look, at the cold end of the temperature spectrum, banging deuterium nuclei together using a quirky attribute of palladium metal and discovering a previously unknown cross-section resonance.

All nuclear interactions, from neutron capture to fusion, are probabilistic events, with the probability expressed as the effective size of the target particle, expressed in barns. In a situation in which an interaction probability between a moving subnuclear particle and its target is very high, the target is "as big as a barn." The probability of two deuterons fusing is extremely low, due to the fact that both are charged electrically positive and they repel each other. They simply cannot get close enough to use the strong nuclear force to hold them together and fuse into a heavier element.

Exceptions, however, have been known to occur. The primary example is fission by slow-moving (thermal) neutrons. In 1934, Dr. Enrico Fermi, coinventor of the nuclear-fission reactor, had measured the fission probability of neutrons hitting uranium-235, and had found that when the incoming neutrons were slowed to room-temperature speed, the probability of fission increased dramatically.[92] By pure

92 Fermi had discovered nuclear fission, but he did not realize it. He thought that neutron bombardment had activated uranium into two new radioactive elements, which he named ausonium and hesperium. The importance of his discovery was that when neutrons were slowed down by colliding with the hydrogen nuclei in a wooden tabletop, they were much more likely to interact with uranium. He only misunderstood the nature of the interaction. This phenomenon was noticed when one of his activation experiments was transferred to a wooden bench from a marble-topped bench. The heavy marble top did not slow the neutrons down, but the wooden top, made of hydrocarbon compounds, did. The fission action was finally sorted out in January 1939 by Lise Meitner and her nephew, Otto Frisch, and Fermi's discovery of the fission resonance in uranium-235 finally made sense. Otto Hahn of Germany was given credit for the discovery, and he was awarded the Nobel Prize in chemistry for it. The fission cross section of U-235 actually increases on a gradual slope as the speed of incoming neutrons increases, but at the lowest level of energy, there is a narrow blip of enormous fission cross section, called the "fission resonance."

chance, his research team in Rome, Italy, discovered a sharp, high upward blip on the graph of neutron speed versus fission probability, a resonance, at the bottom end of the speed scale. The best way to get uranium to fission was not to use fast neutrons but to use very slow ones. Who would have predicted this outcome? This discovery led to the design of the first reactor using slugs of uranium distributed throughout a matrix of pure graphite, used to slow fission neutrons down to thermal speed. Had Stan and Marty discovered an analogous phenomenon in the fusion realm? I wasn't sure. The rest of the interview wandered off to discuss acid rain and the greenhouse effect. Fleischmann indicated that their paper describing their experiments would probably be published in May. The two scientists ended the interview with a vague warning to anyone who would dare replicate their experiment, implying a danger factor.

Having watched the taped interview, Darrell, Rick, and I now knew as much about cold fusion as anyone else in the world, with the possible exception of Stanley Pons and Martin Fleischmann. Rick and Darrell caught fire and wanted to commence a confirmation experiment immediately. I could see the need, but I would have to think about it. Unfortunately, the journalist had not known the right questions to ask. It was clear that the palladium electrode in the cold fusion cell was loaded with deuterium atoms using simple electrolysis. But for electrolysis to occur, there had to be something dissolved in the heavy water to make it conduct electricity—an electrolyte. What electrolyte were they using? Darrell thought sulfuric acid. What purity was their heavy water? Mixing a little sulfuric acid into heavy water would introduce a contamination of non-heavy hydrogen. Would this inhibit the fusion reaction? What voltage were they applying across the fusion-cell electrodes? At what level of electrical current did it operate? We and everybody else were forced to wait until sometime in May to get the details needed to replicate the experiment and confirm their wild-assed claim. One thing the three of us agreed on: Pons and Fleischmann either had a lock on the Nobel Prize for physics or they were crazy. There was no middle ground.

On that day, March 24, 1989, the Exxon Valdez, a 987-foot-long oil tanker, crashed into the Bligh Reef in Prince William Sound, Alaska, spilling eleven million gallons of crude oil into the delicate subarctic environment. A worldwide energy angst had been growing for a long time, and this particular incident kicked it into orbit. People were ready for a positive change and a release from a nagging worry about the atmosphere, oil supplies, and imagined piles of radioactive fission-dirt. The cold-fusion announcement was timed perfectly, punching a human resonance.

On Friday evening, the Pons and Fleischmann cold-fusion experiment was the second-biggest story on *NBC Nightly News*, right behind the big oil spill in Alaska. It was in the newspapers on Saturday, but there were no further technical details.

A couple of things about the announced phenomenon bothered me greatly. Both Pons and Fleischmann had claimed over and over that their apparatus was producing "tremendous" heat and "measurable" neutrons. On the news, they were shown hovering over their operating fusion reactor, closely examining the apparatus as the oxygen and deuterium bubbled. That didn't make any sense. It should have been the other way around—tremendous neutrons making measureable heat. If they were, in fact, making tremendous heat, then they would have been killed by the heavy neutron flux and secondary radiations broadcasting out of the unshielded deuterium-deuterium fusion reaction. Maybe the problem was their definition of "tremendous?" The cooling water in which the fusion cell was suspended wasn't boiling or even misting up their glasses.

The concept of a thin electrode made of palladium holding deuterons together with a force 10 times atmospheric pressure was hard to grasp. It seemed like an energy conservation problem. What was doing the work of forcing the deuterons together, and why did the palladium cathode not explode from the internal force?

We, Pons, and Fleischmann did not realize it at the time, but there was nothing new about cold fusion using electrolysis-loaded palladium. History was repeating.

The year 1926 was a turning point in Germany. The Germans had been on the losing side in World War I, and they had been forbidden from rearming by the Treaty of Versailles. This restriction included rigid, hydrogen-filled airships, which had been used as bombers to harass London and Paris, and this really crimped the economy in Friedrichshafen, home of Luftschiffbau Zeppelin GmbH. Since 1917,

Zeppelin had been building airships in the United States with Goodyear Tire and Rubber, and the German engineers were very impressed with the Americans' use of helium instead of hydrogen as the fill-gas. The hydrogen was extremely flammable, but the helium was completely inert and would not burn under any condition. They wanted helium, and they wanted to build airships back in Germany.

On December 1, 1925, the Locarno Treaties were signed in London, giving the Germans permission to, among other things, build rigid airships, and by early 1926, Zeppelin was in production. Zeppelin wished for helium, but the only industrial-scale production was in the United States, where it was considered a strategic material and could not under any circumstances be exported.

Two Austrian chemists, Friedrich "Fritz" Paneth and Kurt Peters, working at the University of Berlin, thought they had the answer to the helium problem. Although the neutron, a component of the atomic nucleus, had yet to be discovered, and the structure of the element-defining nucleus was still murky, the chemists thought that helium could be synthesized by fusing hydrogen nuclei together. The combination medium would be metallic palladium, known to all chemists as having a strange affinity for hydrogen. A block of palladium, just standing on its own, will absorb 900 times its volume in hydrogen with no encouragement from compression. Palladium was well-known as a "hydrogen-leak window," used in high-vacuum systems to introduce hydrogen, one atom at a time, to be electrically accelerated or otherwise used in several physics experiments. A section of an otherwise gas-tight system made of palladium would admit hydrogen as if it were a hole in the wall.

Thinking that a 900-times compression would surely make hydrogen atoms fuse together, Paneth and Peters made a thin capillary out of palladium, heated it red-hot to expand the distance between palladium atoms, and directed hydrogen gas through the center. Cooling the palladium, they expected the additional compression given by the shrinking of the crystal lattice of the metal to ensure fusion of the hydrogen absorbed under heated conditions.[93]

---

93 This was Paneth's and Peters's fourth attempt to synthesize helium. They had also tried submitting hydrogen to an electrical discharge in an ozone-making apparatus, putting hydrogen in a Geissler tube with aluminum electrodes and bombarding certain salts with cathode rays. The heated palladium tube gave positive results, but a better-controlled experiment followed, in which "paladinized" asbestos was heated.

They were not disappointed. Spectroscopic analysis detected helium mixed with the excess hydrogen flowing out the end of the palladium capillary tube. The amount was small, but in this first experiment they had demonstrated the synthesis of an element, not by radioactive decay on the heavy end of the periodic chart of the elements, but by combination, on the bottom end of the chart. Several sources of possible error were considered and eliminated. The chemists composed a detailed description of their groundbreaking experiment and its results, "Über die Verwandlung von Wasserstoff in Helium" (The Transmutation of Hydrogen into Helium), and sent it to the German chemistry journal, *Berichte der Deutschen Chemischen Gesellshaft*. The interesting paper was received on August 17, 1926, and was published in the September issue, volume 59. The distinguished English journal *Nature* published a full account of the experiment as a news article in the October 9, 1926, edition.

Follow-up experiments were disastrous. Paneth and Peters found, to their ultimate embarrassment, that helium floating around in the lab had contaminated the experimental apparatus. In fact, all of their glassware showed helium inclusions. They had discovered a problem that would bedevil experiments, particularly those involving high vacuum, for decades to come. The fact is, with every breath you take, you inhale atoms of helium. It seeps out of the ground and most building materials, as the slight inclusion of uranium or thorium in most minerals decays slowly, emitting alpha particles. An alpha particle is nothing less than a helium atom, once it readily acquires a couple of stray electrons, and helium will diffuse slowly through the best glassware seals. The chemists immediately submitted retracting papers to the German and English publications. The retraction with explanations of the helium detection errors in *Nature* appeared in the May 14, 1927, issue, volume 119. It was the right thing to do.[94]

94 Friedrich Paneth fled Germany in 1933 when Adolf Hitler assumed full control of the country, and he became a professor of chemistry at the University of Durham in England. In 1943, he was appointed head of the chemistry division of the joint British-Canadian atomic bomb project in Montreal. He returned to Germany after his retirement from the University of Durham in 1953 and became director of the Max Planck Institute for Chemistry until he died in 1958. The mineral panethite and a crater on the Moon are named for him. Kurt Peters remained in Germany during the war and worked at IG Farben, working on catalysts. After the war, the American military government appointed him as trustee for what was left standing of IG Farben. After that, he returned to academia in Vienna, Austria. He died in 1978.

Dr. John Tandberg, humorist, radio personality, and industrial chemist at Electrolux in Sweden, read Paneth's and Peters's first disclosure paper in 1926 concerning hydrogen fusion catalyzed by palladium. He was excited by the idea of fusion and was correct in thinking that the fusing together of hydrogen nuclei to make helium was a change in nuclear structure and similar to nuclear decay. As such, it would release a million times more energy per fusion than any chemical reaction would release per reaction. Chemical reactions twiddled with weakly held, outer-orbital electrons in atoms, while the slightest change in the atomic nucleus involved much more powerful forces and would result in greater net energy release. He therefore saw the fusion of hydrogen as a clean, compact energy production scheme, and helium synthesis was a side product.

In the Electrolux laboratories in Stockholm, Tandberg and his collaborator, Torsten Wilner, set up a unique fusion experiment, using electrolysis to load the palladium. In a glass beaker filled with water he connected a piece of palladium to the negative terminal on a battery and a platinum wire to the positive terminal. Under the electrical influence of the battery voltage, the water disassociated into hydrogen and oxygen, with the hydrogen gas bubbles piling up on the palladium electrode and the oxygen bubbles on the platinum electrode. The palladium sucked up all the hydrogen, presumably subjecting it to enormous pressure within the crystalline lattice of the metal.

Tandberg reasoned that simply loading the palladium with hydrogen was not going to produce fusion. He decided to subject the preloaded palladium electrode to a high-voltage discharge, giving the metal a temperature jolt to trigger a cascade of fusions. Calculating that the energy release from one high-voltage arc across the palladium would be the equivalent of setting off one kilogram of dynamite, he bravely sent Wilner home, instructing him to tell the authorities exactly what had obliterated the Electrolux laboratory once the smoke cleared.

The experiments produced the deafening bangs of an overdriven electrical arc, white-hot flashes, transformer smoke, and rattling glassware, but they didn't produce any detectable radiation, helium, or net energy.

Never discouraged, Tandberg soldiered on. He ignored Paneth's and Peters's retraction of their experimental findings, apparently believing that they had been bought off by the American helium cartel. On February 17, 1927, Tandberg applied for a Swedish patent for "A method for production of helium." The patent examiner found Tandberg's description of the method hard to follow, and the application was denied.

In 1932, deuterium was discovered as a contaminant in ordinary water. Seeing its value as something more likely to fuse, Tandberg begged a purified sample of it from Dr. Niels Bohr in Copenhagen. With deuterium oxide substituted for the water used in his original setup, Tandberg's improved cold-fusion technique was functionally identical to that used by Pons and Fleischmann.

Energy production from the palladium setup was difficult to pin down, and Tandberg's exploits in cold fusion, unknown outside Scandinavia, faded away. Tandberg advanced to the head position in the Electrolux chemistry lab, worked on a home refrigerator design, and retired in 1962. He died on November 3, 1968, in Lund, Sweden.[95]

Ideas for fusion, hot or cold, kept rising, but they would pop and submerge. Fusion and its promise of pollution-free power never lost its allure, but it remained beyond the grasp of science and technology, as the quest for the hidden trick, the undiscovered cross-sectional resonance, continued.

A bright spot flared in 1947, and it would eventually lead a crooked path to an odd convergence in Utah in 1985. In 1947, after wartime service in the Air Ministry, Sir Frederick Charles Frank, FRS (Fellow of the Royal Society) was employed to do research at the University of Bristol physics department in Bristol, England. Intrigued by a new cosmic-ray particle discovered by the head of the department, he hatched an idea.

The new particle was the muon, which is very similar to the electron, only it is 207 times heavier. It has the same negative electrical charge as the electron, and it can orbit the proton at the center of a

---

95 I once gave a lecture on cold fusion at a Mensa convention in Atlanta. My lecture was canned by that time, and I always would mention that Dr. Tandberg had written a book describing his cold-fusion experiments. It was only rumored to exist, and nobody had reported finding a copy, so I jokingly said that anyone having a copy of the book should see me after the lecture. As I turned off the projector and unclipped my microphone, a woman from the back of the audience came up and plopped a copy of Tandberg's book in my palm. I was stunned. Unfortunately, it was written in Swedish.

hydrogen atom just like an electron, but being heavier, it rides 207 times closer to the nucleus. The problem with fusion was always the stand-off distance between nuclei, caused by the fact that, stripped of the electron cloud by the extremely high temperature required for fusion, the protons repel one another. What if you keep it cold and retain the electrons? The size of the atom, due to the radius of the electron orbit, still keeps the protons (nuclei) apart. What if you could shrink the orbits? With muons replacing the electrons in hydrogen or deuterium atoms, they could snuggle up 207 times closer than normal, greatly increasing the probability of spontaneous fusion.

Papers were written discussing muonic fusion in theoretical terms, including one by Andrei Sakharov, father of the Soviet atomic bomb, in 1948. In 1956, Louis Alvarez at University of California, Berkeley unaware of the previous theories, found evidence of muonic fusion in a bubble chamber photograph.[96] He was briefly exhilarated, believing that all the fuel problems of mankind had been solved. However, when the excitement calms down, there are always two problems with cold fusion using muons: a muon has a mean lifetime of 2.2 millionths of a second, and it takes more energy to artificially generate muons than can be recovered from muonic fusion.[97] Hope is diminished, but not eradicated.

Leonid I. Ponomarev of the Soviet Union announced a possible deuterium-tritium (D-T) muonic fusion in 1977, and in 1982, there was a new push for muon-catalyzed cold fusion at the Los Alamos National

---

96 A bubble chamber is a container full of a transparent fluid, heated to just below the boiling point. A high-speed particle zipping though it will leave a trail of tiny bubbles, and a magnetic field running vertically through the chamber helps to determine the electrical charge on a particle. Negatively charged things curve right and positively charged things curve left, and given the speed of the detected particle, the radius of the curve indicates the mass. Electrons and muons both curve left, but the muon moves with a much longer radius of curvature. Watching as the long-curve particles disintegrate in the midst of a bubble track, the half-life is determined. The first fluid tried in a bubble chamber was beer, soon replaced by liquid hydrogen.

97 The mean or *average* life span of a muon is 1.44 times its half-life. Muons are like every other decay-prone object in the nuclear world in that they decay exponentially, and the relative decay speeds are usually expressed as half-lives. After one half-life, half of the original crop of particles has decayed. In this case, the average time that a muon has to work on hydrogen fusion is more important than its half-life.

Laboratories in New Mexico. Involved in the experiments was Dr. Steven E. Jones from Brigham Young University (BYU) in Provo, Utah.

The team achieved something that had never been accomplished before. They achieved eight D-T fusions using one miserable muon. Artificially made muons directed into a liquid deuterium-tritium mixture would replace the electrons in two close-together hydrogen isotopes. If the isotopes happened by chance to be deuterium and tritium, which together have an unusually high fusion cross section, there was an enhanced chance of spontaneous fusion between them. As soon as the fusion happened, the muons would spin off and find other hydrogen isotopes to engage, and on the average, one muon could infect eight successful fusions before it disintegrated into an electron and two neutrinos. To achieve break-even fusion, in which as much energy is generated as it takes to make it happen, an average muon would have to participate in about one thousand fusions. In any other endeavor, this would be seen as a possible failure to succeed, but in the optimistic world of fusion research, this was a breakthrough.

Steve Jones managed to keep the muon fusion concept alive for another five years with Department of Energy (DOE) contracts, but by 1985, the fire was dimming out and he turned to another idea. Fellow physicist at BYU Paul Palmer had been interested in the gases trapped in volcanic ash.

Helium is in the noxious mixture of gases that escape from a volcano, and helium is temporarily trapped in the lava and rocky debris that come up from beneath the surface in an active volcano. Helium gradually diffuses out of any material, into the air, where its light weight causes it to rise to the top of the atmosphere and be blown away by the solar wind. This is the common helium-4 isotope and is understood to be constantly produced by alpha-particle emissions from radioactive nuclides in the Earth's crust. Palmer was finding helium-3 as well as helium-4, and this seemed a little strange.

In the nucleus of helium-3 are two protons and one neutron. Four billion years ago, the Earth was probably contaminated with a measurable helium-3 content, left over from the supernova debris from which rocky planets are formed, but in those billions of years it was thought to have all escaped into outer space. Something underground had to be making new helium-3, and an explanation was fusions of deuterium nuclei. The deuterium-deuterium fusion (D-D) results in one helium-3, one neutron, and 3.2 MeV of energy.[98]

There are 156 deuterium atoms per million water molecules in and on the Earth. The chances of two being next to each other and encouraged to fuse, even in the high temperature and pressure involved in geologic processes, seems vanishingly small. Why would there be D-D fusions inside the Earth? Jones concluded that hydriding, the tendency for certain metals to absorb protons into their crystalline structures, could encourage deuterium nuclei to fuse. Nickel is a hydriding agent, and there is a lot of nickel in the earth's core. In 1986, Jones started trying to make fusion by loading up palladium, which is a more enthusiastic hydriding metal than nickel, using electrolysis.[99]

Incredibly, at the University of Utah (UU), 45 miles away, Pons and Fleischmann had the same idea and began work on the same experiment hoping for the same results at the same time. As Peter Dehlinger, patent attorney for the university, would say some years later, "You'd think the gods would've put them a little farther apart." The two secretive projects bumped into each other by 1989, and the two factions, BYU and UU, reached a handshake agreement: they would each submit their papers describing their cold-fusion discovery to the journal *Nature* on Friday, March 24, shipping by FedEx from the same depot at the Salt Lake City airport, and may the better team win.

---

98 As it turns out, there are natural sources of helium-3 that have nothing to do with fusion. It is in constant production by high-speed cosmic rays hitting nitrogen-14 nuclei in the atmosphere, the decay of tritium, and lithium spallation. Interplanetary dust collecting on the ocean floors, having sifted down through the atmosphere, is estimated to contain 1,200 metric tons of helium-3, and as much as 7 percent of the primordial helium-3 is still in the Earth despite the length of time it has had to diffuse away. As much as 7 percent of the natural gas you may burn in your home furnace and water heater is helium, and a few parts per million of that is primordial helium-3.

99 The element with the highest affinity for hydrogen isotopes is uranium. The second greatest affinity is shown by palladium. Unfortunately, alpha-phase uranium immediately collapses into a white powder upon hydriding. The reader is challenged with finding the hydriding characteristics of beta-phase uranium. In the early days of hydrogen-bomb development, when the fission and fusion components were thought best colocated, a beta-phase uranium deuteride/tritide was considered for the active ingredient of the device code-named Alarm Clock. This research remains classified SECRET.

The potential fame and fortune was too great to leave to chance, and the UU administration convinced Pons and Fleischmann to jump Jones by having a press conference on March 23. This stratagem had not occurred to Jones. His cold-fusion experiments using palladium to hydride heavy water produced, at best, barely detectable neutrons and no heat worth measuring. This seemed a lot more reasonable than Pons and Fleischmann's claim of tremendous heat, but the fortune potential in Jones's findings was a lot smaller.[100] This was the point at which we and the rest of the world intersected cold fusion on that Thursday evening in 1989.

By the time I got to work Monday morning, Rick and Darrell were finalizing the confirmation experiment plan. I had to admit, I was curious about this out-of-left-field claim. We could set it up on a tabletop in one afternoon and get this over with.

Rick's excitement was contagious. As I entered his office, he swung around in his chair. "You won't believe," he began, "how difficult it is to find palladium. It's not like buying aluminum, or steel, or copper. It's a precious metal. You buy it like you buy gold, only it's not as common as gold."

I blinked. "So, you've found some? Where is it? Where is the money coming from? Do we have a project number?"

"It's in Chicago, and they can sell us, get this, one ounce. Only one ounce. It seems that there was a run on palladium this morning, and the price went up 20 percent in a couple of hours, and they don't know why." Rick slapped the desktop and laughed. "They are mystified. The price of palladium has been stagnant for years, and suddenly it takes off."

That sounded ominous. I had to sit down. "So, you're saying that we aren't the only ones who want to try this experiment?"

---

100 To put Steve Jones the scientist in perspective, note that in 2005 Jones presented research indicating that the collapse of the World Trade Center buildings in New York on September 11, 2001, was not caused by airliners crashing purposefully into the sides. These disasters were, according to his findings, caused by previously installed thermite bombs placed with engineering precision against the vertical support structures in the buildings. They were set off by the United States government, with the destruction sequence perfectly synchronized with the sacrificial airliner crashes. This absurd and inflammatory accusation, presented as science, put him under review at BYU, but he resigned before the investigation was completed. His further work involves the study of archeological evidence of Jesus Christ having visited ancient Mayans in Central America and radiocarbon evidence of pre-Columbian horses in the Western Hemisphere, both in support of the Book of Mormon.

"Yep. The metal brokers don't have a clue. Guess how much an ounce costs."

"I have no idea."

"A 99.9% pure ingot, drop-forged in a single block and weighing one ounce, costs a hundred and four dollars. They will have it here tomorrow by plane. We have to pick it up at a bank, over in Vinings."

"And how are we paying for it?"

"We're not buying it."

"What?"

"I've been talking to Billy Livesay. Do you know Livesay?"

I looked at the ceiling. "Uh, yeah. I know him from the Nuclear Safeguards Committee. He works over in the Baker Building, in the Electromagnetics lab?"

"Right. Well, I've been talking to Livesay about hydrogen coupling in metals. Turns out, he's a world-class expert on hydrides. Back during President Carter's alternative energy push, he studied hydrides as a hydrogen storage method."[101]

"Makes sense."

Dr. Billy Livesay from Texas had a PhD in metallurgy, I thought. His location in the Baker Building, which was one block west of us, was a hotbed of earth-science projects. Or, at least it was a *warm*bed. Carter's vision of a hydrogen-based energy economy had seen better times, and the money had dried up years ago.

". . . and," Rick added, "Livesay wants us to skip the palladium and use lanthanum-nickel 5. He says it's a much better hydriding agent than palladium, and he has a sample we can use."[102]

"No."

---

101 If you are wondering about the accuracy of this dialogue after a quarter of a century, I made detailed notes at the time, kept an hour-by-hour timeline, and, by the end of April, had begun writing a book about the experience. It was still fresh, and the word exchange memories had not had time to drift and improve with age. I've condensed it down a bit so that the chapter won't drag, but this dialogue is about as authentic as remembered dialogue can be.

102 We didn't realize it until March 29, Wednesday afternoon, but Steve Jones had already tried lanthanum-nickel in 1988, along with titanium, aluminum, and iron as hydrides. He used both electrolysis and pressurized gas to load these hydrides, and found "tantalizingly positive" results using electrolysis on titanium.

"What? Why? We wouldn't even have to buy any palladium."

"You're jumping ahead, Rick. This is an extremely dubious claim. We have to confirm that the process exists before we can try to improve it. Besides that, buying the palladium is the first test. If we can't buy palladium, then the experiment is not meant to be."

"Okay. I've been thinking about that. If we can get five people to throw in twenty-one dollars apiece . . ."

"No. We're not doing this in your kitchen. We're doing it at Georgia Tech. The institute pays for it, or it doesn't count. Let me know when you've wrangled a charge number. I've got to go see Darrell."

I wanted to talk about electrolysis, and Darrell was the closest thing we had to a chemist in our building. He had worked as a lab assistant in the Department of Biochemistry at the University of Louisville from 1964 to 1972, while pursuing a degree. I began, "How many volts will it take across the electrodes to disassociate water, with sulfuric acid in it?"

Darrell sat back and twiddled his thumbs. "Oh, I don't know. Action will probably start at two or three volts."

"But why didn't they mention any electrolyte in the interview? Does that bother you?"

"Not particularly. The electrolyte has no active role in the process, other than making the water conduct electricity. Just a few drops of sulfuric acid will do, and I can't see it being much of a contaminant. Do we have any heavy water lined up? Don't they have plenty of that down at the reactor?"

"About twenty-three thousand gallons, give or take a gallon. But it doesn't belong to them. It's on loan from the DOE, and besides, it's heavily contaminated with tritium and helium-3. I think I know where I can get a small bottle of it. Over in the Baker Building. On a shelf in the laser lab."

"How pure is it?"

"I don't know. How pure does it need to be?"

"I don't know." Darrell was starting to squirm as the unknowns formed into a pile. "How pure is a palladium ingot?"

"Ninety-nine point nine. Is that pure enough?"

"I don't know."

I tried to bring it back down. "Well, look. We're not trying to make energy. We're not measuring heat. The D-D fusion results in neutrons and helium-3s, in equal measure. No use trying to detect helium-3s. There's too much outside explanation for that. It's the neutrons we want. Establish a background count, and turn on the electrolysis cell. If any neutrons over background are produced, then we've got something. If any neutrons, if one neutron, is produced by a glass of water with three volts across it, then that, sir, is a Nobel Prize for doctors Pons and Fleischmann. That's all there is to it."

The following day, Tuesday, March 28, at about ten o'clock, our cold-fusion quest quietly slipped over the border, a dimensionless line separating that which cannot be done from that which cannot be stopped.

Rick popped into Darrell's office, where we continued to pile up unknowns and loose ends. He announced that his division chief, Hugh Denny, a member of the infamously tight-fisted management, counting pennies and keeping his domain afloat in the storms, was paying for the palladium ingot. "I just danced into his office and asked him, right up front. He just looked up and said, 'Shit. Buy it with petty cash, and leave me alone.' "

My excuse for not trying cold fusion had just been jerked out from under me. Rick placed the order. Our palladium would be at the bank tomorrow morning.

On Wednesday, early, I drove the three of us to the bank in Vinings, 13 miles north of the campus. Darrell put it on his MasterCard, and the teller slid a small, round-cornered slab of hydriding agent into my hands. I took it out of its plastic sleeve, and we stared. The face of the ingot was finely coined, with a beautiful polish on the slightly dished surface. Written on the face in raised letters was "Deak-Perera, since 1928, 99.9 PALLADIUM, ONE TROY OUNCE, 5631." On the back was a repeating, octagonal pattern, slanted at 45 degrees, containing the word "ENGLEHARD."

"It looks like . . . platinum," I mumbled.

"No, not really," said the teller. "Here's platinum." She produced a shiny, round token, coin of the realm for the Isle of Man. I could see the difference. The platinum looked colder, and not as attractive. I passed the ingot to Darrell.

"Well," he said, fondling the metal, "We're in deep now. This object cannot be resold, given away, discarded, or destroyed. It's property of the State of Georgia. It can only be used for the purpose for which it was purchased, or it can be surplused."

On the way back, we talked work assignments. Darrell would find some glassware and some sulfuric acid. Rick would clear out a space in Livesay's storage room in the Baker building. I would get the nuclear instruments and a power supply. We all knew that this experiment would have to be done quietly, with an informal secrecy. The chances of it working were so slight, it seemed too silly to be trying. There were more important, pressing matters at GTRI, and this was just something we wanted to try. We scattered as soon as we got back.

I walked down the hill to my old stomping ground, the research reactor building. Years ago, there had been a big nuclear presence in GTRI, which had then been named the Engineering Experiment Station, or just "the Station." The reactor and its research facilities were another laboratory branch, and were owned not by the school but by the Station. As nuclear research lost its reason to live in the late 1970s, ownership of the white elephant and everyone who worked in the building were transferred to the School of Nuclear Engineering. I was probably the only researcher with nuclear specialty left at GTRI.

I went through the front door, breezed through the security checkpoint, turned left, then right, and down the back hall to the health physics office. "Jerry," I blurted, "I need a portable neutron detector."

Jerry Taylor looked up from what he was writing and decided to start from the top. "Hi, Dr. Mahaffey. How's it going?"

"Just awful. Look, I need one of those little Eberlines."

"How's Carolyn doing? We haven't seen her in a long time down here."

"She's fine. You know, the handheld BF-3, with a built in moderator."

"Come to think of it, we don't see much of anybody."

"And a set of fresh batteries. And a Geiger counter."

"You say you need a detector?"

"That's right."

"Sure. What for?"

"Huh?" I started to sweat.

Jerry, by this time, had changed the slouch-angle in his chair. "What's it for? Just curious. Do you have any neutrons?"

"No. None at all." Oh God. He's going to make me fill out a radiation work permit. Please, Jerry, just give me the damned thing. I hate filling out forms.

"You don't have any neutrons at all, but you need something to count them with?" he asked, beginning to catch on.

"No! I just want to be absolutely sure I don't have any neutrons."

A little smile of recognition came over Jerry's face, and he nodded. "Ahhhh. You're going to try cold fusion."

"Yeah, that's right. But, look, don't tell anybody. It's just something we've got to try. You know, to see if it works."

Jerry laughed, got up, and started sorting through his keys to find the one for the equipment cabinets. "Sure. Let me know if it works. We've been talking about it for days down here, but nobody's doing anything about it. At least you're working on it."

He handed me an Eberline PNC-6 portable neutron counter, normally used by health physics to check for unusual neutron activity around reactor experiments. It looked very similar to a Geiger counter, except that clipped under the chassis was an aluminum box filled with paraffin. A little brass tube, about an inch in diameter and connected by a cable back to the chassis, fit into a hole bored in the front of the box. The tube was sensitive only to thermal neutrons, the type that would escape a nuclear reactor and be floating around a reactor like gas leaking from a cracked hose on a grill. To detect fast neutrons, as were made by fusion, the tube stayed in its hole, and the paraffin slowed them down to thermal speed.

In my other hand I carried away a Ludlum survey Geiger counter. I could not predict any need for a Geiger counter, which would detect gamma and beta rays, but it seemed foolish to set up an experiment possibly involving atomic nuclei without one. In the vanishingly remote possibility of something happening, I would hate to have to explain afterward why I had not thought to monitor for ionizing radiation. For this experiment we would be scratching around in a totally black unknown. I signed for them, begged a variable DC power supply and a set of leads from the electronics shop at the end of the hallway, and walked over to the Baker Building.

The rest of the conspiracy was already in place and fussing over the apparatus. Billy Livesay couldn't make it. He was in meetings all day. I connected up the power supply and positioned the PNC-6 to pick up neutrons spraying out of the palladium ingot. It was hanging by an alligator clip in a beaker of 100 cubic centimeters of deuterium oxide, 99.8% pure, with a platinum wire circled around it in a spiral. Both electrodes, the palladium cathode (negative) and the platinum anode (positive) were supported by a ring stand. Rick set up a video camera to record the event on tape, I adjusted the power supply to deliver three volts across the electrodes, and Darrell measured out three drops of sulfuric acid into the beaker.

"Are we ready?" I inquired.

"Ready here," replied Rick, looking up from behind the camera.

"I can't think of anything else," said Darrell.

"Well," I said, "who wants to turn it on? Darrell?"

"I think I'll just stand back here, out of harm's way."

"Rick?"

"I'm working the camera."

"That leaves me," I reasoned. It was 2:20 on Tuesday afternoon. The amplified speakers attached to both radiation counters, the Geiger and the neutron detector, were making random popping noises about two every minute. That was the background, caused by the occasional cosmic ray or one of its daughters having made it all the way down the atmosphere from outer space, a gamma ray being flung out of the concrete floor, or just a random electronic spasm somewhere in the long paths from detector tube to sound reproduction. I flipped on the power-supply switch.

The power supply hummed slightly in the otherwise dead silence, with the occasional click from a detector. I could hear Rick's tape running. After a few minutes, the novelty of the situation became stale. I spoke up. "Maybe we should look at it."

Darrell responded. "Are you sure you want to get close to it?" By that morning, we had been advised by the rudimentary but active electronic media why Pons and Fleischmann had advised caution: Sometime in the autumn of 1984, one of their cold fusion setups had self-destructed. An electrolysis cell, similar to the one we were presently watching, had been running for seven months when, over one night, the one-cubic-centimeter palladium cube being used as the cathode apparently vaporized, leaving the setup disheveled. The information was sketchy.

Mark Pellegrini, the one who had brought up this subject originally, had taken the rumor in an oddly serious way, and that morning he had almost convinced Darrell that attempting the cold-fusion experiment was an immoral and dangerous act. Putting three volts across a glass of water could kill everyone on campus, and if we persisted, then we should at least broadcast a warning campus-wide so that people could choose to go home. Pons and Fleischmann interpreted the overnight palladium disappearance as definitive proof of principle. Obviously, to them, the palladium had suddenly reached a temperature far above the boiling point of palladium metal. I took it as a possible theft of their cathode.[103]

With the challenge so formally posed, I went over to examine the setup. "Nothing on the Geiger counter. Serious nothing on the neutron counter. Scintillator shows background."

As a backup gamma-ray detector, I had brought my trusty PRI scintillator from my office. Unlike the other counters, it just showed the rate of gamma-ray hits on its detection crystal on the meter, with no event-marking clicks on a speaker. It was one hundred times more sensitive than the Geiger counter. It could find radiation in a common red brick. It was indicating nothing unusual. The electrolyte appeared to be boiling. Oxygen was streaming vigorously off the platinum helix, as expected. There was a faint rushing noise, a white noise, caused by the tiny bubbles breaking at the surface of the heavy water. There were no corresponding bubbles at the palladium anode. One would expect there to be two deuterium bubbles for every oxygen bubble. "We're not getting any D-2 at the cathode," I reported.

Rick offered, "That's because the palladium is hydriding. It's soaking up the deuterium as fast as it can electrolyze it. Are you sure there's no radiation?"

---

103 Accounts differ, but it is agreed that young Joey Pons, Stan's son, was in charge of that fusion cell. He had lowered the electrical voltage and gone home for the night. Upon returning the next morning, he checked in on the cell, saw that the setup was modified, and asked his father to take a look. Later work found that if subjected to electrolysis for many months, the surface of the palladium electrode collects silicate corroded off the inner walls of the glass enclosure, and this greatly increases the resistance of the electrolysis cell. Using a voltage-controlled power supply, the current will rise accordingly, the resistive heating in the cell increases, and the water can start to boil. This effect is not caused by any nuclear reaction.

"Positive. No gammas, no betas, no neutrons."

And so it boiled. We turned the power up. We turned the power down. We checked the batteries in the instruments. We lifted out the palladium piece and looked at it. We thought. We speculated. We debated, lectured, and stirred the heavy water. We zoomed the camera in and out. We shifted the positions of the electrodes. We took the neutron probe out of the paraffin box and waved it around. Nothing made neutrons. Nothing.

We had not proven the existence or the nonexistence of a Pons and Fleischmann cold-fusion effect. All we had proven was that there was more to this experiment than had been revealed in that press interview. If the guys in Utah wanted to play rough and on the sly, so be it. It would not stop us or any of the hundreds of curious researchers around the world. We had all stopped what we were doing and with grim determination were diving headfirst into cold fusion. The race was on.

CHAPTER 4

Good News and Bad News

"Thanks for having generated so much fun."

—physicist Frank Close, February 1992,
above his signature in my copy of
Too Hot to Handle: The Race for Cold Fusion

THE YEAR 1989 WAS A turning point in world history. By March, a wave of revolutions and rumors of revolutions had begun to sweep Eastern Europe, beginning in Poland when the United Workers' Party voted to legalize Solidarity, a shipbuilders' trade union. It was the beginning of a crack in the Soviet control of workers in an important satellite of the USSR, and in a couple of years, the entire Evil Empire would collapse.

Back home, IBM was still a dominant force in the personal computer revolution. It would do to typewriters, airmail postage, and draftsmanship what Solidarity was doing to the Kremlin. There was no Windows operating system. We used the operating system DOS 4.0, and the data-transfer medium of choice was the 3.5-inch floppy disk. There were no smartphones, and digital photography hadn't really caught on. I listened to Enya singing "Storms in Africa" in Gaelic on the cassette player in my car. George Herbert Walker Bush was the new president of the United States.

In March 1989, there were two cold-fusion reactors on Mars. They were part of the soil analysis equipment in the Viking Project lander vehicles, used to activate elements in the Martian dirt by spraying it with neutrons and then identify the resulting radioactive isotopes by the characteristic energies of the gamma rays they emit. One of these reactors is about the size of a fountain pen and consists of a miniaturized Cockcroft-Walton accelerator, throwing deuterons at a tritium target. Each resulting tritium-deuterium fusion releases one helium-4 and one neutron, clocking out at 14.1 MeV. These reactors operate at room temperature, which on Mars is rather chilly. Unfortunately, the energy required to run the Cockcroft-Walton accelerators far exceeds the recoverable energy from the resulting fusions.

This basic problem had been vexing scientists ever since the discovery of fusion in 1932. Stanley Pons and Martin Fleischmann had leapt over this problem with an elegant solution. Instead of crashing hydrogen isotopes together at high speed, they exploited the tendency of palladium metal to gather hydrogens together in a very small space. It seemed so simple, and yet by Wednesday, March 29, 1989, nobody had announced confirmation. The task of finding if cold fusion works outside Utah was still up in the air.

That morning, I wrote a proposal to the Senior Technology Guidance Council (STGC) at GTRI, requesting internal research money to make a serious stab at replicating the Pons and Fleischmann experiment. The STGC, headed by Dr. Hans Spauschus, would very carefully dole out money to research engineers or scientists for start-up development of a promising new concept. Cold fusion was about as new and promising as a concept could be, and I did not have to emphasize that the first labs to confirm the effect would be covered up in externally sponsored work for the rest of the century. I was down as principal investigator, and the team was listed as Rick Steenblik, Darrell Acree, and Billy Livesay, with each salary paid for four weeks. There were $2,450 of materials and supplies, fringe benefits, overhead, a mainframe computer charge, and zero travel. The budget came to $24,991.

By that afternoon, I had a charge number, E-904-031, breaking the record for speed of STGC approval by several days.

At 2:00, we had the first project meeting, behind the padded door of the Quiet Room. We hoped to keep it on the down-low, drawing no attention to the fact that we were doing this crazy thing, but word travels fast in the otherwise secretive environment of technical research. A fairly new hireling who worked in Hugh Denny's division, Gary Beebe, had heard about it, and he wanted in on the experiment. He crashed our private meeting, and he was passionate about wanting to participate. I regretted to tell him that he wasn't in the budget, and we could not pay him. That was no concern. He would work for free.

Beebe was not from around here. He was born and reared in Michigan. He had quit a fine job as an engineer at Motorola just to come down and work at GTRI, and in his strangely aggressive, Yankee way, he wanted to be part of our team. Adding him to the team was as near as we would get to diversity.

Beebe did not understand the popularity of foreign cars around here. To him, it was Detroit iron or nothing, and he drove a beloved albeit ragged-out, salt-eaten, headliner-drooping Pontiac that he had named Monte. He showed up late for work one day, all upset about having been broadsided on the way in by an inattentive driver, but we could not discern any change in Monte's looks.

At this point, there was too much that we didn't know about the Pons and Fleischmann experiment to even start a confirmation experiment. This was noticed worldwide, and Stanley Pons was under enormous pressure to reveal the secret ingredients. He disconnected his home phone, and he and Fleischmann vanished. Nothing was coming out of Utah, and the alt.fusion newsgroup on the infant Internet was in hysteria mode. Darrell and Rick were trying to keep up with it and plumb it for any news or a new reveal. I was phoning around and keeping an ear to the rumble of the rumor mill.

We were trying to decide on discreet laboratory space where we could set up without drawing a lot of attention. We had just about decided on the "penthouse" on the roof of our building. It was originally built as the top of the double elevator shaft on the building, but at the last minute, the building project ran out of money, the elevator doorways were bricked up, and the rectangular structure on top was forgotten. It was certainly obscure, but unfortunately, it didn't have running water. Might we need cooling water? We didn't even have a power figure for a Pons and Fleischmann fusion cell, so this question and others were still up in the air.

When we broke up the meeting and went back downstairs, I noticed a fax in the INCOMING box at the division office. It was from the Los Alamos National Laboratory, Theoretical Division (T-11). It had come in at 2:35. It was a pirated copy of "Observation of Cold Nuclear Fusion in Condensed Matter," by Stephen E. Jones and seven other people. This was golden. It was the paper that Jones had sent to *Nature* in its raw state, naked of any peer review. A referee must have leaked it, bless his mortal soul. This was the first written description of the cold-fusion process that we or anyone had seen. We devoured it.

The paper seemed to wander all over the subject of cold fusion, covering everything from electrolysis experiments to volcanoes, with a lot of discussion about Jones's new neutron spectrometer. We finally had a documented electrolyte, but it was a mixture of "various metal salts" with no explanation. It looked like an electrolyte mixed up by a deranged ninth-grader, containing nine chemicals, everything from 0.2 grams of lead chloride to "a very small amount" of gold cyanate, and a lot of plain hydrogen contamination from the hydrated salts. It made no sense. I put it aside.

Three pages of figures for the paper came over the fax two days later, and they included Jones's best neutron-count data, plotted on a spectrum graph. The fusion neutrons were just barely there, slightly over the noise. It didn't look as though he was fooling Mother Nature, and there was no sin of technical hyperbole. There was no energy production to crow about. His claim for such a slight effect did not look unreasonable.

Our project meeting on Thursday, March 30, started at 1:45. Rick wanted to talk about piezo-fusion, having by this time read about it in Steve Jones's paper.[104] The concept sounded bogus to me, like trying

---

104 Jones and Paul Palmer had begun their version of cold-fusion research assuming that the anomalous helium-3 found in volcanic vents was due to deuterium traces in water trapped between rocks under extreme geologic pressure. They were considering an experiment in which they could simulate continental plates pressing against one another using a diamond synthesis press. Palmer wrote it up in a paper from March 28, 1986, *Experiments in Cold Fusion*.

to catch two deuterium atoms in a weak moment between an anvil and a sledgehammer.

It still bothered me that the electrolysis cell producing "tremendous heat" in Utah was making barely detectable neutrons, but I latched onto an idea. The fusion cell was made from blown Pyrex glass. Pyrex is a boron glass, and the boron-10 isotope component, which was 20% of natural boron, had a monstrous thermal neutron absorption cross section of 3,840 barns. The fusion neutrons, moderated down to low speed by bouncing around in the heavy water in the fusion cell, would be scavenged like june bugs in a duck pen.[105] The electrochemists may not have considered this. We would make a cell out of pure quartz with no neutron absorbers in it.

By Friday, March 31, the plans were beginning to take form. Darrell ordered a flame-sealed vial of 50 grams of deuterated sulfuric acid from Sigma Chemical. There would be no question about ordinary hydrogen contamination if we used D_2SO_4 for the electrolyte instead of H_2SO_4. Rick ordered two more palladium ingots and 200 milliliters of heavy water. Darrell hiked up to the Boggs Building (chemistry) and persuaded the glassblowers to build us a two-inch-wide beaker out of pure quartz. They were intrigued by the idea, but the hydrogen torch was broken. They couldn't blow quartz with anything else, but they would try something.

That evening, Pons and Fleischmann were shown again on *NBC Nightly News* hovering over the open tops of their hot-running fusion cells, cooled by briskly running water and obviously making a lot of power. I watched in awe. The way they were standing, there wasn't any boron glass or moderator in the way of the high-speed neutron stream. Why were they not dead from the neutrons boiling their blood? Pons should have been screaming and scratching at his empty eye sockets as the vitreous humor from his eyes, split open by the pressure of the vaporizing liquid, rolled down his cheeks. Fleischmann wasn't even tugging at his collar, saying "Is it getting hot in here or is it just me?" Why not?

---

105 No, I did not steal the duck-on-june-bug metaphor from *Gone with the Wind*. I stole it from the same place Margaret Mitchell got it: Augustus Baldwin Longstreet's *Georgia Scenes*. Published in 1835, it is a birthplace of some basic southern writing characteristics.

The team came to work Saturday, April 1. We finalized the experiment design, and for the first time, we discussed the control experiment which must accompany the cold-fusion effort. If, by wild chance, we found that the Pons and Fleischmann setup released neutrons, any neutrons, then we would have to run exactly the same setup a second time, but this time the heavy water would be replaced with tap water. If the neutron production disappeared with this substitution, then it would indicate that the deuterium was indeed fusing. A straight hydrogen-hydrogen fusion was theoretically impossible. We saw no reason to attempt calorimetry, trying to determine the rate of energy production. There were just too many traps to fall into, too much uncertainty to justify the trouble, and no realistic chance of producing measureable power. We did, however, plan to surround the electrolysis cell with a water bath and monitor the temperature of the water. Similarly, there was no need to measure helium. There were too many other ways for helium to show up besides fusion. We could measure tritium production, with a background count first. Tritium was detected by its characteristic low-energy radiation, and as was true of any radiation measurement, we would have to keep track of the background count. The key measurement would be the neutron count. Anything above background would be interesting and probably a confirmation.

On Monday, April 3, our vial of deuterated sulfuric acid arrived from Sigma, just as we received word of the first confirmation of the Pons and Fleischmann palladium-catalyzed cold-fusion effect. We had expected a possible hit any day now from one of the national labs. There were rumors that Brookhaven, a well-established national lab on Long Island, New York, was teaming with Yale and would announce soon. The triumphant wording, however, came from the Lajos Kossuth University (now the University of Debrecen) in Hungary, a peripheral Soviet republic, and it stung that they had apparently beat anyone in the United States to confirm. Their setup consisted of a U-shaped glass tube with a platinum wire in one end acting as the anode and a thin-walled palladium thimble in the other end as the cathode and the neutron source. The cell ran at 20 to 45 volts, consuming about 50 milliamperes of electrical current per square centimeter of area on the palladium. They used table salt as the electrolyte, and they had no idea of the purity of their heavy water. The cell was immersed in cooling water running at 20 to 25 degrees Celsius. They detected neutrons at four to six times background using three US-made helium-3 detectors.[106] They had scavenged the palladium thimble from a dusty, unused piece of equipment in a storage room. It sounded believable, but it was from a Soviet satellite, beyond the reach of the coveted industry research and development funding, and therefore it did not count. The race timer was still ticking.

At noon, Darrell picked up our quartz cell at the Boggs Building, a 2-inch black rubber stopper, a stopper drill, tubing, clamps, glassware, and a coil of platinum wire. Getting a torch as hot as he could, the glassblower managed to fuse a quartz disk neatly onto the bottom of a 4-inch-long piece of 2-inch-diameter quartz tube, and he flared the other end for the stopper. It looked good. That afternoon, I got a call from the secretary of Dr. Ratib A. Karam, head of the Georgia Tech Research Reactor. Would I please come to his office? He wanted to have a chat.

Ratib Karam was born in Lebanon. He came to the United States for a physics degree, elected to stay, and worked at the Argonne National Laboratory in Illinois before coming to Georgia Tech to profess nuclear engineering. He would be the last in the line of White Elephant directors, and he would eventually see it torn down. I had to admire him for stepping forward and taking the job, as the research budget was nonexistent, the aluminum coolant pipes in the epithermal bismuth shield were leaking water, and the Nuclear Regulatory Commission was looking to replace the fuel with lower-enriched uranium with which nobody could build a bomb. His plate was full, but not with anything edible.

He knew a great deal of nuclear physics and technology, but he was famous for being unable to budge from a questionable position. He, for example, knew for a fact that a reactor could not be controlled using

---

106 The use of three helium-3 detectors was a good way to detect neutrons. The detector is a hollow stainless steel cylinder, an inch or less in diameter, built like a Geiger counter tube with a thin stainless steel wire running down the center. A high voltage, about 1,200 volts, is established across the cylinder and the center wire, and any ionizing event in the helium gas that is pressurized within will cause an electrically detectable disturbance in the voltage. In this case, the event occurs when a neutron wanders into the helium-3 and is immediately captured. The disturbed helium-3 nucleus blasts apart into a tritium nucleus and a flying proton. The proton ionizes a track in the helium, causing a momentary short across the wire and the inner wall of the tube.

mechanisms outside the core, even with the diagrams of the SNAP-10A reactor laid out in front of him.[107] I wasn't sure what to prepare for. I knocked on his door frame.

"Jim!" he exclaimed. "Come in. Come in. How's it going at GTRI?"

"Just awful. How are things here?" We shook hands. Back when this place was built, they really knew how to make a director's office. It was beautiful oiled wood, floor to ceiling. The doors covering the bookcases on the opposite wall slid on ball-bearing tracks, and the lights were recessed into the ceiling. He had a full picture window behind his desk, the construction of which had killed a tree somewhere in the Amazon jungle.

"We're surviving. Do you need anything to drink?"

"No, I'm fine." Back when I was a PhD candidate, he would spot me wandering through the reactor building, sipping from an aluminum beer stein. I had been unable to convince him that it was powdered Nestea dissolved in water with a few drops of liquid nitrogen to make it cold and not an alcoholic beverage. I did crave a Dr Pepper out of the drink machine downstairs, but I kept that to myself.

"Hey, I hear that you are doing the cold-fusion experiment!"

I flopped down in a piece of blonde-wood Scandinavian furniture. "Uh. Yeah. Word gets around, doesn't it?"

"Yes, well, I got a call from Savannah River."[108]

---

107 SNAP-10A (Systems Nuclear Auxiliary Power Program) was the first and only nuclear reactor that the United States ever put into orbit. It was controlled by four neutron reflector drums, turned by electric motors to face or look away from the core, located outside the core and the cooling jacket. SNAP-10A suffered an irreversible scram incident, but it is still up there, kicked into a high orbit after the electromechanical failure shut it down. You can make a ball of subcritical plutonium, a minimum nuclear reactor, go supercritical just by walking past it. You are reflecting thermalized neutrons back into the core.

108 The Savannah River Site was built in Aiken, South Carolina, beginning in 1950, as a mirror of the Hanford Site plutonium production complex built during World War II in Washington State. While the Hanford Site was populated with eight water-cooled graphite reactors for plutonium conversion, the Savannah Site had five heavy-water moderated reactors built for the same purpose. By 1989, only one reactor, K Reactor, was still in operation, and it was shut down in 1992. In 1956, the neutrino was discovered by Fred Reines and Clyde Cowan using the P Reactor as a neutrino source. The new Tritium Extraction Facility was not completed until 2005. The tritium for nuclear explosion initiators is now made off-site using Tennessee Valley Authority commercial reactors. Savannah River's stockpile of neptunium-237 was quietly shipped to Idaho beginning in 2005.

"Who at Savannah River?"

"Uh, Ray Sigg from the Analytical Development Division. They are extremely interested in this project."

"Oh?" I brightened.

"You know, they're in the process of shutting down all the production reactors. It looks like we've got enough plutonium to last a few hundred years, so they are shutting down Hanford and Savannah River for good."

"Makes sense."

"Yeah, but the trouble is, they've still got to have tritium for the bomb initiators. It decays with a pretty short half-life. After twelve years, half of it's gone, you know."

"Yep," I nodded.

"They won't be able to make tritium by activating lithium-6 in the production reactors anymore, and they are desperate for a new way to do it. Something simple. Non-messy. This cold fusion, if it works at all, will be perfect for that, even if it doesn't make power."

"Hadn't thought of that."

"Well, here's the thing. They want for you to conduct the experiment here, in this building, using Savannah River heavy water."

"Anything special about Savannah River heavy water? I mean, all the coolant here in the reactor is Savannah River heavy water, isn't it?"

"Believe me, it's an important piece of politics to use this water. It will put the SR plant on the map if you can prove cold fusion using water that they purified. Yes, and I know what you're thinking, but we can pre-count the tritium before you start, and we are the only place around here that can count any tritium."

He was right about that. The counting room on the back hallway had a special, very expensive scintillation counter system made specifically for finding tritium. You could not detect tritium with a common Geiger counter, because of the extremely low energy of the beta ray. It was too weak to make it through the wall on a gas-tight detector tube, no matter how thin you made it. The sample had to be dissolved in the scintillation fluid and put in front of a bank of photomultiplier tubes. The feeble glow generated in the fluid by the decaying tritium was detected, amplified, and used as a measure of tritium content in the sample.

"We can set you up with multiple neutron counters," he continued. "We've got lead bricks, we've got neutron shielding. We can make any parts you need in the machine shop. You need us, and we need the publicity, and so does Savannah River."

"Well, I . . ."

"Let's go look at the lab. You can have RM127. It's unoccupied. You can have it for your experiment."

What could I say? We got up and walked to the largest laboratory space in the building, on the main hall and across from the classroom. The lab looked spotless. The gray linoleum floors were even polished. There was a heavy wooden table on the back wall that was perfect for setting up the electrolysis cell, and the tritium collector could be set in the nearby vent hood. Karam was a skilled persuader, and I was hooked.

"This is perfect! I'll need keys to the building, dosimeters, and radiation work permits for the team. Four people, aside from me."

"That will be arranged. Just send them down here. Oh, and we'll have to take urine samples to get a baseline for tritium. Check with HP."

The only radioactive contamination to be measured would be tritium, and the only way to tell if someone was contaminated was to take a sample before and after the contaminating event. Tritium was in trace amounts in the Earth's water supply, and water was constantly being taken into the human body and being lost. The easiest way to measure the tritium load in a person was to count the tritium atoms, one at a time, in the urine. Everybody has just a little bit of naturally occurring tritium moving through their body at a given time. I already had the key and the dosimeter, but I would have to pee in a cup with the rest of the team.

I turned to stride out the door and Karam rushed to his office to make a phone call. I took two steps and almost collided with Dean McDowell, the senior reactor operator.

"Mahaffey!" he enthused. "I hear that you're going to do the cold-fusion experiment here!"

"Word gets around . . . fast, doesn't it?"

McDowell had gotten his initial tritium load aboard the USS *Nautilus (NS-571)*, the world's first nuclear submarine. He had served in the first crew under Vice Admiral Eugene "always enter a harbor at flank speed" Wilkinson, starting in 1955. He was from Texas, and if I ever needed to operate an oil derrick, I would call Dean.

"Heh. Yes, it does. Look, I was thinking. Why don't we get some catalytic converters at a junkyard and use the beads to soak up the deuterium? The beads are covered with palladium, you know."

"Uh, no. It's an interesting idea, but you're jumping ahead. We have to find that the effect exists before we start improving it."

"Oh. Okay. Well, what do you need for your setup?"

"Two or three neutron counters with scalers. What have you got?"

"How about those old Hamner scalers?"

Old? They were, in fact, the Hamner N-221 units previously owned by Lockheed Georgia at the Georgia Nuclear Aircraft Laboratory. An N-221 was a high-voltage power supply, timer, pulse amplifier, digital counter, and pulse-height discriminator built into one box. Vacuum tubes, neon lights, and a mechanical timer, but extremely stable and reliable. The N-221 could survive a lightning hit. I had used them when I was a grad student hand-loading the "sub-critical assembly" in the masters-level course, NE611, Nuclear Engineering Laboratory.

"Good. BF3s to go with them, plus high-voltage cables, a DC power supply with long clip leads, a thermocouple and readout, a rack for it all, a Ludlum survey meter, a portable BF3, about fifty lead bricks, paraffin slabs, a pallet of boron blocks, an HP frequency counter, and a milliammeter."[109]

"Is that all?"

"For now. Oh, wait. I need a block of nuclear-grade graphite with a 2-inch hole bored through it."

---

109 The BF3 is another species of neutron detector tube. A reliable workhorse for detecting neutrons, these devices have been popular since 1942, when they were used to monitor the uranium fissions in the world's first nuclear reactor at the University of Chicago. A BF3 is configured similarly to other ionizing-event detectors. It is a seamless, gas-tight brass tube with a thin wire running down the center. It is filled with isotope-purified boron-10 trifluoride gas, and the boron is extremely likely to capture a slow neutron that happens to wander in through the brass walls. Upon neutron capture, the boron nucleus immediately destructs into an alpha particle and a lithium-7 nucleus. The 2.78 MeV event causes a momentary ionization in the tube that is both energetic and unique, making it possible to discriminate it from other events, such as gamma rays, cosmic rays, or electrical noise. The resulting pulse caused indirectly by a neutron is amplified and evaluated for its pulse height. If it looks like a neutron event, then a digital counter is bumped forward by one.

"Hmmm. Well, we've got the graphite reflector blocks out of the old 104.[110] What's it for?"

"Well, we've had this cell made, and it's very thin quartz. I'd like a graphite support around it, acting as a neutron moderator."

"Gotcha. We can do that. Bring the cell down to the shop, and we'll hand-fit it to the hole in the graphite."

Having thoroughly loaded McDowell with a set of tasks, I trudged back up the hill on Atlantic Drive to my office, thinking, "This is going to be a long week."

On Tuesday morning, April 4, I started working on the setup in RM127, but I had a nagging feeling that the cold-fusion concept was going to turn out to be another dead end or worse, a hoax. The problem was, nobody in the western hemisphere was jumping up and claiming a confirmation. Maybe the palladium-loading was very slow, and it took weeks to accomplish a fusion-causing density? There were vague hints from Utah that seemed to indicate a loading delay, and even veiled indications that in a large array of fusion cells, only a few would start generating power. Pons and Fleischmann were now the most famous scientists in the world, and the latest news film showed them holding a larger cold-fusion reactor, about a foot tall, built to take the place of a home water heater. They had made the leap to commercialization.

Peter L. Hagelstein, the one who had designed the X-ray laser for Edward Teller's late Strategic Defense Initiative fiasco, was a leader in the vast rumor mill that was now substituting for scientific literature in this brightly burning topic, and he posed an interesting problem. It is true that the density of deuterium in palladium deuteride is impressive, just as Fleischmann had pointed out, and making deuterons bunch very close together is what causes fusion; but Fleischmann's claim

110 The "old 104" was an AGN model 201 nuclear reactor, serial number 104, made by Aerojet General Nucleonics in San Ramon, California, in the late 1950s. It was a small research reactor, designed to be used, believe it or not, to teach nuclear technology topics in high schools. The expected demand did not materialize, and only a few were built. This one had been operated in the high-bay of the biology building for several years, but it had been decommissioned and was now sitting on the tarmac in back of the Nuclear Research Center. Its cylindrical core, made of 20 percent enriched uranium mixed into polyethylene disks, had been removed and was in the fuel-storage vault, separated into four subcritical pieces.

was for *average* density. In reality, only one deuterium atom could fit in the center of a cubic lattice element in palladium metal. With every cube filled in a block of palladium, the volume of deuterium gas was certainly compressed, but the deuterons could not touch one another. Access was blocked by palladium atoms forming the cubic matrix. The *local* density was actually quite small, in terms of causing fusion. In fact, the two deuterons attached to an oxygen atom in heavy water are a lot closer together than any two deuterons in palladium deuteride. Why does heavy water not burst into fusion and boil away? Hagelstein went on to speculate that two deuterons could be occupying the same space in their interstitial location in palladium. While not utterly impossible on a quantum mechanical level, it begged the question of what was forcing these unlikely pairings to occur.[111]

With such questions churning around in my mind, I built a castle of lead bricks on the wooden table. When dealing with radioactivity, even if there is a tiny threat of radioactivity, one cannot be too cautious, and it is not unreasonable to anticipate even impossible circumstances. If there were a runaway reaction or "neutron burst" in the small quartz container, such as Pons and Fleischmann suggested, the lead walls surrounding the cooling bath on five sides would dampen any explosion, send it straight up onto the ceiling, and keep shrapnel out of my eyes. It would also shield us from any ionizing radiation.

Neutrons were another matter, and to protect against a neutron blast, I lined the castle with "boron blocks," which are cardboard boxes, slightly smaller than bricks, filled with a mixture of 20 Mule Team Borax and paraffin. No neutrons could make it through that wall, but they could go straight up and start to diffuse in the air. In a neutron runaway, the safest place would be under the table, but a hard sprint for the door would work. The event to look out for is when all the detectors, clicking merrily from the harmless background radiation, suddenly stop working, jammed to overload by a deadly excess.

Inside the boron blocks was a small Styrofoam cooler, designed to keep one beer can on ice, and it was lined with paraffin blocks,

---

111 Paneth and Peters had correctly anticipated this problem back in 1925, and their solution had been to heat the palladium block, expanding the sizes of its cubic lattice locations so that more than one hydrogen atom could fit in one location. To squeeze them together and cause fusion, they then cooled the palladium block. It wasn't a bad idea. We should have thought of it.

meant to be a solid-state neutron moderator, as if we needed one. The graphite block, machined for the quartz beaker, sat inside the cooler, and it was surrounded by water acting as a coolant and a liquid neutron moderator. Any mega-electron-volt neutrons would be slowed down to detection speed by the graphite, water, and paraffin.

By Thursday, April 6, we had yet to absorb any more configuration tidbits from Pons and Fleischmann, but Rick had been thinking about palladium hydride, and he had an idea. Just sitting around in the atmosphere, a block of palladium had probably loaded up with non-fusion hydrogen long ago, finding an occasional nascent hydrogen atom wandering around in the air. It would take time to flush out all the hydrogen atoms and back-fill the matrix with deuterium atoms, and that could explain a fusion starting delay being predicted in the alt.fusion newsgroup. Why not heat the block to incandescence in a vacuum to clear out the matrix? Billy Livesay, our hydride expert, thought it an excellent idea. By 4:30 that afternoon, Rick and Billy had sawed off a section of the palladium ingot, induction-heated it in a vacuum, swaged a platinum wire into the top, and immersed it into a beaker of heavy water for safekeeping.

On Friday morning, April 7, we were close to T minus zero, but a bombshell dropped. Someone had gotten a copy of Pons and Fleischmann's paper, the one they sent to *Nature* that we would not see until sometime in May, and leaked it to the news group. Finally, we and the rest of the scientific world knew their electrolyte. It wasn't sulfuric acid. It wasn't acidic at all. It was lithium deuteroxide, a base. We could not have predicted it, but Pons and Fleischmann were the electrochemists, and they knew more about the electrolyte than anyone in the world. Within 60 seconds of reading it, Darrell was on the phone to Sigma, ordering three bottles of lithium deuteride with overnight weekend shipping. There was, of course, a massive run on lithium deuteride occurring, and soon the phone line to Sigma Chemicals would jam. Normally, there is not much of a market in lithium deuteride.[112]

---

112 Lithium deuteride is the active ingredient in a thermonuclear warhead. It comes in a very small, 3-gram bottle. Mix it with heavy water and you have lithium deuteroxide. We used 0.1 mole lithium deuteroxide, or 0.324 grams of lithium deuteride dissolved in 360 milliliters of deuterium oxide (heavy water). I have an unopened bottle of lithium deuteride on my bookshelf, just to remind myself never to do this again.

Aside from the electrolyte, there was not much in the cold-fusion paper that we hadn't anticipated. Their descriptions of their nuclear measurements did not give me a good feeling about the validity of their findings. This paper was sure to be sent back to them for clarifications before publishing, but it was the most complete description of the cold-fusion experiment that existed.

Why a lithium compound? Interviewed, Fleischmann gave a sly smirk and alluded to a "nuclear reason" for using a lithium-based electrolyte. Was he implying a deuterium-lithium-6 fusion reaction? The lithium-6 reaction had an unusually favorable cross section, but what did it have to do with palladium? The rumor mill went wild with speculation. Although it was a military secret, if you bought lithium deuteride in the United States, it was pure lithium-7 deuteride, with the 7.6% lithium-6 removed by isotope separation at Savannah River for use in H-bombs. Did Pons and Fleischmann buy their lithium deuteride from a supplier in Canada?

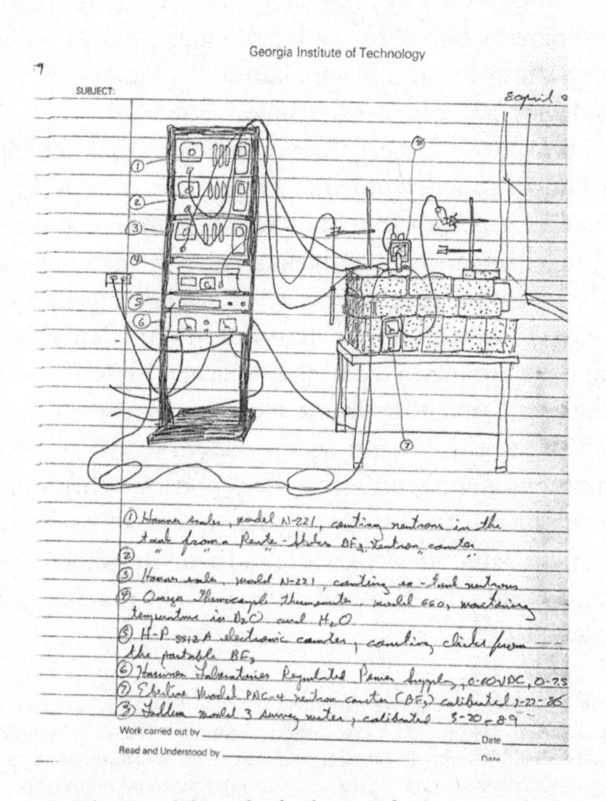

A page from my lab notebook, showing the experiment setup.

We were in hyperdrive until that afternoon, when the day's second bombshell dropped. Pons and Fleischmann had been interviewed at a scientific conference, at which time they announced that they had finally had time to run their control experiment. They had set up another fusion cell, identical to one that was producing heat and neutrons, but this one had tap water substituted for the heavy water. From the control experiment, they were excited to announce that they were getting the same heat and neutron action. Their cold-fusion apparatus worked just as well with ordinary water as it did with heavy water.

Upon learning this, Darrell and I sat quietly for a long time in his office, staring at the wall. I was stunned. Finally, I spoke. "Well, they're crazy."

Darrell looked up and sighed. "Yep. They are certifiably insane. No Nobel Prize for these wing nuts. All this work for nothing. We'll have to hang it up."

I looked back at Darrell. "We can't. We've already spent too much money. We've got to see it through, write the report, and return to life as we know it. They've just saved me the trouble of taking this experiment too seriously."

"How are we going to break this to Rick? He's really excited about cold fusion, you know."

"We? You do it. Don't hit him too hard. Use a soft hammer."

Early Friday, April 8, Darrell and Rick stood in front of the Electronics Research Building in the morning frost at 5:30, waiting for the FedEx truck to show up with our lithium deuteride. I was in RM127 in the reactor building, trying to coax our neutron detectors to work. We had four detector tubes, designated A through D. A, B, and C were Reuter-Stokes RS-P1-0809-101 BF3s, and D, which was our least antique and most sensitive detector, was a top-of-the-line Nancy Wood G-10-5 BF3.[113] It was a beautifully polished brass tube, an inch

113 Nancy Wood was a graduate student at the University of Chicago the day that the first nuclear reactor achieved criticality there, and she had participated in designing and building the boron trifluoride detectors used to measure neutrons in the groundbreaking experiment. She began manufacturing neutron detectors for the Manhattan Project, and in 1949 she started the N. Wood Counter Laboratory on East 53rd Street in Chicago, building special detectors for nuclear research around the world. In 1989, her daughter, Marjory Wood Crawford, was general manager. The label on our detector said it was a G-10-2, but it was an error. I am positive it was a G-10-5.

in diameter, filled with boron trifluoride gas under pressure. It was soldered shut at the detecting end, and the center electrode came out the other end, set in a blown-glass insulating seal, capable of isolating several thousand volts of electricity. The sensitive volume was 5 inches long at the sealed end of the tube, and at the other end was the SHV high-voltage socket, soldered to the center electrode and supported by another 1-inch brass tube, fastened to the detector with three flush-mounted screws. It was basically the same tube they had built since World War II.

As is often the case, the high-voltage cables, carrying up to 2,000 volts, were giving trouble. The problems seemed to be at the interface between the SHV connectors and the cable.[114] Give it a little twist, and a rash of rogue pulses would send the pulse-counting electronics into a frenzy. I swapped out and tried various cables from a pile of candidates, finally settling on the least bad. It was best not to disturb any cable runs once things settled down. I started taking 1-hour background counts. The most stable detector was the Nancy Wood, tube D, giving counts from 98 to 112 neutrons per hour. The counts were reasonable for neutron background and were most likely due to cosmic rays.

Three tubes, B, C, and D, were set up with the Hamner scalers, and tube A was placed outside the lead castle and connected to a Hewlett-Packard 5512A counter to constantly give a background level. I put the Nancy Wood tube in the best position, sitting with its nose in the cooling water next to the graphite block, and adjusted the operating voltage for the middle of its operating plateau, at exactly 1,175 volts. The other two BF3s were also in the water, near the paraffin. One temperature probe was in the heavy water, inserted through the top stopper in the quartz cell, and one was in the cooling water outside the graphite block. An Omega model 660 readout instrument continuously tracked the temperatures. With neutrons wandering off in every direction in the water, one detector tube would be lucky to register one out of a thousand fusion neutrons coming from the palladium, as if there were going to be any.

To run a one-hour neutron count, one would set the timer by spinning the big, black knob in the center of the clock mechanism and

---

114 SHV means "safe high voltage."

setting it to 60 minutes. Press a small lever at the bottom right corner of the timer and the neutrons start registering on the three-decade indicator as the timer runs. Each decade is a vertical column of ten neon lights, counting from zero to nine from the bottom up. After an hour, the timer shuts off with a loud snap, and the counter remains frozen until it is reset.

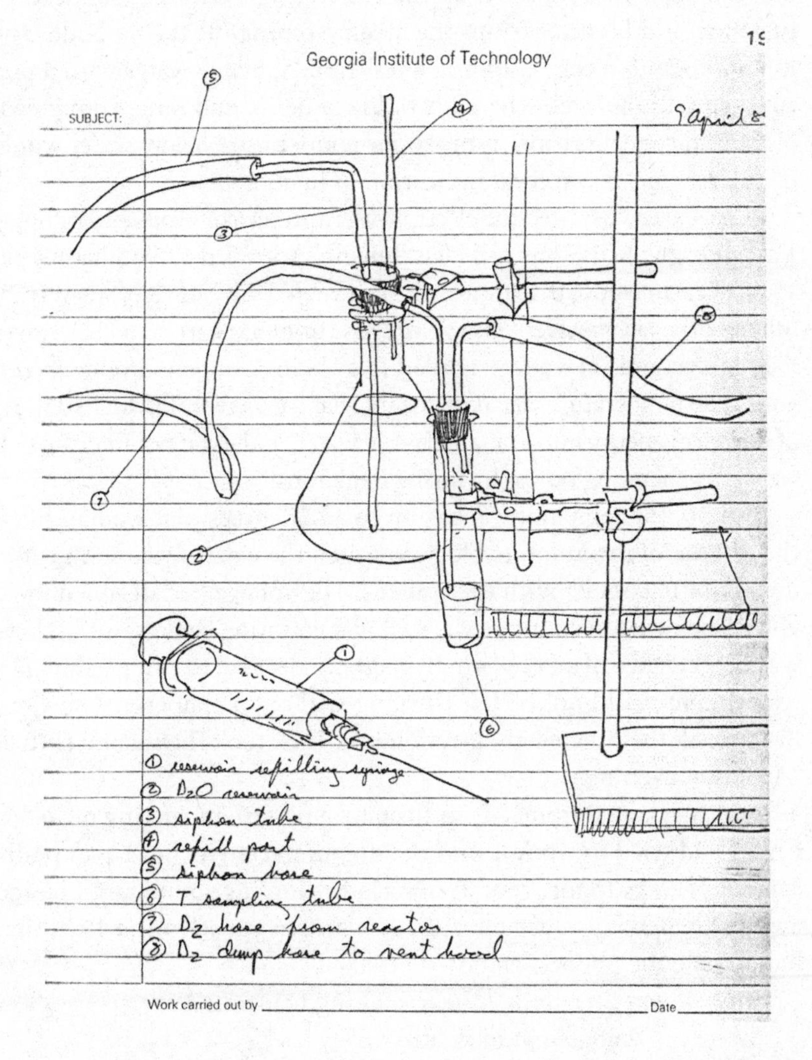

A page from my lab notebook, showing the tritium sampling tube and the deuterium replenishing flask.

While I was sorting out the instruments, the team was at work on the bench. Darrell mixed the lithium deuteroxide, and Beebe built the stopper assembly, drilling holes and inserting various tubes and wires. Rick set up the tritium collector under the vent-hood, using a tube leading back to the cathode to scrounge excess hydrogen isotopes boiling off, and a deuterium level maintenance system, as we expected deuterium to be consumed by the electrolysis. Tritium produced by fusion would be mixed into the steam coming off the cathode, and it was piped into the bottom of a test tube of heavy water with a pre-counted tritium level, where it would condense and mix. An increase of tritium concentration in the room-temperature heavy water would be interpreted as a second indication of fusion.

At 4:24 P.M., we started the electrolysis, running at 48 volts and drawing 1 ampere, giving us Pons and Fleischmann's specified 125 milliamps per square centimeter of palladium. The cell vented oxygen vigorously. In 10 minutes, the heavy-water temperature rose from 83 degrees to 112 degrees Fahrenheit, and the voltage needed to sustain 1 ampere dropped to 36 volts. The temperature rise only meant that we were unloading 40 watts of electrical energy into a small glass of water, and it did not indicate that we were generating power by fusing deuterons.

By 4:50, the temperature was up to 132 degrees, but we had lifted the cell out of the graphite block to watch the electrolysis action, and it was not in contact with the coolant. The voltage had dropped to 27 volts. The neutron counters were all just counting background. There was no evidence of any fusion. By 6:00, we decided to call it a day. The experiment could not be left running without a chaperone, so Gary Beebe took the first watch. I went home. Rick took the second shift at 11:00 that evening.

I was awakened from a deep sleep by a persistent ringing noise. At first, I had tried to work it into the dream narrative, but I eventually ran out of ways to incorporate a periodic jingling sound and I bolted awake. Somebody must have died! Carolyn turned on a light and handed me the phone. "Hello?" I croaked.

"Jim!" It was Steenblik. "We're making neutrons!"

"We're . . . doing what to neutrons?"

"It's cold fusion! The neutron flux is rising. The count rate is up to 60 times background!"

"What? No, Rick, that can't be happening. Something's out of adjustment. The high voltage must have drifted . . ."

"No! It's really working. You've got to come down here and see it. Pons and Fleischmann were right. It's working, and it lit up after a few hours of electrolysis. We've beaten everybody to the confirmation because we cleared out the palladium matrix."

"No, Rick, we can't be . . . What time is it?"

"It's one o'clock in the morning. You've got to come down here right now."

"Something's not right. Are you sure you haven't jiggled the cables? I'll check it out when I come for my shift. G'bye." I handed the phone back to Carolyn, rolled over, and passed out.

I walked into RM127 at 7:00 in the morning, Sunday, April 9, wide awake and ready to find out what the hell was going on. Rick was floating on a cloud of ecstasy for which I could not really find fault. He had, after all, found his Grail, or so it seemed. I found him at the bench, drawing graphs of the neutron count increase. The takeoff had started immediately on Beebe's shift, making a steep climb from 63 to 957 counts per hour on tube D. Tube B went from 25 to 172 counts, and tube A increased from 10 to 90 counts per hour in Beebe's 8-hour shift.

The counts from the three internal detectors did not agree very well, but all counters are not created equal. The best tube, the Nancy Wood, gave the most dramatic results, and the neutron count did seem to have taken off sometime last night and continued at a much higher rate. I glanced at his charts as Rick babbled on about the neutron production, but I was more interested in examining the counting equipment. I looked at the volt meters. Each Hamner N-221 had a big volt meter on the right side showing the level of high voltage given to the tube at the end of the cable. Everything looked correct, and the voltages were still spot-on with the values that I had so carefully measured and dialed in the day before. An increase in voltage could have moved out of the operating plateau and stumbled into the running-wild region in which a tube starts counting nonevents. The decade columns on each Hamner would jump at unpredictable intervals. Sometimes a counter would make a two- or three-count leap, and sometimes it would be quiet for an extended period. It all looked normal, in the utterly random world of nuclear events.

What else could it be? There could be neutrons streaming out of the reactor adjacent to the building and skewing the data, but it had been dormant for the past week. Nobody had come in and started it up last night. Was there an ongoing cosmic-ray storm? Not impossible, but the two gamma-ray monitors, scintillator and Geiger, did not show any unusual background. My best guess was that the cable ends were acting up, but no matter how I shook the tubes, everything seemed rock solid. At any rate, it would be easy enough to prove that there was no cold fusion happening.

"Here's what we're going to do, Rick", I pronounced. "I will remove the inside detectors and position them outside the shield. I'll make a 1-hour count, then put them back in the water and make another one-hour count. I'll cycle this over and over throughout my shift. Darrell will take the next shift, and he will continue my cycling, and so on for the rest of the experiment. Outside the boron blocks, down low, there's no chance of any fusion neutrons being counted, and all we should get is background. So, if the count rate doesn't change when we shift the positions of the counters, no matter what is causing the counts, it will prove that cold fusion has nothing to do with it."

Rick, still sparkling, agreed with my strategy, encouraged me to keep adding to his charts, and left for a well-deserved sleep.

It was dead quiet in the lab, and it was a good place to think, as I tried to come to grips with the conflicting information stream. At 8:15 A.M. the counters snapped to a stop. On tube D, the count rate had dropped from over 600 down to 102 counts per hour, which was background. In my notebook I wrote, "Putting the 2 counter tubes outside the boron did apparently lower the neutron count. We'll see! No gammas on the scintillation counter."

I put the tubes back in the water and started again. After one hour, I read off the counts and entered them in the log. Tube D now recorded 625 neutrons. I was not convinced. I configured another outside count, this time immersing the counter tubes in beakers of water, thinking that water vapor in the SHV connectors had to be causing spurious counts. In this condition, the count rate on tube D fell to 69 counts per hour. There was no water-vapor problem. Tube C registered the same phenomenon, but with a smaller contrast. Tube A seemed only slightly aware of the changed venue, but it was the only tube that I had not pretested using a californium neutron source in the scramble on Saturday morning.

I put the tubes back in the water and ran it again. The neutron counts jumped back up to 533 counts per hour on tube D, the Nancy Wood. I was convinced. I was seeing a nuclear phenomenon occurring in a glass of water with 27 volts across it. There was no chemical or any other process at this temperature that would cause deuterium to give up a neutron out of its nucleus. I was looking at a nuclear process under conditions that cannot produce a nuclear process. It was a slight reaction, and not one that could produce measureable heat, but producing neutrons, producing just *one* neutron, in that glass of water, was miraculous.

We had taken and stored tritium collection samples at the end of each shift, and I would have them counted Monday morning. An increase in tritium, another signature of deuterium-deuterium fusion, would tie a bow on it. I called Dr. Spauschus, our sponsor, at home to tell him what we had found. By 12:30, he was standing in the room seeing the miracle, and he was properly enchanted. We rounded up all the nuclear physics experts we could find on campus, brought them to RM127, and carefully explained our findings of neutron production, asking for another explanation of the neutron production in that small vessel filled with heavy water. They came, pondered, looked quizzical, twisted a few cables, studied the graphs, and generally failed to find fault. It was far from the definitive test of the experiment's validity, but in the unusually compressed time line of cold fusion, it made us feel confident enough to proceed.

At 3:00 P.M., Darrell took the shift, and I went home to fetch a bottle of champagne, Freixenet's "black bottle bubbly," from the bottom bin in my refrigerator. At 4:15, the team was gathered in the classroom across the hall from RM127. Finding that we were missing champagne flutes, we, of course, substituted with urine-sample cups and drank to Pons and Fleischmann. Darrell produced a box of cigars, and a few of us blew smoke out an opened window while we talked strategy.

Normally, we would just coauthor a paper and send it off to a journal of our choosing, but these were not normal circumstances. Pons and Fleischmann had set the bar as low as it would go for this line of research, and it was in mob-rule, last-lab-standing mode. We eventually agreed, more or less, that the proper interface was a press announcement.

Georgia Tech made press announcements all the time, trying to rouse interest in what we were doing here, and most were roundly ignored. This was not a normal announcement. It could not sit on someone's desk for a week. It had to be released on an emergency basis, and the only way get it done was for me and Billy to take it directly to the top, to Dr. Thomas E. Stelson, vice president for research of the Georgia Institute of Technology and in the running to become the next president of the institute. We would do it in the morning, after I had the tritium counting results and before I had time to adequately ponder possible outcomes. Billy had the graveyard shift that night, until 7:00 in the morning, and he would be gelatinous. I would do the talking.

Monday morning, April 10, I got the tritium counter results. The tritium concentration in the samples had doubled every 8 hours, and that backed up our neutron results. When deuterium fused with deuterium, half of the reactions would result in neutrons and half would result in tritium.[115] Rick was taking two shifts and would be in the lab for the next 16 hours, alternating the detector positions on an hourly basis. I showed him the tritium results, which pleased him immensely, and rounded up Billy. We set out on foot for the other end of the campus, to the Wardlaw Center for Communications and Public Access. Stelson's secretary had warned him that we were looking for him.

Stelson was in the middle of a huge fund-raiser or a coronation of some sort, and he was dressed for the opera. He had taken his nosebleed position at Georgia Tech in 1974, coming to us from Carnegie Mellon University, where he was head of the civil engineering department. We had a bit of history, but I hoped that he had forgotten it by now. He and Dr. Demetrius Paris, director of the School of Electrical Engineering, interrupted what they were doing and came to meet us, waiting in an adjacent room. I started hammering them with results of our cold-fusion experiment while spreading our hand-drawn charts and cryptic tritium-counter readouts on the closest surface, the top of an

115 The products were evenly distributed between tritium plus ordinary hydrogen and helium-3 plus a neutron. The tritium and the neutron were both detectable in small quantities and were clear signatures, while hydrogen and helium-3 were harder to pin down as having come from fusion.

exquisitely polished grand piano. Stelson squinted at the papers, nodded, rubbed his forehead, grabbed a telephone, and made a quick call.

I could not make out what he was saying, but Stelson was off the phone in what seemed a few seconds. "Go see John Toon," he commanded. "He's in the Centennial Building. Second floor." I glanced at Billy. We were launched.

Young, tall, and serious, John D. Toon, director of GTRI Communications, did his part, hustling to write up our experimental results in words that would interface cleanly with the news-hungry public and distributing it with electronic speed. With that we could relax and twiddle our cold-fusion setup at our leisure until the funding ran out.

Or so we believed. In reality, the entire world was wound up like a spring, waiting for a solid confirmation of the cold-fusion effect. Our brief press release hit the journalistic community at precisely the optimum time for maximum impact, somewhat like unloading a squadron of Mitsubishi dive-bombers on Pearl Harbor. Within an hour, the telephone switching circuitry at GTRI overloaded, the Reactor Building was covered up with journalists, and GTRI found it necessary to convene a sudden, extemporaneous press conference in the classroom across the hall from our gently fizzing fusion reactor.

It was 4:00 P.M. I sat at a hastily configured desk next to Dr. Donald J. Grace, director of GTRI from Stanford University, behind a microphone hive and never actually having spoken in public before. The room was tightly packed with newspeople, camera operators, people wearing headphones, and journalists from afar. An army of lenses was aimed right at my face, and I had no prepared statement. Grace started off, announcing that we had tried to replicate the Pons and Fleischmann cold-fusion experiment originally performed in Utah, and we had apparently confirmed the newly discovered effect of dissolving deuterium in metallic palladium. The implications were large. He then turned to me, saying that I would describe what we had done in more detail.

I started rolling off details, being a lot more specific than necessary, sort of like this chapter. Finally, we opened for questions. I, being in shock at the time, can only remember the last question. Someone in the back asked, "And what did this experiment cost?"

"About twenty K," I answered.

Silence. "That's twenty thousand dollars, in engineer-speak," added Dr. Grace.

The next morning, Wednesday, April 11, my phone was ringing when I opened my office door. It was a guy at the Brookhaven National Laboratory. He wanted to know what we had done differently that had given us results so quickly.

"I thought that you guys were about ready to announce," I said.

"Announce? Hell, we don't have anything to announce. Our fusion cell is dead as Teller's mother," he replied.

That dead, I thought. Just then, I noticed about twenty missed-call notes stuck all over my chair, started reading them, and felt a twinge of dread. Every nuclear lab, from Los Alamos to Harwell, wanted to know how we were getting neutrons. We were good, but we weren't *that* good.

That evening, we were on the televised national news, and the new CNN 24-hour news network, headquartered across 10th Street from the reactor building, spread that news conference across the globe. On Tuesday morning, April 12, I started answering an endless string of phone calls. Magazines, such as *Newsweek* and *Paris Match*, called wanting interviews.

Our first offers of funding came early that morning, from Babcock & Wilcox Research and Development Division and Combustion Engineering, wishing to get us under contract as quickly as possible. Oak Ridge National Laboratory called, wanting to send a team down to see our setup. All manner of conspiracy theorists and pseudoscience practitioners contacted me, one claiming to have evidence that the German navy had perfected cold fusion during World War II for submarine propulsion. Pons and Fleischmann were elated at the news.[116] Scientific groupies, who I previously did not know existed,

116 The same day, almost at the same minute, Texas A&M University in College Station, Texas, announced that they confirmed that their Pons and Fleischmann apparatus produced net energy using calorimetry, but they had not tried to detect neutrons. Their experiment looked a bit strange on the news, as electrical arcs and flashes were shown coming from the setup. Shortly after having made this announcement, TAMU went quiet, and it was hard to get confirmation of their confirmation. At Georgia Tech, we knew Texas A&M by its alternate name, the Sam Houston Institute of Technology—or its acronym.

camped out on the lawn behind the reactor, just to be close to what was going on, and undergraduate students showed up at the door, volunteering to take a shift in RM127. By lunch, the division secretary took the phone off the hook and went home to calm her nerves. *Business Week* and *U.S. News & World Report* sent photographers to take pictures of Billy and me standing at the experiment table. My parents sent flowers.

The next day, April 12, I had a commitment, made weeks prior, to give a talk at the American Nuclear Society meeting in Birmingham, Alabama. I was too exhausted to drive to Birmingham, so my wife, Carolyn, drove as I cranked back the passenger seat and tried to rest my eyes. At the meeting, the only thing the nuclear engineers wanted to hear about was the cold-fusion experiment, and my planned remarks on computer-driven emergency response systems went unspoken.

We got back late, after 11:00 P.M., but I wanted to stop in at the reactor and see how the experiment was going. As I entered RM127, I knew immediately that something was not right. Gary was on the phone and looking grim, Rick was worrying over graphs, and Darrell was holding a BF3 tube in his fist. "Uh, what's wrong?" I inquired.

Darrell looked over at Rick. "Tell him, Rick."

Rick put down his pencil and turned to me. "Well," he began, "we got an urgent phone call from Lawrence Livermore."

"Okay," I said. "And . . ."

"And they have a cold-fusion setup that's practically identical to ours. They've got the same electrolysis cell, the same instrumentation, the same cooling water that's running over a hundred degrees, and they put their counter tubes right in the water like we do. They are even using the exact same Nancy Wood BF3 that we're using. Turns out, they are getting the same neutron increase that we've got, and they even relocated the tube outside the coolant just to make sure."

"But?"

"They were getting really excited, and then one of them happened to be holding the Nancy Wood in his fist as they were getting ready to put it back into the coolant. For some reason, they had the counter clock going, and he could see the count rate climb by a factor of ten."

"Oh. No." At that moment, the gravity of what was found at Livermore washed over me, and I could see my career sliding away gracefully into the dark abyss.

"Yes. That's right. The BF3 tube is temperature sensitive. He told us to put ours in a beaker of warm water and watch it count neutrons, just like we started to get when the coolant warmed up."

I had to sit down and loosen my tie. "We're screwn," I declared. But why were the counter tubes so sensitive to temperature, and why had I never noticed this before? As it turned out, the problem was with the glass-to-brass insulator at the back of the tube, where it connects the center wire to the SHV connector. When the brass gets warm, it expands. The glass does not, and the diameter of the hole through which the glass insulator connects grows in size, very slightly. It's not enough of a difference to let any pressurized boron trifluoride gas escape, but the crack is just big enough to let water molecules diffuse into the tube, probably one molecule at a time. There is a strong ionizing chemical reaction between the water molecule and the boron trifluoride. It looks, to the electronics, exactly as if a neutron was just captured, and the counter trips by plus one. This was never noticed before because it is such a slight effect, but if you are looking for very slight neutron activity, which is seldom the case in work with fission reactors, then this small bit of noise becomes very significant.

"We've run the counters over and over," Rick continued. "I've subtracted the temperature effect from the data. I've reworked the data every which way. The neutron counts from every tube cancel down to background. Fusion is making no neutrons."

"But, Mr. Jim," Darrell offered, "what about the tritium counts? Doesn't that mean something?"

I shrugged. "Not without neutrons, it doesn't. The tritium signature was always weak. Karam talked me into using that damned Savannah River water, which is heavily contaminated with tritium."

"But, we subtracted that as background."

"Yeah, but don't you see? We were loading the test batch with whatever was evaporating off the cathode, which was preferentially soaking up deuterium over tritium. The sample tube couldn't help but get an enhanced tritium concentration. I should have just assured him we were using the SR water and used the Sigma product instead. That was my fault."

Darrell looked discouraged, Gary looked beaten, but in his eyes, Rick looked as if the world's energy salvation had died in his arms. He was staring at the wall. "What are we going to do?" he asked.

"We've got to un-announce it. We've got to have another press release. A negative one of the same intensity as our positive one, so that the two press releases will cancel each other out."

"Absolutely," said Darrell.

It was a logical fix, but it seems that press releases do not operate with any mathematical precision. The next morning, Thursday, April 13, I visited John Toon and spilled everything. He agreed we should issue another immediate press release, but instead of admitting that our confirmation announcement was utter crap, we should more gently mention that inconsistencies in our data have been found and that further study would be required before we can definitely confirm the Pons and Fleischmann cold-fusion effect. I left it in his hands and went back to my office to break the news to Babcock & Wilcox, Combustion Engineering, and hopefully the boys from Oak Ridge before they took off for Atlanta.

Incredibly, no one found it easy to believe me when I told them that we had been mistaken, the neutron detectors had fooled us, and there was no reason to draw up a contract or come to Georgia Tech to see the experiment. The Oak Ridge delegation showed up anyway. I gave them the tour, explaining what had happened, and discounting all our data. In their opinion, I was wrong the second time, and we were simply misinterpreting the findings. I wasn't sure how to counter that without causing physical harm.[117]

It was Friday, April 14. If the first press release went over like a bombing raid, this one was a 6.9 earthquake followed by an inundating

117 The Oak Ridge researchers had improved the Pons and Fleischmann experiment by using a very large palladium cathode. It was spherical, hollow, and the size of a basketball. It exploded and destroyed the lab one night, right after the researcher tending it had left the room, and this they interpreted as confirmation that cold fusion was extremely energetic. It was more likely the result of deuterium compressing into the air space at the center of the sphere by hydriding action and suddenly igniting, exploding the sphere like a grenade.

tsunami wave. The division phone lines again jammed, this time by calls from people who were certain that our research team had either been bought off or threatened with death by the government, the energy cartel, or General Motors, and we were thus artificially suppressing our confirmation. Another press conference had to be arranged, but this would be the big one. We could not hold it in the reactor classroom. This one would be in the ballroom in the Centennial Research Building.

I met Don Grace, apparently on a prescribed tranquilizer, in the lobby, and together we walked the gauntlet into the Valley of Death. The hallway was lined with spectators. Most looked fearful, foreseeing death by journalism and an end to research as we knew it. Some wore an unmistakable smirk. Dr. Paris was there, wearing a look of irrational, eye-bugging terror. He jumped out, latched both claws around my left arm, and hissed directly into my ear, "Don't tell them anything!" I felt my legs wobbling, beginning to fail, and my mouth went dry. The full realization of what a complete fool I had been, not running the control experiment as we had planned, was starting to crush me.

There were now a hundred cameras pointed at me.[118] All I had to do was lift an eyebrow or change my stance slightly and I could hear a sudden rash of shutters firing. Grace opened with, "It certainly has been a busy week for us, as maybe it has been for you as well." He went on to remind them of our previous press conference and indicate that we had modified our position on the validity of cold fusion due to newer measurements. I was up.

I started with, "First of all, for those of you who may have read the *Constitution* this morning, I think it was probably reckless for me to say we were reckless." Laughter. The previous evening I had taken a call from Hal Straus, a reporter for the *Atlanta Constitution*, and I had spilled everything, admitting that it had been reckless to release the experimental finding so quickly. I discovered in the morning edition that a reporter will always take the worst thing you say instead of

118 I don't really know how many cameras were pointed at us, but it seemed like about one thousand, so I trimmed that number down to one hundred as a more realistic estimate.

your best phrasing and make a headline out of it.[119] When I made it to work, I endured a stern lecture from Toon about talking to anybody off campus. To the assembled journalists, I went on to describe our experience with the neutron detectors in gruesome detail and promised to report back after a complete revision of our experimental setup and an attempt to arrive at believable results.

As I stepped down and started to walk away, a man threw me one last question. "Will you identify the maker of the faulty neutron detector?"

"No," I replied.

The next day, Saturday, April 15, Billy Livesay bailed out of the cold-fusion project with a formal memo. If only it were that easy, I would have done the same. We did go through the motions of reconfiguring the experiment with all new equipment, helium-3 neutron detectors, gamma spectrometers with intrinsic germanium detectors, computerized isotope identifiers, and a control for every experiment, but we knew it was all for show. There was no cold fusion to discover and nothing to confirm. Confirmations were coming out of Europe and the Soviet Union, but all were bogus, usually falling for the similar low-count neutron detection problems that had bugged our fiasco. As Rick pointed out, every neutron detection that was reported seemed to be on the edge of the equipment's level of perception. In Italy, they had a monster BF3 so big it could detect as few as one neutron per day. They reported finding the production of one neutron per day from the Italian cold-fusion cell. Babcock & Wilcox representatives came to my office to discuss teaming. I unloaded my studied opinion that there was nothing on which to team.

---

119 Take this as a lesson learned: The reporter who interviews you will be extremely sympathetic and understanding, and your interaction will be cordial and relaxed. You will have made a friend, but the reporter, although he or she must be treated as such, is not your friend. The reporter is looking to make a sensation, and has no particular regard for your welfare. The worst case of this was a crew from the BBC. They came to film me for a *Horizon* show on British television. It would be repackaged as *Nova* on American public television. I made one slip, talking about the conspiracy theorists who thought we had been paid to suppress our cold-fusion findings and joking that we might have considered doing that but we were never offered. That was the part of the film they showed, and they reported on *Dr.* Gai at Yale and *Mr.* Mahaffey at Georgia Tech. It was subtle but effective in making their twisted point.

We had another press conference on Tuesday, April 26, in which I tried to put cold fusion to bed, and I forced myself to write the final report. As I read it over one last time, I found that I had misspelled the word "neutron." I chuckled. It seemed appropriate, so I left it in. Pons and Fleischmann disappeared, never admitting to have been mistaken. Given time to think it through, Billy Livesay, who had worked in palladium hydriding for several years, pointed out that if a fully loaded chunk of palladium in electrolysis is removed from the water or is allowed to run dry, exposing the electrode top, then bare deuterons will start boiling out of the surface of the metal. These nascent deuterons will combine into deuterium-deuterium molecules, and this will produce heat. With heat generated on the metal surface, the release of stored deuterium will increase, until it starts to react with oxygen in the air and ignites into a hot, invisible flame, which eventually sets the palladium on fire. Palladium, when started, burns like magnesium and will spall its way down into concrete, just as once happened in the laboratory at the University of Utah. It was this oxidation breakdown that firmly convinced two chemists that they had broken the fusion code and would give nuclear power to the world.

Cold fusion had erupted and then crashed with such blinding speed, it would still take a while for the scientific papers, expense reports, retrospections, and recriminations to catch up. There had already been planned and scheduled a Workshop on Cold Fusion Phenomena for May 23 through May 25 in Santa Fe, New Mexico, before the concept melted down, and Steenblik had wrangled a registration fee and a paid trip. The meeting was scheduled to start exactly eight weeks after Pons and Fleischmann had announced their discovery.

His plane touched down in Albuquerque at 6:35 P.M. He rented a car and drove to Santa Fe, noting the countryside "blazing with fusion," as he characterized the bright sunlight, and seeing for the first time why Georgia O'Keeffe had been so fond of it. The workshop started the next morning, with five hundred registrants, mostly from the United States and Canada. There were 110 papers to be given and 75 poster presentations, and the entire conference was to be broadcast by satellite on the University Net. Topics included Integrated Experiments, the Physics of Fusion Reactors sessions I and II, Neutron and Gamma

Ray Spectrometry, Calorimetry, Applicable Condensed Matter Physics, Applicable Electrochemistry, and Analytical Chemistry of Applicable Products. There were teams of researchers and observers there from Texas A&M, the Department of Energy, Massachusetts Institute of Technology, Chalk River Laboratories from Canada, AT&T, the California Institute of Technology, the Los Alamos National Laboratory, the Oak Ridge National Laboratory, Sandia National Laboratories, the University of Alabama, the Gran Sasso National Laboratories in Italy, the Brookhaven National Laboratory, Princeton University, Stanford University, the University of Rome, Argonne National Laboratory, the Colorado School of Mines, Washington State University, Aarhus University in Denmark, the State University of New York, and the Ford Motor Company research lab at the University of Michigan. Expertly conducted, it was the largest scientific symposium held in the twentieth century dedicated solely to a phenomenon for which no one in attendance could offer evidence of its existence.

The International Conference on Cold Fusion has been held yearly since 1989 in Salt Lake City in Utah, Washington, DC, Cambridge in Massachusetts, Italy, Japan, Monaco, Canada, China, France, India, Russia, and South Korea. In 1995, the meeting was chaired by Stanley Pons, and in 2003, Peter L. Hagelstein of the X-ray laser fiasco was cochairman. The topic is considered excessively controversial and is avoided by professional scientists fearing career damage, but the conferences, resembling religious revivals, are attended enthusiastically by true believers, pseudoscientists, crackpots, frauds, and a few sociologists.

Fleischmann went back to England, and Pons moved to France to spend a generous research grant from Toyota in peace. The United States Patent Office assumed a policy in which they will never accept a patent application that states or implies the existence of cold fusion. A proto-blog developed on the news groups: "PERSONAL FUSION: The News Magazine for Desktop Nukes."

Our failure to confirm cold fusion had been so spectacular, at least in the press, and so perfectly timed, it alone was enough to drive a stake through the heart of the cold-fusion phenomenon. It died. Eleven years later, the Modern Physics Collection of the Smithsonian Institution in Washington, DC, requested and was given my lab notebook, the Georgia Tech cold-fusion cell, and many pounds of documentation. Pieces are incorporated into a display from time to time.

The aftershocks and faint echoes seemed to never go away. Several times I was required, for the purposes of public relations, to sit and hear a person's theory about cold fusion, sometimes going as far as to explain how a "fusion" reaction can generate energy without resulting in any change in a nucleus. I would try calmly to get it across that there is no need for a theory to explain a phenomenon that does not exist, there was no ongoing research, and no organization outside Japan was funding any studies, theoretical or experimental, ever.

There were times when it was all I could do to sit and stare at the wall, contemplating the soul-crushing gravity of it all. I would have to put it behind me, find new work, publish or perish, and dive back into the furnace of technology innovation. Cold fusion echoed for a long time. Like the microwave background relic from the Big Bang creation of the universe that was still echoing, cold fusion seemed as if it would never go away. Every few days, I would get a call. There was a rash of new books that tried to unscramble the cold-fusion phenomenon and make a coherent story of it: *Bad Science: The Short Life and Weird Times of Cold Fusion* by Gary Taubes; *Cold Fusion: The Scientific Fiasco of the Century* by John R Huizenga; *Cold Fusion: The Making of a Scientific Controversy* by F. David Peat; and *Too Hot to Handle: The Race for Cold Fusion* by Frank Close. I talked to many writers. There was a time when I wouldn't buy a book unless I was mentioned in it.

Rick Steenblik left Georgia Tech and started a phenomenally successful business. One of his inventions now appears on every one hundred dollar bill rolling off the presses of the United States Bureau of Engraving and Printing. Darrell Acree quit and opened a violin shop in Norcross, Georgia. Billy Livesay retired and lived the good life. Gary Beebe drifted away and died in a fall on a Caribbean Island. I soldiered on.

CHAPTER 5

The Lost Expedition to Mars

> "The Earth is the cradle of mankind, but one does not live in the cradle forever."
>
> —Konstantin E. Tsiolkovsky,
> a Russian schoolteacher who designed
> interplanetary rockets in 1903

It is odd that a subject so unlikely and so ultimately inconsequential as cold fusion would hold the world transfixed for weeks, grasping for details and engaging in spirited armchair speculation, while a subject as important and serious—capable of changing the course of humankind—as nuclear-powered spaceflight would have trouble generating a headline. An expensive, long-running effort to develop nuclear engines for use in interplanetary travel was almost unknown among taxpayers.

Perhaps it was the fact that this work was government sponsored and involved nuclear fission. In the early days, throughout the 1950s, most nuclear research was either secret or wasn't widely published. Through the 1960s, the hard-core government research was still centered on secret nuclear weaponry, but even nonmilitary topics were not given adequate exposure in the press. It was a challenge to explain nuclear physics to the larger public audience, which was thought likely to misinterpret progress, such as nuclear rockets exploding on the test stand, and cause a stampede of negative feelings.

There was actually an unbroken stream of government-funded nuclear rocket motor development projects for eighteen years, starting in 1955 and ending in 1973, often with more than one effort running in parallel. The ultimate goal of building exotic, high-performance reaction engines was to send human beings to Mars and back. It would have been an amazing feat, using lessons learned in the successful lunar landing effort of 1969 plus advanced and highly developed nuclear technology, to push us farther out into space and contemplate living somewhere off the planet.

How did this exciting concept die such a quiet and uneventful death? It didn't go down hard with wailing and gnashing of teeth. It just disappeared. There wasn't any one, central reason. There were several, adding together.

The serious idea of going to Mars probably started with H. G. Wells's novel of 1898, *The War of the Worlds*, in which obviously superior beings from Mars blast off from the surface of their planet, travel one hundred million miles to Earth, and crash-land in a field, making a big impact crater. Being technical wizards, they of course survive the landing, build war machines in situ, and proceed to lay waste to civilization. It was an exciting read, and it got a sixteen-year-old boy in Worcester, Massachusetts, Robert H. Goddard, thinking about ways to leave the ground and fly to Mars. He would eventually become the father of liquid-fueled rocketry, pioneering the technology that built German ballistic missiles in World War II, sent Soviets into orbit in 1961, put Americans on the Moon, and lofted global positioning satellites into orbit.

Goddard completed his PhD in physics from Clark University in Worcester in 1911. Still thinking about how to get to Mars, he reasoned correctly that a self-contained rocket engine would work in the vacuum of space, pushing a vehicle using the reaction caused by high-speed exhaust out a nozzle in back. Solid-fuel rockets had been in military and amusement use for centuries, starting with the Chinese back in 1045 C.E. Solid rocket fuel, which is usually black gunpowder, is burnable fuel mixed with an oxidizer, and it operates without air, burning explosively even in a complete vacuum. Goddard saw rocketry as the way to Mars, but solid rockets are woefully inefficient, and there is no way to control the thrust, throttle it up or down, or even stop the engine on demand. For these important features, Goddard invented the liquid-fuel rocket engine.

In Goddard's invention, an oxidizer, such as liquid oxygen, and a fuel, such as kerosene or alcohol, are carried in separate tanks in the vehicle. Upon launch, the two liquids are pumped into a combustion chamber, mixed together, and ignited with a spark. A continuous explosion results, controlled by the amount of oxidizer and fuel allowed to flow into the chamber, and energetic detonation products leaving through a nozzle opening in the bottom of the engine throw the rocket upward. Goddard's first demonstration of this new mechanism was on March 17, 1926, on his Aunt Effie's farm in Auburn, Massachusetts. The engine produced power for a modest 2.5 seconds, and the rocket reached an altitude of 41 feet. Further experiments in liquid-fuel rocketry continued until 1941.

Even though he spent his life working on chemical liquid-fuel rocket propulsion, Goddard realized early that it was not going to send him or anybody else to Mars. For all its efficiency over crude solid-fuel engines, the amount of releasable energy per ton of rocket was simply not going to send a heavy, manned expedition to another planet. A rocket's burn time was measured in seconds before it ran out of fuel and oxidizer. A Mars shot would require hours of burn time. While it is true that most of such a space voyage would be spent coasting in the frictionless vacuum of outer space, to reach the objective with human cargo and slow down for the landing in reasonable time was going to require energy that no chemical reaction can produce.

As a college sophomore in 1906, Goddard studied the newly found energy locked in the atomic nucleus, millions of times greater per unit mass than anything previously known. Radium had been discovered and found to convert mass directly into energy by nuclear decay. He calculated that one gram of radium would release enough energy to lift 5,000 tons 100 yards high. Interplanetary travel, he reasoned, would begin when control of nuclear energy release was achieved. All attempts to publish these thoughts were rejected out of hand, and the scientific community disappointed him repeatedly.

Others had similar theories concerning the use of the energy in radium for space travel, but a problem with this particular nuclide, radium-226, is that while the energy in a gram of it could indeed lift the British navy a thousand feet in the air, the rate of energy release is terribly slow. To derive that lifting power out of a gram of radium would take 160,000 years. In 1912, a French spaceflight theorist named Robert Esnault-Pelterie solved that problem by suggesting an engine using not one gram of radium but 400,000 grams of it. Emitting radiation in parallel, this large number of grams of radium would suffice to power an interplanetary flight.

There were a couple of problems. The radium energy release could not be throttled down or turned off. While sitting there, idling, the engine would have to shed a full blast of energy in some nonproductive way. As a practical note, by 1940, after decades of strenuous, worldwide extraction, there were only 900 grams of radium-226 in existence. Spaceship designs using radium as interplanetary fuel were still being bandied about in 1936, when Philip Cleator, founder of the British Interplanetary Society, brushed radium aside as impractical and predicted the discovery of controlled nuclear disintegration, fission, which would supply the ideal propulsion for a trip to Mars and back.

In July 1945, when the successful nuclear reactor and atomic bomb were still American top secrets, two members of the British Interplanetary Society began studies and designs of nuclear-powered rocket engines. Working independently, A. V. Cleaver and Leslie R. Shepard outlined solid-core graphite-to-hydrogen heat exchangers, an idea that ten years later would be rediscovered. The first public article describing this concept was published in November 1946 in the *Journal of the British Interplanetary Society.*

Meanwhile, under a strict information lockdown, scientists at the Los Alamos National Laboratory, such as Stan Ulam and Frederic de Hoffmann, drifted off-task and discussed nuclear rocket engines as early as 1944.[120] Unfortunately, there were wars to be won and complicated weapons to be perfected, and talk of space travel was shelved until the world had calmed down. There would be no active research into nuclear rocketry until an armistice ended the Korean War in 1953.

It was not a good time for planetary exploration enthusiasts. The United States government had correctly ascertained that nowhere in its Constitution was there a mention of needing to blast citizens off the surface of Earth and send them to Mars. There were in the document, however, stated obligations to protect people from belligerent foreign powers, and this could be interpreted as a need to send 10-megaton thermonuclear weapons halfway around the world and land them within five miles (the estimated crater width) of Moscow, capital of the Soviet Union.

City-killer hydrogen bombs started out very heavy. In 1952, the "wet" version, the Ulam-Teller bomb, using liquid deuterium/tritium, weighed about 100,000 pounds, but it was eventually reduced to 1,500 pounds by using a dry chemical, lithium deuteride, as the active ingredient. To throw it 5,500 miles in a semiorbiting trajectory and hit the aiming point with adequate precision would require dramatic advancements of all phases of chemical rocket technology. The development contract for the liquid-fueled launch vehicle, code-named Atlas, was awarded to Convair of San Diego, California, on January 16, 1951. The Atlas missile became the highest-priority development project in the U.S. Department of Defense (DOD), but the number of technical innovations required to make it perform as specified was daunting. A backup technology was needed in case it did not work.

In 1953, the Oak Ridge National Laboratory made an attractive offer to Robert W. Bussard, fresh out of Princeton University with a physics

---

120 Stanislaw Marcin Ulam was a Jewish mathematical genius from Poland who escaped his native country on the eve of World War II, taught math at the University of Wisconsin-Madison, and was invited to join the Theoretical Division at the Los Alamos National Laboratory to work on the atomic bomb project. After the war, he and Edward Teller invented the thermonuclear weapon, or hydrogen bomb, but his most prized idea was for a unique nuclear spacecraft-propulsion system, code-named Orion. Frederic de Hoffmann was a Harvard-educated nuclear physicist from Austria who met Ulam at Los Alamos in 1944. He was assigned to work for Teller, which was something of a baptism by fire, and in 1948 he received his PhD. In 1955, he founded the General Atomics Corporation in San Diego, California, and was its first president. The first defense project at General Atomic was to develop Ulam's idea for interplanetary flight, Project Orion.

PhD, to work on the Aircraft Nuclear Propulsion program. He agreed to come aboard, but only if he could also work on designs for nuclear rockets. Since the age of seven, he'd dreamed of a Mars expedition, and nuclear power was obviously the way to get there. He had read the Shepherd-Cleaver reports from Great Britain, and he wanted to expand on their design for a solid graphite-core nuclear-fission engine. Oak Ridge allowed it.

At night he worked on the nuclear rocket, during the day he labored over the nuclear jet engines, and in between he got two hours of sleep. He published the results of his nuclear-rocket studies, showing that the concept is vastly superior to any chemical rocket, in December 1953. The January 1954 journal in which it appeared was classified secret, and it could be read only by people having the proper security clearance. Few people were able to find the article and understand it, but one of the impressed readers was John von Neumann. Von Neumann was a physicist, mathematician, polymath, computer pioneer, and unusually well-dressed inventor of everything from game theory to a hydrogen bomb. He would become an Atomic Energy Commission chairman in March 1955, but when he read Bussard, he was a highly influential proponent of the ICBM strategy. He pushed the nuclear rocket engine as the perfect H-bomb hauler.[121]

The basic nuclear rocket engine consists of a fission reactor, held at criticality and making heat energy at a steady rate of about one billion watts. A liquid, such as ammonia or liquid hydrogen, is pumped into one end of the reactor. Moving forward through open channels in the hot reactor core, the liquid is heated to thousands of degrees, expanding with tremendous pressure and exiting from the end of the reactor through a nozzle. There is no burning, oxidation, or mixing of two chemicals involved. The cold liquid stored in the fuel tank is suddenly transformed into a hot gas by the power of the nuclear fission,

---

121 There is a touch of irony here. The expatriate Hungarians in the Manhattan Project, von Neumann, Leó Szilárd, Edward Teller, Theodore von Kármán, and Eugene Wigner, among others, were collectively known as "the Martians," because their native language was incomprehensible and did not trace to any other language, modern or ancient. These individuals were therefore obviously not from this planet, or so it was conjectured, and most likely had come from Mars. They formed a core of unusual intellectual power without which the atomic bomb would have been delivered later and with greater confusion. There must have been something in the water in pre–World War I Budapest.

as if it were continuously involved in an explosion, and the fact that initially cold liquid is moving rapidly through the core prevents the nuclear reactor and the back end of the vehicle from melting.

The advantage of a nuclear rocket over a chemical rocket, even a liquid-fueled chemical rocket, is in the efficient use of the fuel, as designated by its "specific impulse" (SI). SI, which is expressed in seconds, is the "hang time" of a rocket, or the maximum number of seconds it can accelerate at 32 feet per second per second, balancing the pull of gravity and hanging still above ground. It is analogous to the miles-per-gallon rating of an automobile and is the most important measure of rocket performance, particularly for long voyages to other planets. The SI depends on many factors, such as the weight of the fuel that must be carried and the speed with which the rocket exhaust leaves the engine.

The type of fuel is also a critical factor in rocket performance. As the superheated gas escapes through the rocket nozzle, the recoil pushes the space vehicle forward. The faster the flying gas exits the nozzle, the more reaction is derived, and the speed of the gas particles is due to their temperature and the light weight of each particle of gas. The absolute lightest possible gas particle is hydrogen, which is why the perfect propellant for a nuclear engine is hydrogen. A particle of hot hydrogen is just one proton. For a chemical rocket engine, the lightest possible combustion product is steam, the result of oxidizing liquid hydrogen. A steam particle weighs eighteen times that of a hydrogen particle.

Solid rocket boosters, such as were used in the Space Shuttle launches, have a maximum SI of 290 seconds. It doesn't matter how big and powerful a solid rocket is. After 290 seconds, the best one will run out of fuel while trying to stay hanging in the air. The enormous F-1 engines used in the Apollo 11 Moon shot Saturn V booster in 1969 were the best that could be achieved with engines burning kerosene in liquid oxygen. They had an SI of 350 seconds. The absolute theoretical limit to the SI of a chemical rocket is achieved using a liquid-hydrogen/liquid-oxygen mixture (LH2/LOX), as was used in the upper stage of the Saturn V and later in the Space Shuttle main engines. Its SI is 450 seconds, and there is no way to make it higher. The SI for a nuclear rocket starts at 900 seconds and can increase, in theory, into the millions of seconds.

A Rocketdyne F-1 engine, the most powerful chemical rocket engine ever made, can run for as long as 165 seconds. You can only start it once, and its only throttle setting is full thrust. A NERVA nuclear rocket engine runs for 10 hours, can be stopped and restarted sixty times, and it throttles down from full thrust to barely moving. Those are the basic differences between a chemical rocket and a nuclear rocket, and it is all due to the efficiency of using the extreme power of nuclear reactions versus the relatively feeble power of chemical, or "atomic," reactions.

In October 1954, two national labs began work on nuclear rocket engines for an enhanced Atlas missile, even though the design for the Atlas XSN-16A had just been frozen and production was supposed to start in January 1955.[122] Herbert York, director of the Lawrence Livermore National Laboratory in California, formed the Rover Boys committee to study nuclear rockets, while Darol Froman, deputy director of the Los Alamos National Laboratory in New Mexico, put together the Condor Committee to do the same thing.[123]

Lawrence Livermore was heavily involved in thermonuclear weapons development, and Los Alamos was hungry for a new mission. By 1955, Los Alamos had established the N-Division specifically to develop nuclear rocket engines, and the first official design was

---

122 The Atlas was a complicated beast, and production of test rockets produced twelve units, named the X-11s, by June 1957. Two out of three of the first test launches resulted in spectacular explosions on the launchpad, but eventually the Atlas was refined to the point where it was cleared for manned spaceflight, and ten of them were used to launch Project Mercury space capsules. John H. Glenn became the first American to orbit Earth, launched by an SM-65 Atlas missile, and variants of the Atlas were used by NASA to launch scientific probes to Mars, Venus, and Mercury. The last SM-65E/F Atlas missile was launched on March 24, 1995. The Interstate Highway System, being planned when the Atlas was in development, did not use the clearance height of tractor trailer trucks as a specification for the height of overpasses. The overpasses were all built to allow an Atlas missile carrier to pass underneath.

123 *The Rover Boys* was a very popular series of juvenile books written by Edward Stratemeyer between 1899 and 1927. The Rover boys, Tom, Sam, and Dick, were involved in thirty stirring adventures, always getting into trouble with various authority figures as well as criminals and running into Germans, Italians, Irish, Chinese, and African Americans, all of whom were portrayed as having degrading speech patterns. A typical title was *The Rover Boys Under Canvas, or The Mystery of the Wrecked Submarine* (1919, second series, populated by the sons of the original boys). Herbert York probably read these books as a young man and so named his gallery of young, eager engineers, ready to take on a very daunting mission. Froman named his project for the largest bird in the Western Hemisphere that can fly, the California condor.

Dumbo, based on using tungsten as a high-temperature structural material. A nuclear rocket is at its most efficient when it is running at just below the melting point of the engine, so building it out of material that can stand to be at thousands of degrees was essential.

A big question was which fissile nuclide to use as reactor fuel. Plutonium-239 was ruled out, because none of its compounds can withstand the high temperature. Uranium-233 forms high-temperature compounds, and its fission properties would result in a reactor core that is one half the size of a uranium-235 core with the same performance. Unfortunately, there was no U-233 being produced, mainly because of the high gamma radiation present in certain impurities. That left uranium-235, which was readily available in any desired purity.

Five reactor core designs with power ratings ranging from 600 to 2,000 megawatts (2 billion watts) using graphite as a neutron moderator came out of the Condor project. Their aggressively odd code names were Uncle Tom, Uncle Tung, Bloodhound, Shish, and Old Black Joe.[124]

Old Black Joe, designed to run at 1,200 megawatts of power, was approved in November 1956 for continued development. The design was upgraded to 2,700 megawatts, and plans were to use it for a range-extending second stage for the Atlas missile. This "Super Atlas" would be capable of parking a heavy H-bomb in geostationary orbit, 22,236 miles above the equator in the Eastern Hemisphere, ready at a moment's notice to pounce on Moscow. It would be 9.6 feet in diameter and 96.6 feet high, and to the delight of the air force, it would seem an even better idea than carrying missiles around in submarines. The bad news was that the estimated cost of developing the nuclear rocket for it was an eye-watering one billion dollars.

Coming back down to Earth, in March 1956, the Armed Forces Special Weapons Project decided to invest $100 million in test facilities

---

124 The use of African American characters in two of the names, Uncle Tom and Old Black Joe, may have been due to the use of black-as-you-can-get graphite for the high-temperature fuel assembly structures. In the early 1950s, racial sensitivity was not an agendum. The Uncle Tung design was a combination of Uncle Tom and Dumbo, which was based on tungsten core structures. The reasons for Bloodhound and Shish are not obvious. Shish did not use fuel assemblies that looked like shish kebobs. They were hollow tubes. "Soda Straw" might have been appropriate. The Bloodhound did not have fuel channels resembling the turbination in a bloodhound's nasal passages, but it did have a big can of heavy water dominating the center of the core.

and nuclear rocket engines to be tested over the next three years. The money was to be split between Lawrence Livermore and Los Alamos. Los Alamos jumped at the opportunity, renamed Old Black Joe "KIWI-1," and planned to make it fly in five years. In Washington, DC, where the money came from, Senator Clinton P. Anderson from New Mexico stood in front of the Senate and expounded on the future benefits of a nuclear rocket program, for the first time on record citing potential goals of sending men to the Moon and eventually colonizing the rest of the solar system. Livermore's Rover Boys project was immediately transferred to the Los Alamos lab, where it combined with the Condor project and became Project Rover. Livermore was given Project Pluto, which was a dead-ended effort involving a nuclear ramjet propelled cruise missile carrying an H-bomb in the nose.

At this point in 1956, no nuclear rocket engine had ever been built, and the technology consisted of designs on paper and some computer simulations.[125] The mind-numbing list of impossibilities didn't seem to bother the engineering climate of the time. The fuel pump would take a frozen hydrogen slush at -259 degrees Celsius, which was 14 degrees above absolute zero, and push it at a rate of 70 pounds per second into the top of a nuclear reactor running at two billion watts. No such pump existed.[126] No nuclear reactor had ever run at that power

---

125 In 1956, more than 90 percent of all electronic computer calculations in the United States were devoted to simulations of nuclear processes. For all the expense of running large computer jobs, over and over to correct and improve, it was still a lot less expensive than building and running a reactor or bomb design on speculation. Most electronic computers of the era, which were vacuum-tube-based money pits and often one-of-a-kind units, were owned by national laboratories and the Atomic Energy Commission. The Monte Carlo method and neutron diffusion theory calculations used in reactor design were of such extreme complexity, there was no way to do the arithmetic other than using high-speed electronic computation. Without a burgeoning need for computers in nuclear science and engineering in the 1950s, there wouldn't have been enough demand for commercial development, and life would be different now.

126 Rocketdyne was awarded the contract to come up with this exotic turbo-pump because it was the only company with any experience. In 1956, Rocketdyne worked on a parallel program for Lockheed to develop a fuel pump for the top-secret CL-400 edge-of-the-atmosphere reconnaissance airplane, code-named Suntan. Its turbojet engine, operating at Mach 2.5, 98,000 feet above the ground, had to use the most efficient burnable fuel in existence in the rarified oxygen environment of the edge of space. That fuel was liquid hydrogen slush, pumped into the jet engine at 5 pounds per second. The KIWI pump would have to be fifteen times bigger and operate in a lubricant-destroying radiation field.

level. The most advanced power reactor in the world, the S1W in Idaho, topped out at 10 megawatts. In 52 inches, from the hydrogen intake to the nozzle, the liquefied fuel would go from near absolute zero to 2,000 degrees Celsius through multiple, mechanically chaotic phase changes and a severe pressure drop that would try to suck the core out the end of the engine. The fuel, liquid hydrogen, was the most corrosive substance known, and while sitting quietly in the storage tank, it would diffuse into all the metal structures it touched, rendering them even more brittle than simple freezing would make them. Nothing was known about how stray neutrons from the reactor would interact with the hydrogen slush in the fuel tank, whether two nuclear rockets sitting side by side would cause each other to go supercritical by neutron exchange, or how to keep the hot, unsupported end of the reactor, glowing incandescent, from following the hydrogen gas out the rocket nozzle.

Competition for attention and federal funding had to be fought for tooth and nail. The president of the United States, Dwight D. Eisenhower, former five-star general and Supreme Allied Commander during World War II, was opposed to any war and hated to expend effort in anticipation of one. He publicly warned of the existence of a military-industrial complex, while his administration approved the development of the most advanced weaponry in the history of the world, and he could not see why anyone would need to fly around in outer space. In parallel with any nuclear rocket work were devised the B-52 jet-powered strategic bomber, nuclear-tipped ballistic missiles (from intercontinental versions to one carried on the back of a jeep), the Sidewinder heat-seeking air-to-air missile, supersonic jet fighter planes, 15-megaton hydrogen bombs, a real-time computer network linking early-warning radars, extreme high-altitude reconnaissance jets, a personal "pogo-stick" helicopter for soldiers, a hovercraft jeep or "fleep," nuclear-powered ram and turbine jet engines, a miniature nuclear power plant that could be carried around on a truck, and, in 1959 a detailed plan for taking over the island of Cuba with a clandestine force landing in the Bay of Pigs.

Project Pluto was set up for tests at the Nuclear Test Site (NTS) in Nevada, surrounded by two air force test ranges. Rover found a home in Jackass Flats, Nevada, bounded by the Shoshone Mountains, Skull

Mountain, and Yucca Mountain. Neither location was exactly a holiday land of fun and easy living. Keith Boyer, the construction supervisor, installed thirty trailers for workers to live in at Jackass Flats while they labored making the engine test facilities. They called their home Boyerville.

Engines under test were to be mounted with the nozzle pointed straight up, exhausting directly into the boiling hot Nevada air, in Test Cell A. Construction began in the middle of 1957. Electronic equipment used to record and monitor every datum during a test run was put behind a 3-foot-thick concrete wall for protection, the control room was located underground two miles away, and the connecting cables were buried in a trench.

The most impressive structure at Jackass Flats was the Reactor Maintenance and Disassembly Building (R-MAD). It was an enormous hot cell, made for safely working on naked reactor cores that had very recently been run at two billion watts, making them extremely radioactive. The main hall was 250 feet long, 143 feet wide, and 63 feet high. It could swallow a string of train cars holding rocket engines still hot-to-touch and glow-in-the-dark radioactive.

The world's smallest and slowest railroad, the Jackass Flats and Western, linked the R-MAD to Test Cell C using standard-gauge rails. An all-electric train engine (L-1) with radio remote control did the hauling, with a larger diesel-electric locomotive (L-2) that had a lead-shielded cab for the engineer used for heavier loads. A farm of pressurized hydrogen tanks was installed to hold test fuel. A puzzling electrical problem was solved when it was found that rats were chewing the insulation off the buried cables.

Three versions of KIWI-1 were envisioned. KIWI-A, the first baby step, was power-rated at only 100 watts. The reactor was to be contained in a welded aluminum can, 4 feet in diameter and 4 feet high, with the fuel arranged in plates made of a combination of graphite and uranium. The neutron controls were at the center of the cylindrical core with heavy-water coolant to prevent overheating. The liquid fuel would be ammonia or methane, which were much easier to handle than liquid hydrogen.

KIWI-B was rated at a respectable one billion watts, using liquid hydrogen as rocket fuel and a pressure vessel only 25 inches in diameter. Rocketdyne was contracted to design and build the fuel pump and the nozzle. Aerojet General Nucleonics did the plumbing. The reactivity of the reactor core was improved by having a beryllium neutron reflector at the fuel inlet.

KIWI-C would be the flight-test engine, or the "iron horse." It was to have liquid hydrogen piped around the nozzle to keep it from melting loose, and a computer would automatically start it and control it with no human intervention. This engine, with its multitude of untried features, would be a milestone in nuclear engineering.

These were well-thought-out plans with goals and a schedule, and engines would be ready to test by late 1958, when construction of Test Cell A would be completed. For the Atomic Energy Commission (AEC), things were moving too slowly, and the world situation being the way it was, planning for a mission sometime in the 1980s was looking too far ahead. The commissioners proposed a heartbreaking cut to the Project Rover budget on October 2, 1957.

The dread only lasted two days. On October 4, 1957, the Soviet Union put the first man-made Earth satellite, Sputnik 1, into orbit. It was a shock to the American sense of preeminence, and its implications were profound. One month later, the 8,000-pound Sputnik 2 was orbited, this time with a live dog on board. On March 17, 1958, we finally got the Vanguard 1 satellite into orbit. It weighed 3.25 pounds. The weight of Sputnik 2 meant that the Soviet rocket booster had 1,350,000 pounds of thrust. The Atlas missile, the best rocket we had, could do 300,000 pounds of thrust on a good day. Eight out of our first eleven space shots failed to achieve orbit. As a nation, we felt a lack of significance.

There was no more talk of budget cuts, and a year later, Project Rover was handed over to the newly configured National Aeronautics and Space Administration (NASA) in anticipation of nonmilitary use of the nuclear rockets. We didn't seem to have much chance of beating the Soviets any time soon with a bigger, more aggressive space shot. By October 1959, they had sent Lunik 3 to the Moon with a 1,000-pound payload, and it sent back the first pictures ever of the lunar far side. Whatever we did, they had already beaten us to the milestone with a heavier, better-looking projectile. It was time to end-around them and send a crew to Mars. As Washington ran amok with fears that the

Reds would take over outer space, AEC administration reminded the House Space Committee that a KIWI-C nuclear engine could take a 55,000-pound load to Mars and come back with 25,000 pounds. The best chemical rocket could only take 7,500 pounds out and return 750 pounds.

Making use of the moment of panic, Project Rover at Los Alamos proposed a hybrid rocket named Helios. There was, even at this early time, concern that a nuclear rocket firing in the atmosphere would spew radioactive debris and chunks of fuel all over the world. Although the scenario was more realistic than AEC would care to admit, the fear of biosphere damage was definitely overstated.[127] To erase this concern, Helios was designed to blast off with six liquid-hydrogen/liquid-oxygen engines encircling a single nuclear engine. At 100,000 feet, the chemical rockets, out of fuel, would fall away and the nuclear engine would start and complete the mission outside the contamination-sensitive atmosphere. Although it was planned for a manned Mars or Venus mission, Helios could also put 80,000 pounds on the Moon or push an entire, quarter-million-pound space station into orbit in one swat.

On July 8, 1959, KIWI-A Prime was tested. Three KIWI-As had been tested in 1958, and A Prime was a redesign based on lessons learned. There were sticky hydrogen valves, various fires, and component failures in KIWI-A, including a graphite part that failed and blew out the nozzle. It ran 500 degrees hotter than it was supposed to. KIWI-A Prime was definitely an improvement, but still three fuel modules

---

127 This concern about nuclear fallout from nuclear propulsion originating at ground level came up several times as nuclear propulsion enthusiasm waxed and waned. The nuclear technologists had no choice but to acknowledge the fear, and accommodating mission adjustments were always made. However, those who understood the technology were aware that any debris from a rocket engine, whether it is in the atmosphere or in a low-earth orbit, eventually comes down and lands on the ground. Starting a nuclear rocket at 100,000 feet or assembling a Mars mission in orbit and then leaving by starting the nuclear engines will result in almost the same radioactive imprint on the ground as would blasting off from ground zero with nuclear rockets ablaze. In conversations with lawmakers and congressional hearings, it was hard to keep a straight face when fear of nuclear rocket radiation killing spacemen was mentioned. The hazards of high-energy particles zipping through outer space and solar flares make fission radiation dangers seem trivial, and the daily radioactive fallout from cosmic rays hitting Earth's atmosphere will exceed the danger of nuclear rocket exhaust.

broke off and flew away. The corrosive hydrogen from the ammonia fuel, very high temperature, and heavy vibration made design work for the uranium fuel assemblies challenging.[128] New exotic coating materials had to be developed, such as a vapor deposition onto the fuel modules of gaseous niobium carbide diluted with helium.

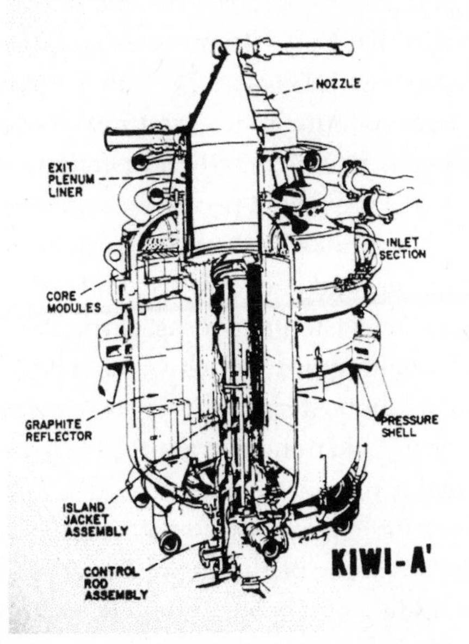

A cutaway diagram of the KIWI-A Prime nuclear rocket engine. The output nozzle points up, but at this early stage of development it was not a true rocket nozzle, and it was using ammonia as a substitute for liquid hydrogen rocket fuel. Conventional control rods are inserted through the bottom of the reactor core.

KIWI-A3, one of the low-power units that had been rebuilt, was tested on December 19, 1959. It ran for 259 seconds on liquid ammonia

128 A component of the nuclear engine vibration problem is that superhot hydrogen gas coming out the exhaust nozzle immediately mixes with oxygen in the air and explodes. This phenomenon also adds to the lifting capability of the nuclear rocket, similar to the afterburner on a high-performance jet engine. In outer space, there is no air to mix with, and to get an idea of how the engine would work with no back-reflection of neutrons from air, a huge vacuum chamber was built to fit around an engine under test. Split down the middle and sealable using rubber gaskets, it was pushed into place on the test stand using a locomotive and came together like a clamshell around the engine.

at 118 megawatts. It did not overheat, and there was less corrosion than with KIWI-A Prime, but every uranium fuel module cracked and one was broken in two.

The United States finally shot a man into space on May 5, 1961. The Soviet Union had already beaten us by twenty-three days, putting Yuri Gagarin into orbit in a mission lasting 108 minutes on April 12. We sent Alan Shepard on a boost just past the end of the atmosphere and back down in a flight that lasted 15 minutes and 22 seconds.

The orbital mission with Gagarin, Vostok 1, was bold and risky. Out of the past twenty-four Soviet satellite launches, twelve had failed, giving a success rate of exactly 50 percent. The mission was secret so that if it failed the Soviet people and the world would not be disappointed. The American shot with Shepard, Mercury-Redstone 3 Freedom 7, was cautious and conservative, and it was covered in real time on all radio and television networks. If the Redstone rocket had blown up on the launching pad, which was not impossible, it would have been a public relations disaster. If we were going to send people to Mars with a nuclear rocket, then it would have to be developed with crew safety as a major design element. This was a rock-hard NASA mandate, but it was a foreign concept to the Los Alamos and Lawrence Livermore laboratories at the beginning of Project Rover. The nuclear labs were used to building a complete system with multiple unknowns, running it until it came apart, doing an autopsy, and building another one with the lessons learned. NASA insisted that before a complete system could be built, the perfection of each component had to be previously verified. The Rover Boys knew that they had to do as much development as possible before NASA came down on it with a hammer. There was no time to waste.

KIWI-B-1A was ready to move to Test Cell A by November 1961. The United States had not put anyone into orbit yet, and Gus Grissom had almost drowned when his Liberty Bell 7 space capsule landed in the Atlantic Ocean after another fifteen-minute pop-up. This KIWI was completely redesigned. It had 1,644 hexagonal uranium-graphite fuel modules, each 52 inches long with seven vertical channels for hydrogen to run through and absorb the heat of fission. The fuel modules were held in place by ⅛-inch steel rods running top to bottom, with one rod for every seven modules. The neutron controls were moved from the interior of the reactor to outside, redesigned into vertical cylinders placed in a circle surrounding the reactor core. Each cylinder was half cadmium and half beryllium. Turn a cylinder so that the cadmium side faces the reactor and it absorbs thermal neutrons, slowing down the fission reaction by discouraging strays from wandering back into the core. Turn the cylinder so that the beryllium side faces the reactor and neutrons are reflected back into the core, improving the fission action and causing the power to increase exponentially.

During checkout in a movable aluminum shed, pressurized, nonliquid hydrogen leaked through a stuck valve and exploded, not quite sending the building into orbit but injuring several workers. This was just the kind of thing that would give the NASA administration nightmares.

The new KIWI engine was under full test a month later. Its reactor scrammed at 225 megawatts due to a large fire caused by a hydrogen leak between the reactor pressure vessel and the rocket nozzle, but in the world of nuclear-rocket testing, it was no big deal. Only two fuel modules cracked, and the new control drums performed well. From the Jackass Flats perspective, the glass was half full.

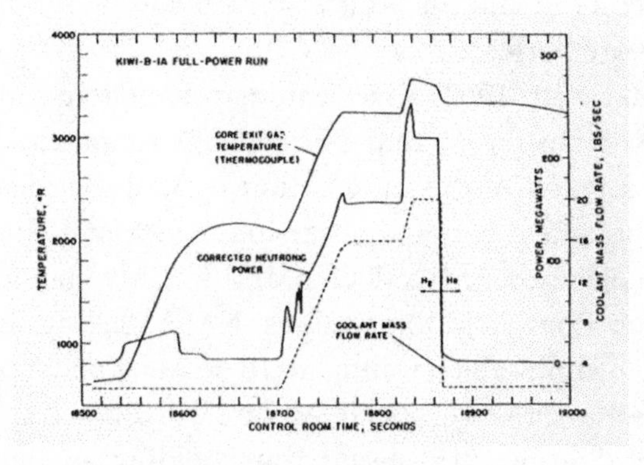

The data chart from the KIWI-B-1A test. Notice how the power curve ("corrected neutronic power") seems to follow the coolant-rate curve. A major finding of Project Rover was that the reactor in a rocket engine can be controlled by throttling the liquid hydrogen flow rate.

The Georgia Nuclear Laboratory (GNL), formerly known as the Georgia Nuclear Aircraft Laboratory, had reason to celebrate after

December 1961. Wernher von Braun, former German "vengeance weapon" designer during the war and now the director of NASA's Marshall Space Flight Center, was given responsibility for the reactor in-flight test (RIFT). It had been decided that the third stage of the Saturn V rocket that would send Americans to the Moon would use a KIWI-C engine, renamed NERVA, the Nuclear Engine for Rocket Vehicle Application. Von Braun chose GNL as the ideal facility to make radiation survivability tests on the hydrogen slush tank that would feed NERVA.

Sitting atop a Saturn V booster in the hot Florida sun, without a lot of high-tech insulation, the hydrogen slush would start to boil in the tank, eventually pop the pressure relief valve, and send the precious fuel away into the air. The insulation was to be on the inside surfaces of the tank, and there was fear that it would break down under gamma radiation streaming out of the fission reactor underneath it, shred, and clog the output pipe. GNL had the perfect setup for the test, but they would have to install a liquid-hydrogen handling facility and a special, all-aluminum railcar to send the tank to within inches of the naked, above-ground reactor. The similarity between the GNL reactor test setup and the close proximity between the hydrogen fuel tank and the NERVA rocket engine reactor could not have been more perfect.

On February 20, 1962, American astronaut John Glenn orbited the earth thrice in the Mercury Friendship 7 capsule, lofted by an Atlas D. As a result of this successful mission, the idea of manned spaceflight became a national obsession, and Marshall Space Flight Center immediately began work on EMPIRE, Early Manned Planetary-Interplanetary Roundtrip Expeditions. NASA, feeling the optimistic afterglow of Glenn's orbits, wanted to fly to Mars and Venus by 1972.

At Jackass Flats, the world's largest liquid-hydrogen storage facility was built by Chicago Bridge and Iron, holding an impressive one million gallons but tending to leak and causing the world's largest liquid-hydrogen explosion later in 1962. Working quickly while public enthusiasm was peaking, the Los Alamos lab built another huge hot-cell building, the E-MAD (Engine Maintenance And Disassembly), and two engine test stands, ETS-1 and ETS-2. ETS-2 was intended for big, going-to-Mars engines.

KIWI-B2 was supposed to be the next engine to be tested, but a design review concluded that its large graphite neutron reflector disk was going to be a problem, and it was put aside. That left the KIWI-B4, but it almost went critical as it was being assembled. The Project Rover reactors had always been hot-rod designs, hanging over the edge of danger, but they had never had one threaten to start up and generate power before the last layer of fuel modules was installed. It turned out that water vapor, even in the very dry Nevada air, had soaked into the fuel as it lay in a storage building. Interstitial water, filling the tiny spaces between graphite-uranium particles, acted as an unplanned neutron moderator, giving just enough enhanced fission environment to kick the assembly over the critical line and wake up the reactor. After this pants-wetting incident, all rocket-engine nuclear fuel was stored sealed in cans filled with inert gas.

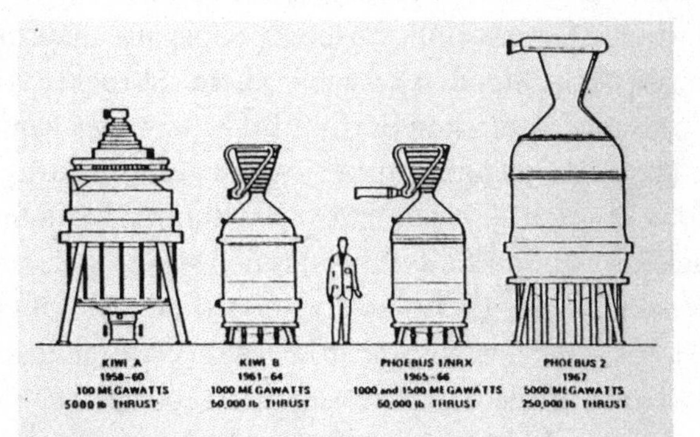

The sizes of Project Rover nuclear rocket engines shown in relation to the height of an adult human being. The Phoebus 2 is the largest, and it is rated at five billion watts.

The project was in a hurry, making hay while the sun shined, and there wasn't time to rebuild the KIWI-B4, so that left the KIWI-B1B, an engine with a questionable core design that was expected to have trouble with two-phase flow through the tiny channels in the fuel. It was bolted to the test stand and started up in September 1962. It was supposed to run for ten minutes at one billion watts.

Start-up was ragged, but the engine settled down at 60 percent of the target power, and it did not seem to surge as had been predicted.

It started throwing pieces of the core out of the nozzle. The control drums rotated to compensate for the reduced reactivity, and power was increased to 965 megawatts, almost to the target. It hung there for 100 seconds, despite the fact that the core was still leaving in chunks. It ran for a few more minutes, with the control drums turning slowly to keep the reactor critical, until a nozzle sensor blew out and a fire started. All in all, it was seen as a successful run.

Back in Washington, details of the KIWI-B1B test were interpreted differently, and key people in charge of the budget had to be peeled off the walls by Wernher von Braun at a hearing. He laid out the post–Apollo Moon landing plans for a lunar base, in which a nuclear rocket engine would be essential. Being able to start and stop many times and running for hours on one tank of fuel, the nuclear engine would power an Earth–Moon shuttle, carrying personnel to and from a permanent Moon base. It would be parked in lunar orbit, and when needed, it would start up and transport people and material to Earth and back again to the Moon. A reusable, chemical-rocket space shuttle would lift passengers into low-Earth orbit, where they would transfer directly to the nuclear Moon shuttle or stop at the orbiting space station for a day of rest and briefing. Even without a Mars mission, the nuclear rocket was important if the manned spaceflight program was going to be something other than an isolated moment of glory.

KIWI-B4 was tested on November 30, 1962. Wally Schirra had made six perfect orbits in the Mercury Sigma 7 capsule. The Soviets had launched an unmanned probe to Mars and had orbited two men in one spacecraft. The KIWI-B4 engine reached 250 megawatts effortlessly and then started making flashes of light at the rocket nozzle. At 500 megawatts, it started throwing pieces of fuel 1,000 feet into the air, a new altitude record, making all colors of the rainbow as the chunks caught fire in the air. Only 44 of the 1,644 fuel modules remained intact, and most had broken in the last 15 inches from the hot end, nearest the rocket nozzle. A careful teardown revealed a seal failure at the top of the engine, allowing hydrogen to leak into the half-inch gap between the reactor core and the pressure vessel that contained the engine. This half-inch void was supposed to allow the core to expand when it heated to 2,000 degrees Celsius without splitting open the pressure vessel. The pressure difference between the hydrogen in the middle of the reactor and the hydrogen flowing through the gap caused enough chaotic vibration to tear the bottom 15 inches off the engine. The "quick fix," a term that caused screaming fits at NASA, was to move the hydrogen seal from the top of the reactor down to the bottom.

President John F. Kennedy and his entourage visited Jackass Flats on December 7, 1962, and were given the royal tour. Serious plans for the Mars missions were beginning to jell. Given a realistic budget of $32 billion, an appropriate space mission after the Apollo Moon landings would be a manned Mars flyby in 1982. The solar flare activity, for which crew radiation shielding would be difficult, was predicted to peak in 1980 and be over by 1982, when an Earth–Mars opposition flight would be ideal.[129] A manned Mars orbiter mission was planned for 1988, and a manned landing would be ideal in 1993. Nuclear-engine specifications were detailed with these missions as the goal. A 4.5 billion-watt engine would need to be flight-tested and man-rated by 1980. A compact reactor running at 4.5 billion watts had never been built or contemplated, but in 1962 nothing nuclear seemed impossible. The president seemed excited by the bold, frontier-pushing attitudes at Jackass Flats and the thought of sending men to Mars.

KIWI-B4D was tested in May 1964. This was the first nuclear reactor ever operated with fully automatic controls. Just press the Start button, and the engine went from cold dead to one billion watts in two very smooth steps. Everything was perfect until the power started fluctuating between 900 and 1,030 megawatts. There were several small explosions, followed by a large liquid-hydrogen fire. The problem was not in the automatic controls. It turned out that oxygen, under the influence of the very cold hydrogen slush, had liquefied

---

129 Both Earth and Mars are in slightly tilted, elliptical orbits, and the distance between them and the resulting trip time varies greatly, from 33.9 million miles to 249 million miles. About every two years, Mars is in opposition mode, meaning that when viewed from Earth it is on the opposite side of the sky from the Sun. At the time of opposition, if at midnight Mars appears in the exact center of the ecliptic in Atlanta, then it is high noon in Kansk, Siberia. Near this time, within a couple of days, Mars happens also to be the nearest to Earth that it has been in the last two years. Scheduling a trip to Mars is complicated, as it involves both the variable distance to be traveled and the radiation level due to possible solar flare events, which are not easy to predict. A massive solar flare on October 25, 2003, would have baked a Mars flight if it had taken advantage of the unusually close opposition that year. On August 23, 2003, Mars was as close to Earth as it had been since 1924. It will be almost that close in August 2050.

out of the air and collected in the spaces between brazed joints in the structure of the engine. When mixed with the liquid hydrogen in a hot-running engine, explosions resulted, with at least one opening up a leak between the coolant tube and the nozzle shell, and this caused the fire. Amazingly, no fuel was damaged or lost.

On August 28, 1964, the next engine, KIWI-B4E was up for a test. It went from zero to one billion watts in 2 minutes, but it ran out of liquid hydrogen in 9 minutes and shut itself down in an orderly fashion. There were no cracks or breaks in the reactor core, but despite some very sophisticated coating work on the fuel, there was corrosion. The same engine was retested in September with very encouraging results, and this time the specific impulse was measured at 740 seconds, beating any possible chemical rocket engine. A question arose. Why were they trying to build such a big rocket engine? The Soviets had sent up everything from an Earth satellite to a probe to Venus using twenty-five World War II–vintage V-2 rocket engines bolted onto one kerosene and one liquid-oxygen tank. Instead of making a huge booster engine, they simply used many tiny engines. A billion-watt NERVA engine wasn't tiny, but why not use five of them clustered together instead of one huge NERVA?

The reason for the single-engine configuration was a fear of reactor cross-coupling. These engines were innocent of any lateral shielding, either gamma-ray or neutron type, and neutrons generated in one reactor could migrate over to another, skewing the carefully balanced neutron count. At a steady power, a reactor has to make neutrons at exactly the rate at which neutrons are lost. One neutron less makes it lose power, and one more makes it gain power at an increasing rate. Some scientists predicted that it would be no problem, and some said it would be a disaster. Computer simulations could be tweaked to make it go either way, so the only way to answer the question was to run two engines side by side.

The effects of reactivity in a nuclear core scale up or down perfectly, so there was no need to run the two engines at liftoff power. They could, in fact, be run at essentially zero power with the same results or lack of results. A low-power engine used to test fuel assemblies, PARKA, was set up in a lab next to KIWI-B4, and both reactors were made critical. Separation distances of first 16 feet, then 9 feet, and finally 6 feet were tried without any reactor runaways. Nuclear engines can be clustered.[130]

In 1964, the Los Alamos reactor experiments officially became the NERVA rocket engine development, and Westinghouse took over the construction of new test engines in Pennsylvania. KIWI-B4 became the NRX-A, and NRX-A1 passed cold hydrogen-flow tests early in the year. By September, NRX-A2 was at Test Cell C and ready for a full-power test. It was run up to 1,094 megawatts (1.094 billion watts) in four steps, with a minute wait at each step for instrument calibration, and it held full power for 40 seconds until the liquid hydrogen ran out. Specific impulse was 760 seconds, the highest SI yet measured. Westinghouse was ecstatic and ran a full page ad in the *Wall Street Journal* with a three-word banner over a photograph of NRX-A2 firing straight up, trying to bury itself in the ground: ON TO MARS!

On a roll, Westinghouse then ran NRX-A2 with the control drums locked in the subcritical condition. Theoretically, the rocket engine reactor could be controlled perfectly just by throttling the liquid hydrogen. Hydrogen, made dense in the liquid state, enhanced the reactivity for an enriched-uranium reactor, as slowing down the fission neutrons by collisions with cold hydrogen would greatly increase the probability of fissions. The density of the hydrogen in the reactor core made a big difference. The fission rate, and therefore the power produced, could be controlled with a throttling valve on the fuel pump. As was always the case, some scientists thought it would work, and some did not. Using only the hydrogen flow as a neutron control, Westinghouse was able to start up the reactor and modulate its power level perfectly. The complexity and the extra weight of the cadmium-beryllium control drums were not necessary at all, and when there

---

130 Excess reactivity in a nuclear reactor is measured in dollars and cents. (An atomic bomb has an initial excess reactivity of about $3.00.) One dollar of excess reactivity means that the reactor is prompt critical, or critical just using the neutrons that are immediately available and not waiting for the delayed fission neutrons to affect the reactivity. A prompt critical reactor is very supercritical, and prompt criticality is a factor in some severe criticality accidents, such as the SL-1 incident in Idaho in 1961. At a separation of 16 feet, each of the two rocket engine reactors had an excess reactivity of $0.03 (three cents) due to neutron sharing, and this degree of excess reactivity was easily erased by a slight rotation of the control drums. At 9 feet, it was $0.12, and at 6 feet it was $0.24, and all of these excess reactivities were easily extinguished with no modifications to the controls.

was no more liquid hydrogen, the fission stopped for lack of neutron moderation.[131]

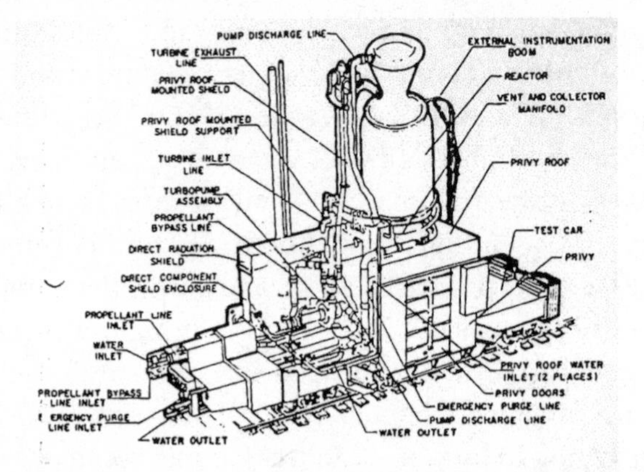

The NRX nuclear rocket engine test setup. If the reactor should go wild and melt, it would fall into the heavily shielded box on which the engine sits. It is named "the privy," bringing to mind an antique sanitation facility, the outhouse.

Meanwhile, on January 12, 1965, Los Alamos National Laboratory decided to end its KIWI nuclear rocket program with a bang. NASA, always safety-conscious, had asked, "What's the worst that can happen to one of your nuclear rockets?" That would probably be toppling off the top of the Saturn V booster, where it was scheduled to be the last of three boost stages for the manned Moon shot. Los Alamos dropped a KIWI 75 feet onto a concrete pad, to see if it would somehow throw the reactor into uncontrolled criticality, which they knew would not happen. That's interesting, said NASA, but it's a 300-foot drop off a Saturn. Los Alamos took the challenge, bolted a KIWI to a rocket sled, and crashed it into a concrete barrier at high speed. Then, to top it off, they put together a special KIWI, cleverly named KIWI-TNT,

---

131 Controlling a reactor with moderator flow was called "mapping." It was not an entirely new idea. It had been considered as early as 1944 in reactor designs at Chalk River, Canada. The Canadian reactor ZEEP controlled the reactivity with actively pumped heavy-water coolant/moderator, but it ran at essentially zero power, and the pump-rate of the heavy water was nothing compared to the aggressive turbo-action used to force hydrogen slush down the throat of the NRX reactors at Jackass Flats.

having controls that could be slammed into the full-on position with pneumatic cylinders, putting the reactor into prompt critical mode all of a sudden. The reactor exploded in blue flash, of course, with a blast the equivalent of 300 pounds of black gunpowder, scattering its remains over a 1,700-foot radius. Over half of it was found and picked up, eventually. The Soviet Union immediately registered an outrage, stating that the United States had violated the Limited Test Ban Treaty by setting off an above-ground nuclear explosion. The State Department fired back, pointing out that it was a reactor test and not a nuclear weapon, while wiping the carpet with AEC representatives for giving the thing such an ill-advised name: TNT, the explosive used in hand grenades.

By March of 1965, it was decided that the NERVA program was not moving fast enough for inclusion in the Apollo Moon shot. A single J-2 liquid-hydrogen/liquid-oxygen chemical rocket would do instead, but a new, more powerful NERVA II would be designed and put atop a Saturn V variant, named the Saturn N. It would be perfect for post-Apollo manned Mars expeditions. There was only one problem. Estimated development cost of the 4.1 billion-watt NERVA II with a specific impulse of 825 seconds was $6.2 billion. President Lyndon B. Johnson would have to think about it. To put the money in perspective, the war in Vietnam was costing us $2 billion per month and growing.

NRX-A3 was tested in April 1965. It ran at full power for 15 minutes, using hydrogen flow as the power control. Stop and restart were advertised advantages of the nuclear engines, so it was shut down, restarted, and pushed to 1,093 megawatts. Three and a half minutes into the second run, the reactor scrammed on a too-much-liquid-hydrogen-flow signal. The power had jumped to 1,165 megawatts, and the entire engine overheated. The steel tie-rods in the core were meant to run at 22 degrees Celsius, but the rod-temperature pen-chart recorder ran off the paper at 822 degrees Celsius. Even with that amount of stress, there was no obvious fuel breaking or cracking, so the engine was immediately restarted and run at 1,072 megawatts for another 13 minutes. There were no overstressed tie-rods, and NRX-A3 ran very smoothly using hydrogen "mapping" for control, showing an ability to throttle down to 95 megawatts.

The Los Alamos laboratory was working in parallel with Westinghouse to experiment with a new class of very powerful engines, named Phoebus. It was an improved KIWI-B with graphite fuel tips, coated with niobium carbide, glued to the ends of fuel assemblies at the hot end of the reactor, above the rocket nozzle. The steel tie rods holding the fuel together were reduced in diameter from one eighth of an inch to one tenth of an inch, and this simple modification alone increased the performance by 5 percent.

In June 1965, Phoebus 1A was pushed to 1,090 megawatts and ran very smoothly for 10.5 minutes. It suddenly ran out of liquid hydrogen, overheated instantly, and blew a fifth of the fuel out the nozzle. During the run, gamma radiation had damaged the fuel level sensor, and it falsely told the automatic control that it was out of liquid hydrogen. Picking up the scattered fuel fragments, which were radioactive from having run the engine at high power, was a good exercise in dealing with what could have been a realistic launchpad accident. At first, a truck with a 200-horsepower vacuum cleaner was tried as a high-tech method of cleanup, but it turned out to be easier and safer just to pick up the pieces with tongs and drop them into a paint can on a rolling dolly. One cleanup worker got a dose as high as 3 rems, but in 1965 the maximum allowable occupation dose for a year was 12 rems. Test Cell C was down for a year and a half for the cleanup and improvements. New 500,000-gallon liquid-hydrogen tanks were installed.

Three new NRX engines, A1, A2, and A3, began testing in February 1966. On the run to Mars, a nuclear engine would have to be started up for the deceleration into Mars orbit after having been cold-soaked for months in the absolute-zero environment of outer space. NRX A1 was cooled down, simulating the Mars trip, then "boot-strap" started, first using reactor heat to vaporize liquid hydrogen to power up the turbine that ran the fuel pump. It was a perfect, smooth start-up. NRX A2 was tested under complete control by liquid-hydrogen throttling. It ran flawlessly. NRX A3 was supposed to test mapping control at high power, simulating course correction or maneuvering in space with a high specific impulse, but a bad liquid-hydrogen flow-rate instrument caused a premature shutdown. After two hours of run time, the fuel had corroded badly, losing 120 pounds of partially spent uranium into the rocket exhaust. Still, after only 18 months of effort, the Westinghouse work was a brilliant success.

By February 1965, President Johnson had made up his mind about the nuclear rockets, and he approved the further development of the next step: NERVA II. Los Alamos had run Phoebus 1B, which was slated to become NERVA II, at 1,496 megawatts for half an hour with the turbo-pump clocking out at 95.4 pounds of liquefied hydrogen slush per second with the core at 2,165 degrees Celsius. A year or two before then, this temperature and flow rate would have been impossible to achieve. In December 1967, the last NERVA I (NRX-A6) was tested, running at full power, 1,180 megawatts, for 62 minutes. Only 40 pounds of fuel disappeared out the nozzle, but the beryllium reflector rings at the top of the core cracked under the intense neutron load. NERVA I had about reached its performance limit, and if more power was required, it was suggested to dial back the power and cluster NERVAs together.

Phoebus 2A was started up in February 1968. It was designed to hit five billion watts, but at half maximum power it started to chug, spitting exhaust with a periodic lunge. This problem was new, but much analysis was done and it was finally concluded that the surging power was due to the Heisenberg Curse.[132] In June it was run again, this time with certified parahydrogen, and power was increased to 4,082 megawatts. That is more than four billion watts. No smoothly operating nuclear fission reactor, before or since, has ever reached anything near that power level. The largest electrical generating plant reactors are run at a quarter of that power, and they are twenty times

132 In 1929, Werner Heisenberg, German physicist and noted quantum theorist, studied the strange properties of liquid hydrogen, particularly the spin isomers of H_2, the hydrogen molecule. In its normal gaseous state, the molecule is orthohydrogen, in which both atoms are spinning in the same direction. In the parahydrogen state, the two atoms spin in opposite directions. Orthohydrogen when liquefied is unstable, it evaporates quickly, and it absorbs neutrons, shutting down a reactor. Liquefied parahydrogen is stable, does not absorb neutrons, and is an excellent moderator. Parahydrogen is made by blowing the gas through a catalyst, such as iron oxide, while it is being liquefied, and apparently that step was missed when the Phoebus 2A liquid hydrogen was processed.

bigger than Phoebus 2A. After running for 12.5 minutes, the reactor started to lose power, and it came to a gentle, complete stop. It was not the fault of the hydrogen fuel. There are theories as to why, but the reason for this odd shutdown has never been determined. The Phoebus 2B test was scrubbed, and the NERVA II development project was suddenly canceled.

The state of Nevada was starting to complain about radioactive debris from upward-pointed nuclear rockets floating around in the atmosphere, contaminating the Mars-like landscape, and the United States Congress was starting to design a national environmental policy. Seeing it coming, the Los Alamos National Laboratory built a new test setup called Nuclear Furnace. The first engine to be evaluated was named Peewee. This rocket was tested firing downward into a gas scrubber that would catch any particles in the exhaust. It was a good idea and just in time to keep the entire effort from being closed down due to environmental concerns. Peewee ran on pressurized hydrogen instead of liquefied hydrogen to save money, and it only needed 11 pounds of uranium. It was a challenge to make criticality with such a small mass of uranium, but if anyone could do it, the Rover Boys at Los Alamos could. The hot test was in December 1968. It ran for an hour with a specific impulse of 845 seconds, the best yet, and a power density 20 percent higher than Phoebus 2A. There was no fuel cracking or corrosion while running at a blistering 2,285 degrees Celsius, and it proved once and for all that there was no need for rotating control drums.

In May 1971, Peewee ran for two hours with a specific impulse of 1,000 seconds, and it made five stops and restarts. At the start of fiscal year 1972, the Peewee program was canceled for budgetary and environmental reasons. The Environmental Protection Agency had been signed into being by President Richard Nixon on January 1, 1970.

In February 1969, the newest NERVA I engine, XE-Prime, ran for 3 hours and 48 minutes, trying new start-up techniques. On January 5, 1973, NASA canceled all nuclear propulsion activities. There were no plans for any manned Mars flights, and no inhabited base on the Moon was to be built. There would be no Saturn N and no mission for it. Apollo Moon landings 18 and 19 were canceled, and the Apollo 20 space station was dropped. The only surviving manned spaceflight program was the reusable Space Shuttle, which was supposed to shuttle people and supplies up to the nuclear-powered Moon shuttle in support of the Moon base. There would be nothing waiting to meet the new shuttle in low-Earth orbit.

What happened to the twentieth century dream of walking on the planet Mars? How did it vanish so completely and so suddenly? There were a few reasons that ganged up, mostly having to do with the federal budget and a lack of public enthusiasm, but the central factor was the resignation of Clinton P. Anderson, the senator from New Mexico, on January 3, 1973, three days before NASA pulled the plug.

Anderson was born in 1895 in hard-living South Dakota. At the age of twenty-two he was dying of tuberculosis, was given six months to live, and the only thing to do in those days was to move to New Mexico, where the dry air would discourage the bacteria in his lungs from procreating. If that was not enough of a challenge, he further suffered from pleurisy, diabetes, a heart attack, prostate and gall bladder surgery, blood clots, and Parkinson's disease, and those were just the ailments that were successfully diagnosed. He tried journalism, made a fortune selling insurance, and got his pragmatically conservative but socially liberal self elected to a seat in the House of Representatives in 1940.

It was old-school government back then, and he attended Speaker of the House Sam Rayburn's "board of education" meetings, which usually ended up in the basement playing cards, drinking whiskey, and smoking cigars with the boys, including Vice President Harry S Truman. When Truman found himself suddenly president in 1945, he made Anderson secretary of agriculture. In 1948, Anderson was elected senator from New Mexico, where his close comrades were Lyndon B. Johnson of Texas and Robert S. Kerr of Oklahoma. Combined, they were a force to be reckoned with. Anderson championed the crash ICBM development in 1954, served as chairman of the Committee on Atomic Energy for two years, and during the most active period in the history of space exploration, from 1963 to 1973, he was chairman of the Senate Committee on Aeronautical and Space Sciences. In this position, he was the key policy maker, and he was in charge of the funds outlaid to NASA. He passionately believed in the Mars mission and in the power derivable from the atomic nucleus, but he was growing old and feeble toward his last days in the Senate, and he could no longer fight the fight. The Bureau of the Budget (BOB) was determined to dismantle his pet space program, which it considered unnecessary, piece by piece. The BOB won, by default.

The NERVA engine did pop up again in 1987, this time as the rocket propulsion to be used in the scientifically questionable Strategic Defense Initiative. Project Timberwind, as it was named, was first part of this very secret Department of Energy effort, but in 1991 it was transferred to the Air Force Space Nuclear Thermal Propulsion (SNTP) program. Three new engines were designed, the Timberwind 45, 75, and 250. Specifications seemed lofty, by Project Rover standards, with a specific impulse of 1,000 and a burn time of 8 hours assumed in an engine that weighed one quarter as much as the old NERVA. High-temperature materials had improved in the years since Project Rover, and the accuracy and detail of computer simulations had grown more than was imagined in 1965. It was time to reengineer the nuclear rockets.

The Strategic Defense Initiative had burned through a $139 million budget without actually building an engine, and the Air Force estimate for a newly built test stand for Timberwind was $500 million, with Environmental Protection Agency design approval, of course. It was a "black program," top secret and above the normal budget oversight, but news of it leaked out and was published, and it was no longer black. This and the collapse of the Soviet threat of a massive missile attack caused the Air Force to lose interest in it, and Timberwind was canceled in 1994.

The Soviets had their own secret nuclear rocket program, and when the smoke cleared after their government went away around 1990, Russians could point proudly to their RD-0410 engine. Their project began in 1965 after clandestine observation of the American efforts since 1955, but it was stopped in 1986 after the Chernobyl power reactor catastrophe. They claimed to have ground-tested the RD-0410 at the secret Semipalatinsk facility, but there is no test stand or engine-handling facility there, and they were never able to master the intricacies of pumping liquid hydrogen or even keeping it in a tank. They also never put a comrade on the Moon. Think of the money they saved.

CHAPTER 6

The Chic-4 Revolution

"We are fighting a thousand-year war with India, and we will make an atomic bomb even if we have to eat grass."

—Zulfikar Ali Bhutto, chief martial law administrator, Pakistan, January 20, 1972, at the Scientific Conference in Multan

WITH THE NUCLEAR-ROCKET ENGINE PROJECTS, NASA and the Los Alamos National Laboratory were honestly trying to develop a nuclear technology with good, nonviolent intentions. The goal was to expand the reach of mankind beyond our immediate orbit. They managed to scatter many pounds of fission product into the Nevada landscape with engine tests, but it was not with malice, and to be realistic, the resulting pollution was insignificant compared to that caused by exploding hundreds of atomic bombs above ground in the same area. Although the rocket projects and the entire span of atomic energy began with grim, military justifications, they morphed into a quest for something better. Once we got beyond the excitement of making multimegaton explosions and watching simulated towns get blown away by the shock waves, the nuclear art took interest in generating clean power, curing cancer, and interplanetary voyaging. The prestige and sex appeal of being a nuclear ordnance engineer dimmed down to zero.

Other countries, smaller than the United States and with less technological infrastructure, have started the atomic adventure but are still in the opening phase, thrilled with making nuclear weapons and scaring the bejesus out of enemies and rivals. By getting on the wagon late, these governments have the advantage of years of essential data collection and experience, costing billions of dollars but provided free of charge, beginning in 1954 with President Dwight Eisenhower's "Atoms for Peace" program.

While Atoms for Peace provided the world with detailed nuclear knowledge, including energy-dependent neutron interaction cross sections for thousands of materials and a chart of all known nuclides in the universe, it was careful not to provide any aspect of atomic bomb design that is not obvious. To know those little details, you were on your own, and the fact that atomic bomb development is difficult kept the world relatively safe from militant theocracies, drug cartels, civil wars, and border disputes. This era of what passed for calm suddenly ended in 1998, when it seemed as if pariah regimes suddenly had the keys to the liquor cabinet of nuclear weapons. The crisis condition exists to this day, as the United States tries to put the cap back on and relock the cabinet. Tracing backward, this trouble may have begun on April 1, 1936, in Bhopal, India.

On that spring day, Abdul Qadeer (A. Q.) Khan was born to Abdul Ghafoor, a retired school headmaster, and his wife, Zulekha. The British Raj, or the period of British dominion over the Empire of India, was undergoing a gradual downward slide as unrest and rebellion began to flourish in the hot climate, and life was not comfortable for Muslims in a predominantly Hindu land. Suffering general exhaustion at the end of the Second World War, Great Britain gave up on the once colony of India and divided it into two countries in 1947. The long-standing clash of the Muslims and Hindus was settled by breaking off the eastern and northwestern provinces. These two sections, separated by the Indian Subcontinent in the middle, became the Islamic Republic of Pakistan.

Unfortunately, millions of Muslims found themselves living in Hindu-Sikh India, and millions of Hindus and Sikhs found themselves living in Islamic Pakistan. The cross-adjustment took hundreds of thousands of lives in riots and retributive genocide. Topping that, the first of three wars between Pakistan and India over ownership of the Kashmir Province began almost immediately upon partition in 1948. Not being able to take it anymore, Abdul Ghafoor, his wife, and his sixteen-year-old son A.Q. made the move to Pakistan in August 1952, settling in Karachi, Sindh. A.Q., an exceptionally bright student, moved to Karachi University from a brief enrollment at the D. J. Science College, majoring in physics, in 1956. He graduated in 1960.

In 1961, he went to Germany to study metallurgy at TU Berlin, transferring to the Delft University of Technology in Holland in 1965. His terminal degree was a DEng (doctor of engineering) from the Catholic University of Louvain, Belgium, awarded in 1972. He learned to speak fluent German, English, and French, and married a Dutch woman named Hendrina. With his new degree, A.Q. joined the senior staff at the Physics Dynamics Research Laboratory in Amsterdam and, quietly at first, began his quest to take over the world.

In August 1954, when the eighteen-year-old A.Q. graduated high school, the Atoms for Peace program from the United States allowed a nuclear technology transfer to any country that pledged not to make bombs with it. Pakistan, a country that could not manufacture a sewing machine needle or a decent bicycle, decided to jump right in, and the oddly named High Tension Laboratory was inaugurated at the Government College in Lahore by Dr. Rafi Muhammad Chaudhry for the study of nuclear physics. Chaudhry was more than qualified to start this effort, having worked at the Cavendish Laboratory at Cambridge University under Lord Ernest Rutherford, but the relatively new Islamic Republic was underdeveloped in most technical areas, starting with the manufacture of soap.[133] A year later, Chaudhry's colleague, Dr. Nazir Ahmed, led the Pakistan delegation to the Peaceful Uses

---

133 For what started out in 1947 as a "Third World" country with many technical disadvantages, Pakistan has turned into a manufacturing economy of notable accomplishment. The first Unilever factory for making vegetable oil was built in Rahim Yar Khan in 1958, and Unilever moved on to make everything from Lifebuoy soap to Lux dishwashing detergent in Pakistan. The Pakistan Soap Manufacturer's Association reported in 2014 that Pakistan now makes 120,000 metric tons of toilet soap per year. That is 60 percent of the domestic consumer demand. Eleven percent is made up for by smuggling of the product by the Afghan Transit Trade, and the rest is imported.

of Atomic Energy International Conference in Geneva, Switzerland, and shortly after Ahmed was named chairman of the Pakistan Atomic Energy Commission (PAEC).[134]

By 1960, Pakistan's ultimate goal of participation in Atoms for Peace was obliquely disclosed by the new head of the PAEC, Dr. Ishrat Hussain Usmani, appointed by President Ayub Khan. Usmani's PhD in physics was from the Imperial College, University of London, in 1939. At a meeting to discuss the national importance of nuclear research and development and provocative intelligence reports from India, he let go with, "If there will be a sixth nuclear weapon state, there will be a seventh one." In two years, he and the government of Pakistan had established nuclear research centers at Lahore and Dhaka. Initial installations were a 14 MeV neutron generator, a subcritical assembly using British magnox fuel, and a sparkling new IBM 1620 computer, named CADET.[135] The same year, India suffered a humiliating loss in a border war with China. These events—the establishment of nuclear research in Pakistan, and war footing between India and China—set the venue for A.Q. Khan's future operations.

---

134 In April that same year, 1955, representatives from Pakistan and China met at the Bandung (Asian-African) Conference, held at Bandung, Indonesia, and the two countries mutually recognized their importance to each other. China wanted Pakistan to counterbalance the west side of India as a nagging threat, and Pakistan wanted technical assistance from China for certain projects. India was a growing problem for China, with Himalayan boundary disputes and an unease in Tibet, thought to be encouraged by India. Dr. Ahmed, feeling bold at the Atomic Energy Conference, asked the United States to sell Pakistan a CP-5 research reactor—for peaceful purposes, of course. "Not for all the tea in China," replied the United States, aware of the meeting in Bandung. The CP-5 used metallic uranium-aluminum fuel, fully enriched to 94 percent U-235, and with a full fuel-core load plus a couple of spares, Pakistan could have its revenge on New Delhi with a 16-kiloton atomic bomb built using the fuel. (The Georgia Tech Research Reactor was a CP-5.)

135 The 1620 was a very small computer at the time, about the size of a large desk, but it was designed for the type of calculations appropriate for nuclear research and development. CADET was an acronym for "Can't Add, Doesn't Even Try," referring to its use of a lookup table instead of a logical adder for additions, a method of processor design that would reappear thirty years later in the Intel Pentium. The 1620 ran FORTRAN II (assuming that PAEC had bought the necessary extra memory option) and a simplified interpreter version called GOTRAN. Programs were composed on paper tape or punched cards, and the cycle time was an unusually sedate 20 milliseconds. For many decades, various versions of FORTRAN were used in the United States and around the world as the language of choice for most nuclear calculations and large-scale simulations.

The start of dangerously bad relations between Pakistan and India can be traced to December 1963, when a sacred relic, said to contain strands of the Prophet Muhammad's hair, was stolen from the Hazratbal Shrine in Kashmir. India's unwillingness or perhaps inability to replace these remnants of the Prophet created a seething rage that was difficult to contain. India tried, by releasing the popular leader Sheikh Mohammad Abdullah, who had been held for eleven years, but violence and widespread chaos remained in Kashmir. Delhi sent in the Indian army. Things would get worse.

At 7:00 Universal Time on October 16, 1964, China set off a 22-kiloton nuclear explosion atop a steel tower at the Lop Nur weapons test range.[136] The psychological shock was felt around the world, and we braced for the second tier of nuclear-equipped nations to complicate things. India wasted no time. Prime Minister Lal Bahadur Shastri approved a new defense expenditure of 28 percent of India's budget and funded a project called Study of Nuclear Explosions for Peaceful Purposes (SNEPP).

Meanwhile, things were still off the rails in Kashmir, and the second Kashmir war with India started in February 1965 when Indian troops evicted Pakistani troops from an old fort named Kanjarkot. Pakistan countered on March 6, ordering the 51st Brigade to hold Kanjarkot at all costs. On April 10, the brigade made it across the border of the Sindh province at the Rann of Kutch, engaging the Indian army in what went down in history as the Battle of the Raised Mound. Pakistan declared itself victorious, and India, taking exception, rearrested Sheikh Abdullah.

This only made things worse, and on July 24, Pakistan started Operation Gibraltar, which was intended to infiltrate Indian-administered Kashmir with disguised Pakistani soldiers. On September 2, 1965, Pakistan went to phase two and initiated Operation Grand Slam to

---

136 If the United States had been ready for this event, we would have named it CHIC-1. We didn't name Chinese weapons tests until CHIC-2 (CHInese Communist). The Chinese named the weapon "596," meaning June 1959, the time when the Soviet Union refused to provide China with a prototype nuclear device. It was a solid-core implosion weapon, very similar to the Fat Man of World War II, only made with highly enriched U-235 instead of plutonium. It weighed a hefty 3,420 pounds, and the Chinese had no vehicle that could fly this bomb anywhere.

capture the choke point at Akhnur. The infiltration failed to materialize, and India attacked four days later. The war ended in stalemate with China solidly on Pakistan's side and feeling emboldened. Seeing that India was now exhausted from the war with Pakistan, China insisted that Indian troops and all military equipment be removed from the Chinese border in three days. The Pakistani alliance had paid off for China, and these events again improved the setting for A.Q. Khan's plans.

In the midst of this war situation, on May 24, 1965, Dr. Usmani of the PAEC negotiated a deal with the Canadian General Electric Company to provide Pakistan with a turnkey 137-megawatt-electric CANDU heavy-water-moderated power reactor, to be built in Karachi. Pakistan had only to promise that they would never use it to do what it could be used for, to make weapons-grade plutonium.[137] It would be built in Karachi, and it would be named KANUPP (KArachi Nuclear Power Plant). On December 21, the 5-megawatt pool reactor, PAAR-1, at Usmani's Pakistan Institute of Nuclear Science and Technology (PINSTECH) in Nilore, achieved first criticality. By 1968, three thousand Pakistani students were studying nuclear science and engineering at home and abroad, and the Ayub Khan government had spent 742 million rupees on nuclear technology development.

---

137 The term "weapons-grade plutonium" is thrown around a lot, but a weapon can be (and has been) made using "reactor-grade" plutonium taken from spent commercial light-water reactor fuel. Weapons grade is defined as 93.6% Pu-239, 5.8% Pu-240, and 0.6% Pu-242, and it is made by low-burnup reactor operations, in which natural uranium fuel in a low-absorption moderator, such as heavy water or graphite, is not used to full efficiency. By pulling the fuel before it is spent, the Pu-239 converted by neutron activation from inert U-238 in the fuel has less likelihood of being further neutron-activated into Pu-240. The Pu-240 is an undesirable contaminant, in that it fissions spontaneously, trying to set off the bomb before it is time and causing the plutonium to at least undergo subcritical fission and generate heat. For this reason, making electrical power with a nuclear reactor and making weapons-grade plutonium using a nuclear reactor are crossed purposes. Reactor-grade plutonium is 2.0% Pu-238, 61.0% Pu-239, 23.0% Pu-240, 10.0% Pu-241, and 3.0% Pu-242. An A-bomb built with reactor-grade plutonium would be about 50 percent heavier and larger than one built with weapons-grade material, and weight is everything for weapons shot using guided missiles. The Canadian CANDU design uses natural uranium fuel, heavy water for the moderator, and is refueled while it is running. This makes it a perfect plutonium converter reactor, if it is run on a weapons-grade plutonium refueling schedule instead of a power-production refueling schedule, which would keep the fuel burning as long as possible and start to turn Pu-239 into Pu-240.

Back in Area D at Lop Nur Test Range, China, an extremely important nuclear weapon test was conducted on October 27, 1966, at 2:10 U.T. Needing a new type of nuclear warhead that could be delivered with the modestly powered Dong Feng (East Wind) DF-2 medium-range ballistic missile, China developed the "548," or the CHIC-4 as we named it.[138] It was their fourth nuclear weapon test. The CHIC-4 was small and lightweight, burned highly enriched U-235, could stand to be roughed up in the launch of a Chinese missile, and gave an explosive yield of 12 kilotons.[139] It also lacked the wide girth of an implosion weapon, so it could even be hung under the wing of a fighter plane.

Despite progress in the Islamic bomb project, Ayub's regime fell apart in 1969 and was replaced by martial law under army chief General Agha Mohammad Yahya Khan. On March 24, 1971, East Pakistan, the wayward part of Pakistan on the other side of India, seceded from the Union. Seeing this as an opportunity to demonstrate the futility of being Pakistan, India intervened militarily and captured ninety thousand Pakistani troops. This was a turning point. The humiliation was so damaging that the rivalry with India turned permanent and severe, and an ultimate form of weaponry was needed. On December 20, 1971, Zulfikar Ali Bhutto became the chief martial law administrator, or president, of Pakistan, and the atomic bomb development became Job One.

Dr. A.Q. Khan, cloistered off in the Netherlands and working at the Fysisch Dynamisch Onderzoek (FDO), made note of this development down in his native country. FDO was a subsidiary of Vernidge

---

138 The Chinese DF-2 (NATO name CSS-1, Chinese Surface-to-Surface) is interesting because it shows a transfer of secret defense technology from the Soviet Union to China in the early 1960s. The DF-2 is a copy of the Soviet R-5 Pobeda ("Victory," or the SS-3 Shyster), which was introduced in 1956 and was considered obsolete by the time the Chinese were building them in 1966. It ran on alcohol and liquid oxygen, with a range of 1,250 kilometers. The first Chinese version, launched in 1962, blew up on the pad.

139 The Chinese name, 548, means August 1954, but I'm not sure what this date means. The Constitution of the People's Republic of China, or "The Organic Law of the Central People's Government of the People's Republic of China," was written in August of 1954 and enacted on September 20. It was the first document to stipulate that China is a Communist country.

Machine Fabrieken, which made things for Ultra-Centrifuge Nederland, a member of the Uranium Enrichment Consortium in Almedo. He was asked to translate reports from Germany concerning their new G-1 and G-2 centrifuges, the latest and greatest hardware used to turn mined uranium into bomb-grade U-235.[140]

All was not well. On March 15, 1972, Dr. Usmani suddenly disappeared as chairman of PAEC, replaced by Munir Ahmad Khan. Usmani had been working for the CIA, seeing that Pakistani International Airlines planes were sprayed with a special adhesive to which dust particles would stick as they flew over China. At the time, the Pakistani airline was the only one that flew on a schedule to China, and the American intelligence organization wanted to monitor Chinese nuclear tests by analyzing the radioactive isotopes sticking to high-flying airplanes. President Bhutto called Chou En-Lai, premier of the People's Republic of China, to assure him that the problem had been taken care of in March 1973.

By 1973, Pakistan had delicately negotiated the purchase of a nuclear fuel fabrication plant for the KANUPP from Canada, but at the last minute, with the shipment in port, pressure from the United States killed the deal. A similar agreement for a heavy-water plant from

---

140 Mined uranium is only 0.72% U-235, and the rest is unusable U-238. To make a bomb, the U-235 isotope must be extracted from the uranium, making "highly enriched uranium." There are several ways to do this, none of which is easy. An early development was the ultracentrifuge, invented by Dr. Jesse Beams at the University of Virginia in 1936. He used it to separate neon-22 from neon-20. It was and still is the most efficient way to separate gaseous isotopes, but it is also the most complicated, and a great deal of further development stood between the separation of neon and the separation of uranium turned into a gas. During World War II, the United States atomic bomb project required as much highly enriched uranium as could be produced, and the ultracentrifuge was tried, but it would clearly take years to perfect the machinery to spin uranium hexafluoride gas at 90,000 RPM, so the ultracentrifuge separation method was dropped in 1944 and much simpler thermo-column, magnetic mass-spectrometer, and gaseous diffusion methods were all employed instead. During the war, Professors Wilhelm Groth and Paul Harteck tried to separate uranium isotopes using ultracentrifuges for the German atomic bomb project without any practical result. After the war, practical versions of the ultracentrifuge were developed by the German scientist Max Steenbeck in the Soviet Union. In 1956, an Austrian mechanical engineer named Gernot Zippe, who had worked on the Soviet centrifuge, was allowed to leave and go back to Austria. His dispersal of information is responsible for the ultracentrifuge design that is now in use all over the world. The ultracentrifuges familiar to A.Q. Khan were "Zippe-type" centrifuges.

Germany was stopped in its tracks at the same time. The United States, at least, was onto them. So was Libya, with different results. The one-man Libyan government, run by Muammar Gaddafi, excited by the prospect of an Islamic atomic bomb, gave Pakistan half a billion dollars in oil-earned cash plus 450 tons of uranium oxide (yellow cake) from Niger and wished them luck.

In March 1973, a team from PAEC went to Belgonucleaire in Dessel, Belgium, to learn how to separate plutonium from spent reactor fuel, doubtlessly with only peaceful applications in mind. Back home in Pakistan, Munir Ahmad Khan jump-started a program to develop explosive lenses for a plutonium-fueled implosion bomb. His team of eager technologists was named the "Wah Group."

The implosion method, developed in the United States in World War II, is about the only way that plutonium can be made to explode. Any Pu-240 contamination in the Pu-239 material causes the bomb to fissile out prematurely unless the plutonium core is assembled into a hypercritical mass with incredible speed, and all plutonium is Pu-240 contaminated. The Pu-240 radiates free neutrons, and they cause fission to start working before the core is completely assembled into the correct configuration. To prevent a premature fission start, plutonium has to change from a subcritical mass to a hypercritical mass in 1 to 4 microseconds. A way to achieve this extremely fast assembly is to use a ball of high explosives to compress a plutonium sphere, located in the exact center of the ball, into a smaller plutonium sphere. When the plutonium ball is made small by the spherical shock wave produced by setting off the explosive, its fission probability is increased by the reduction in distance between plutonium atoms, it becomes suddenly three critical masses instead of just under one critical mass, and the fission rate increases with explosive speed.

To explode a large ball of high explosive, you set it off at the very center of the ball. It starts off a tiny spark and increases in size extremely fast, shooting an expanding, spherical shock wave of hot gas outward. To *implode* a ball of explosive, you have to change the ignition configuration, setting off every point on the outside surface simultaneously. The explosive burn proceeds not from the center, but from the outside, moving inward and creating a shock wave that increases in strength and concentrates the force into a smaller and smaller sphere. It creates enough inward force to crush a metal ball, plutonium, from something the size of a soft ball to something the size of a lime. The only problem with this concept is initiating and maintaining a perfectly spherical explosive shock wave.

The United States Corps of Engineers at the secret Los Alamos Laboratory, consisting of 1,500 scientists, engineering specialists, and precision machinists, solved the problem of spherical shock-wave propagation in 1945 in the Manhattan Project. It was the most difficult task in the entire project, in which highly abstract theoretical physics was turned into a weaponized device. It took twenty thousand test explosions to perfect the technique, and for every test explosion there were probably twenty experimental configurations that were found to be not worth testing. The explosive shock waves could be treated as light waves, and hemispherical shocks starting out at a finite number of detonation points outside the ball of explosive could be bent and turned inside out like light traveling through a lens.

The ball of explosives was designed to be assembled from 32 segments, each an explosive "lens" with a detonator at the surface. The lenses were made by casting liquid explosives in molds, and these metal forms were precisely calculated, complex curves that were almost impossible to machine. It took the best scientists in the world to figure this one out.

The resulting bomb was an extremely heavy thing, 5 feet in diameter, and it was all that a big, four-engine bomber could do to fly it over Japan. It was not the sort of thing that could be carried by a rocket or by a lesser airplane. It took ten years, a few hundred billion dollars, and thousands of computer hours using very sophisticated two- and three-dimensional hydrodynamic computer codes to work it down from 60 inches in diameter to 16 inches, suitable for rocket-carry.

Being realistic, it wasn't likely that the Wah Group was ever going to solve the vast problems of explosive implosion, resulting in a device that could be delivered using a small airplane or a smaller rocket. For one thing, they lacked computer numerically controlled (CNC) metal-working machinery to make explosive molds, assuming that they

ABOVE: Chapter 1. Young Ronald Richter in his research laboratory before the war with his plasma-arc setup arranged on a bench. The equipment configuration looks neat and orderly. BELOW: Chapter 1. Richter's experimental setup in a corner of Kurt Tank's jet fighter factory in Cordoba. The work-space appears cramped and loosely organized.

Chapter 1. Dr. Richter with his wife, Ilse, and his daughter, Monica, in their new home in Bariloche.

Chapter 1. The small fusion reactor under construction. Immediately after construction was finished, Richter had it torn down. This strange action affected his relations with the construction organization and with the military.

ABOVE LEFT: Chapter 1. The first fusion reactor constructed in Richter's large laboratory building. ABOVE RIGHT: Richter's 47-ton electromagnet. That may be his wife, Ilse, standing next to it. BELOW: Chapter 1. Richter at the control desk of his final fusion setup. To his right are two very expensive Tektronix oscilloscopes, and in front is an array of pen-chart recorders.

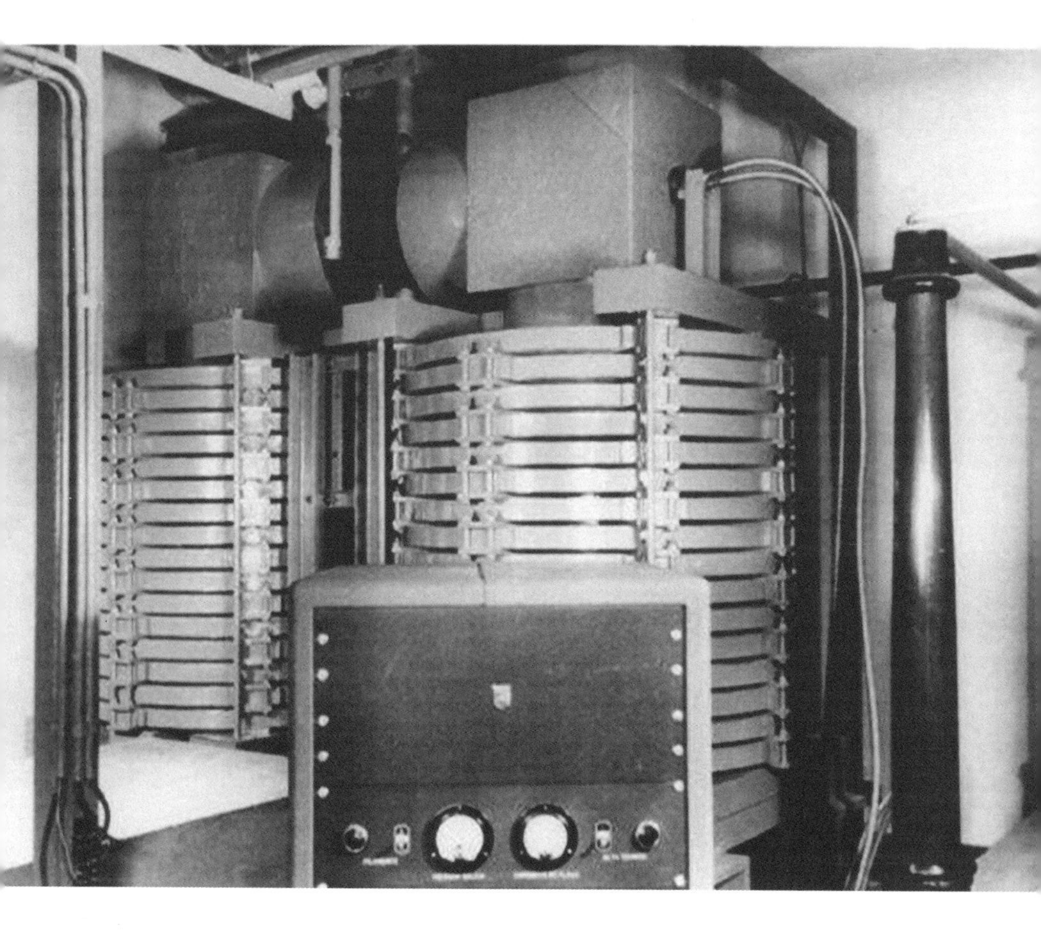

ABOVE: Chapter 1. Two lesser magnets forming a gap and a set of inexplicable high-voltage stand-offs at the fusion experiment. The reason for the extremely tall porcelain insulators is unknown, but there is speculation that Richter simply found them beautiful. LEFT: Chapter 1. This is another angle of the control desk, showing two Speedomax X-Y plotters in the top left corner. To the left of Richter's chair are two pressure cylinders, probably filled with hydrogen gas. The setup is crowded but impressive for the amount of instrumentation. The oscilloscopes, at least, are not connected to anything.

ABOVE: Chapter 1. Behind the control desk and a concrete shield is the spark control circuitry. The big glass tubes on the left are kenotron high-voltage switching-diodes. The setup looks neat and orderly. RIGHT: Chapter 1. Richter and his beloved cat, Epsilon, taken on July 12, 1955 at 5:00 in the afternoon. Two months later his patron, Juan Perón, would be ousted by a military coup.

Chapter 1. Two of Richter's laboratory buildings as they now appear.

Chapter 1. Visitors would stand behind the railing to be dazzled by one of Dr. Richter's demonstrations.

ABOVE: Chapter 2. The main reactor at the REF being installed. Notice the very shiny stainless-steel walls.
BELOW: Chapter 2. Mel Dewar touching something in the CEF (Critical Experiment Facility) reactor with a 10-foot pole.

ABOVE: Chapter 2. The ground-level portal to the underground control room as it now appears. BELOW: Chapter 2. Billy Statham, reactor operator, standing at the control console for the REF reactor. He is changing a roll of paper in a pen-chart recorder.

ABOVE: Chapter 2. The east side of the hot cells as they now appear. All window and manipulator penetrations have been blocked with steel plugs. BELOW: Chapter 3. Our palladium electrode sawed off a one troy ounce ingot of precious metal.

Chapter 4. The team, in RM127. From left to right, Gary Beebe, Darrell Acree, Rick Steenblik, the author, and Billy Livesay.

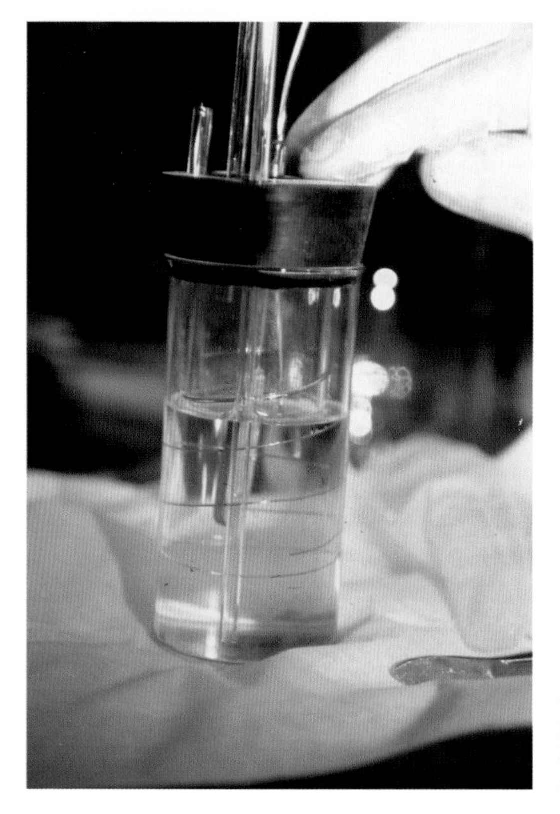

ABOVE: Chapter 4. Billy and Darrell preparing to clear our palladium of any existing hydrogen. LEFT: Chapter 4. The Georgia Tech cold fusion cell.

ABOVE: Chapter 4. The cold fusion cell, before being lowered into the graphite block. Three BF3 neutron counter tubes are immersed in the cooling water. RIGHT: Chapter 4. Gary and I working on the detector setups. In this picture, the lead castle atop a wooden table is clearly visible. A portable neutron detector sits on the boron bricks inside the castle walls.

ABOVE: Chapter 5. KIWI-B4-A set up at Test Cell A, in front of the combination blast and radiation shield used to protect the data-recording instruments. BELOW: Chapter 5. Setting up for an NRX test. This is half of the clamshell vacuum chamber. It clamps around the engine and simulates the operating environment of outer space.

ABOVE: Chapter 7. The 27-inch cyclotron at RIKEN. It is shown with the accelerator chamber racked out from between the poles of the enormous electromagnet that accounts for most of the weight of this machine.
BELOW: Chapter 7. The accelerator chamber of the 27-inch cyclotron at RIKEN, on the bench and ready to be installed in the magnet gap.

ABOVE: Chapter 7. The RIKEN laboratory before the war. BELOW: Chapter 7. Dr. Yoshio Nishina.

Chapter 7. A big piece of a Japanese cyclotron being dumped into Tokyo Bay.

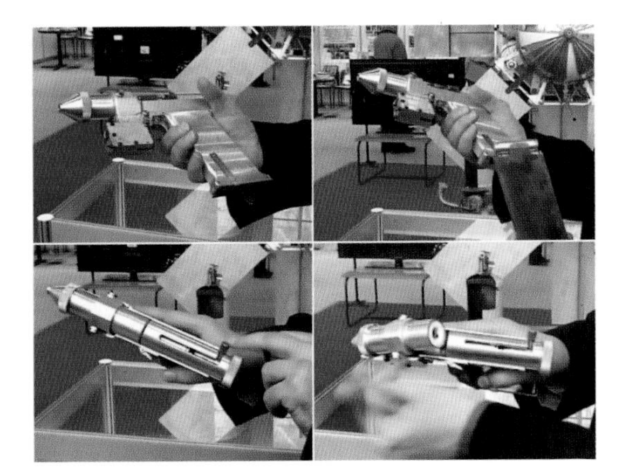

ABOVE: Chapter 8. From the top left, moving clockwise: The prototype Soviet laser pistol to be used for defense in Earth orbit. The ejected magazine containing disposable zirconium flash cartridges. The breech snapped open to eject a spent cartridge. Push the lever to the left to load a cartridge and pull the trigger to fire. BELOW: Chapter 8. The 1K17 Szhatie Soviet mobile laser weapon built at the height of the Cold War in the 1980s. As you can see, it weighs several tons and runs on Msta-S tank tracks, and therefore it may be impractical for criminal use. It may have taken a half an hour to charge up the capacitors for a single shot, so its value as a military weapon was also questionable.

ABOVE: Chapter 8. A cluster of four Soviet radio-thermal electrical generators, filled with enough strontium-90 to run a lighthouse. A small block of strontium-90, which was an inexpensive by-product of nuclear power reactor waste, gets so hot it glows dull red. BELOW: Chapter 9. The original Victoreen Model 247, from 1943. It turned out to be too delicate for use in the beach landing at Normandy in 1944.

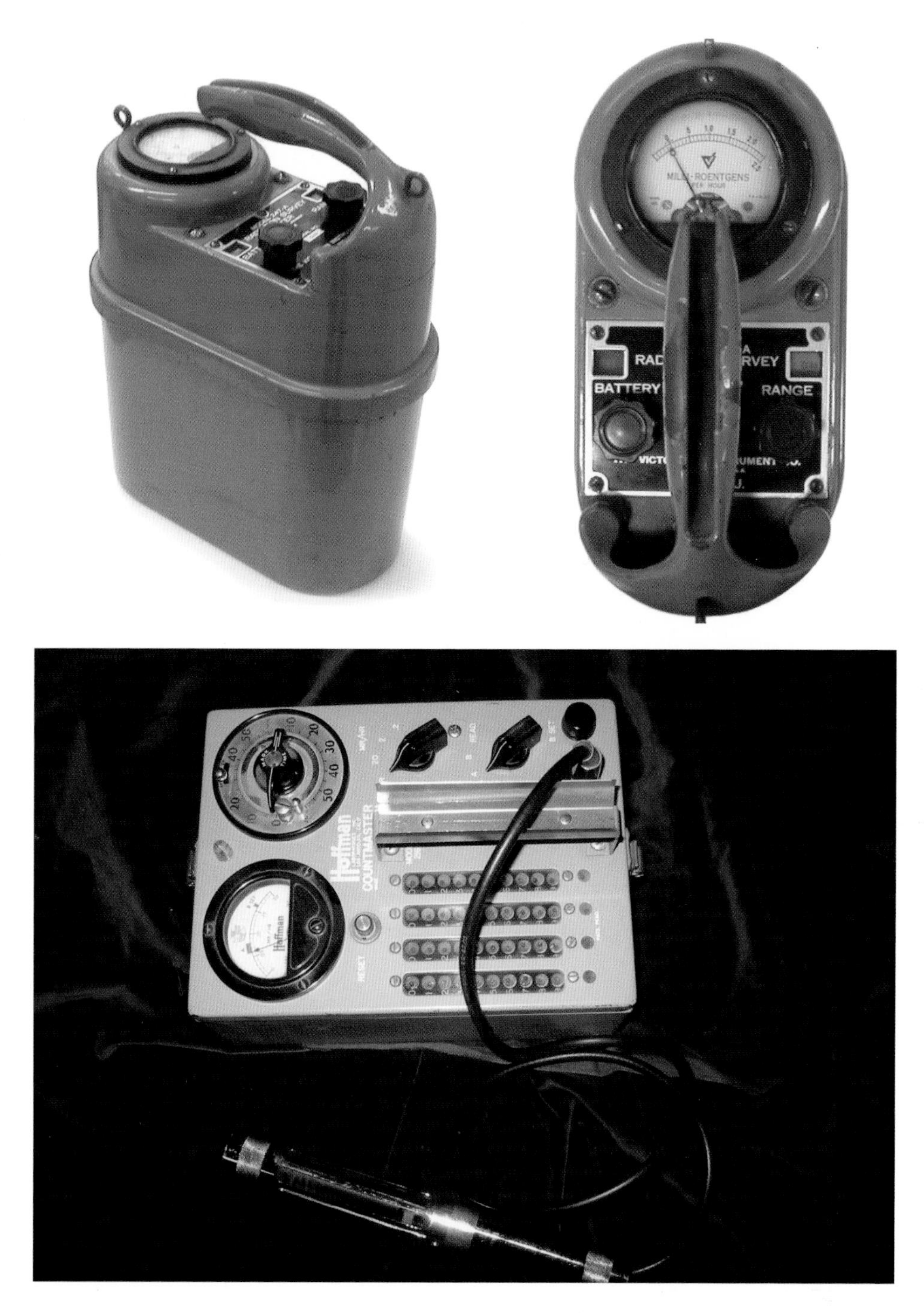

TOP LEFT AND RIGHT: Chapter 9. The Victoreen Model 247A radiation survey meter from 1946. It met all the needs for a rugged, dependable instrument to measure dangerous radiation levels in nuclear weapon tests. BOTTOM: Chapter 9. Costing $299.95 in 1954, the Hoffman Countmaster was one of the most expensive portable Geiger counters available, but it was unique in that it was equipped with a four-decade digital counter having an impressive nanosecond response time. It was appropriate for uranium ore assaying and precision radiation contour mapping as well as field exploration. The countdown clock is mechanical.

Chapter 9. January 1955, Blossom Barton of Claremore, Oklahoma, checks out her mineral-water bath with a Geiger counter before getting in to make sure it's radioactive enough. The general feeling about radioactivity was different back then.

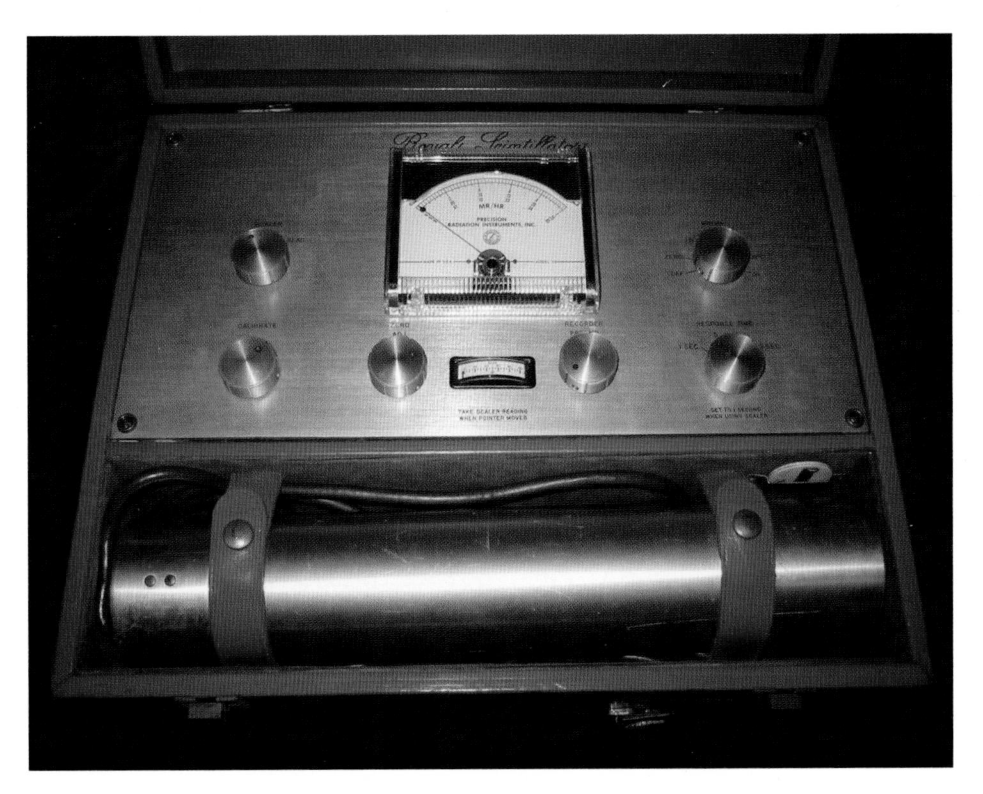

Chapter 9. The author's gold-plated PRI Model 118 Royal Scintillator, tastefully presented in a leather-covered case, is one of two working examples known to exist. It has a counter function similar to the Hoffman Countmaster, but it is analog instead of digital, showing gamma-ray counts taken during an electronically derived countdown period as a needle deflection on the large meter. It can find uranium in a fireplace brick.

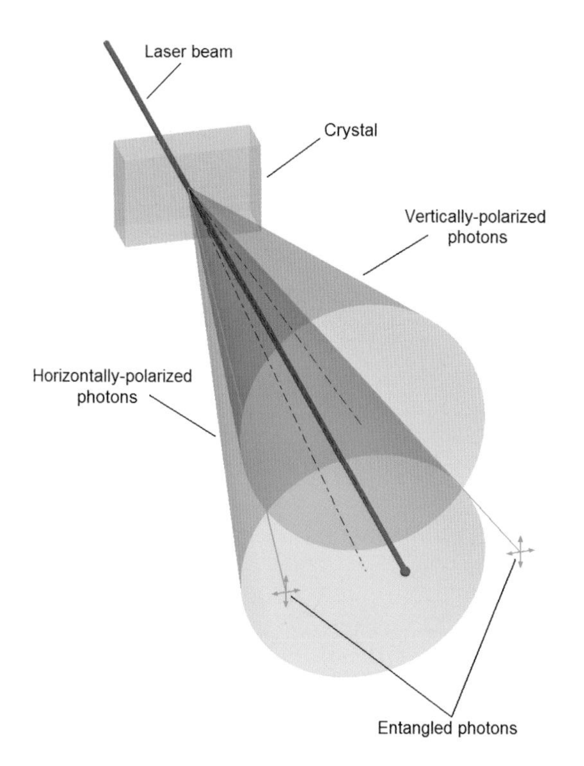

ABOVE: Chapter 10. For the purpose of making entangled photons, a laser beam is directed through a down-conversion crystal. The unused portion of the beam continues straight on, but down-converted photon pairs are segregated into vertically and horizontally polarized examples, exiting the crystal in cone-shaped regions. The lines where the two geometric cones intersect define where to look for the special entangled pairs. BELOW: Chapter 10. The Bussard Interstellar Ramjet, showing the large hydrogen collector in front. The visible exhaust is fusion products heated to incandescence.

Chapter 10. Just to give an impression of the size of the unmanned Daedalus probe, this is how it looks standing next to the Empire State Building. It is very large, and it will be expensive to build.

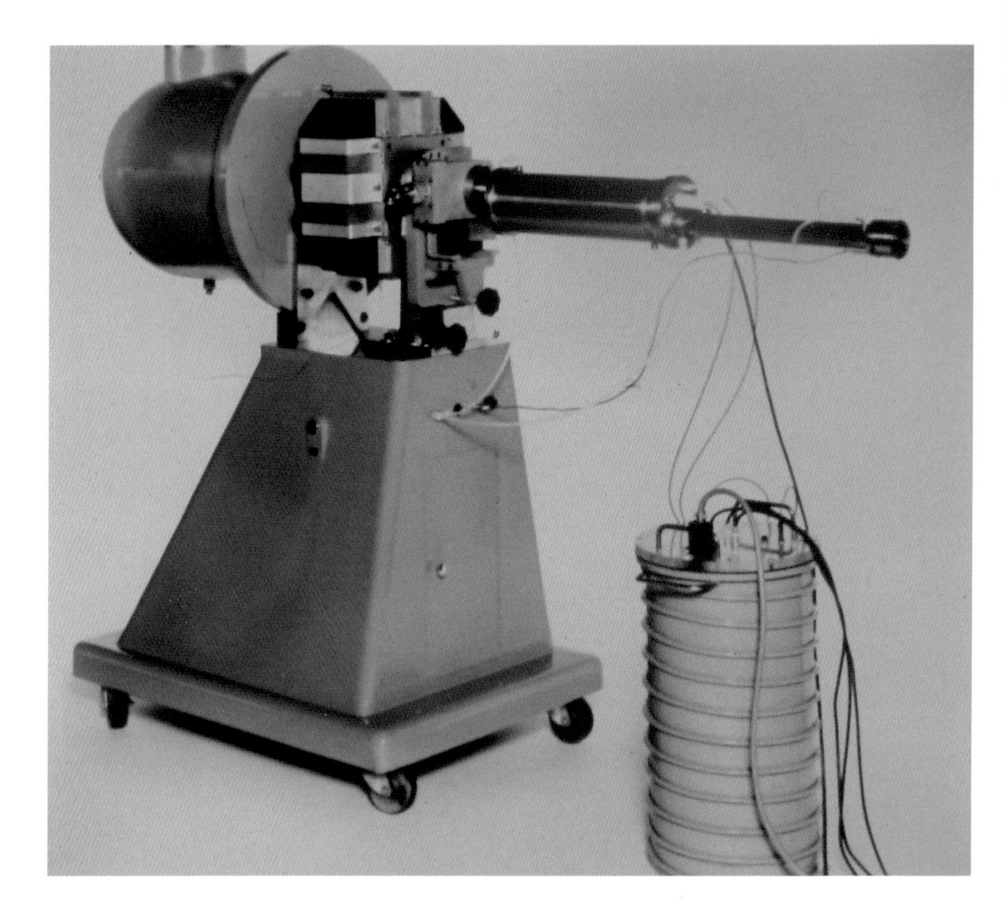

ABOVE: Chapter 11. Judge Ellison's Activitron, built and sold by the Ellison Company, Atlanta, Georgia. The cylinder in front is the high-voltage power supply. Neutrons come out of the nose-piece, on the right. BELOW: Chapter 11. Our Kucherov glow-discharge cold fusion reactor, on display in the Georgia Tech Library.

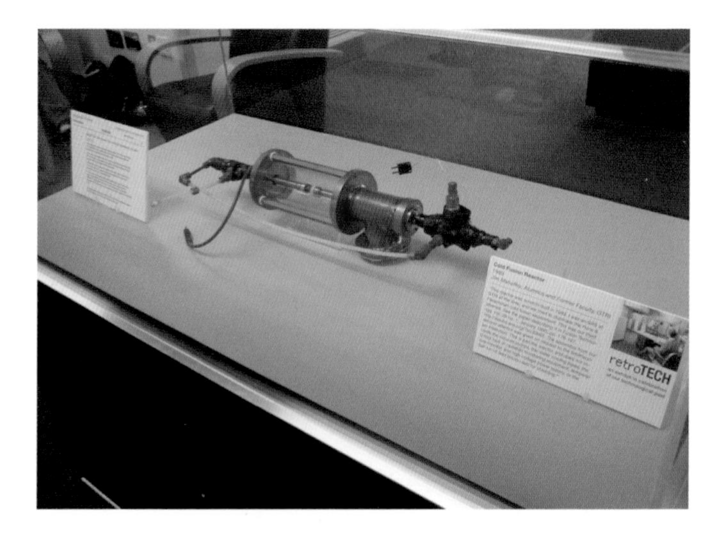

arrived at a design. As Munir Ahmad Khan famously asked his team, "If the Americans could do without CNC machines in the 1940s, why can't we do the same now?"[141]

Things were not looking good in 1974 Pakistan. A poor wheat crop left the economy in dire straits, and on May 8 at 8:05 in the morning, the seismometers in Karachi quivered. Homi Sethna, chairman of the Indian Atomic Energy Commission, phoned Prime Minister Indira Gandhi with a simple, four-word message: "The Buddha is smiling." India had just successfully tested a 10-kiloton atomic bomb.

Things could hardly have looked worse, but on September 17, Pakistani prime minister Zulfikar Ali Bhutto received a letter from A.Q. Khan. The letter carefully explained the advantages of using highly enriched uranium over plutonium in designing, building, and implementing an atomic bomb. They could spend the next twenty years working on an implosion device. Why not chuck that and build a simple uranium bomb? All they needed was an enrichment plant, and A.Q. claimed to know all about how to build one, and he was ready to quit work in the Netherlands, come back to the homeland, and be put in charge of a bomb program. Bhutto was intrigued. He wrote back, telling Khan to stay right where he was and learn as much as possible about this "ultracentrifuge" to which he referred.

The uranium-enrichment plan was named Project 706. With long-distance encouragement and written plans from A.Q. Khan, Bhutto approved a $350 million budget for Project 706 to build a production-level cascade of three thousand centrifuges, capable of

141 The question was incorrectly phrased. The Americans at Los Alamos in 1945 used their AA1 priority to commandeer a Kellermatic CNC milling machine at the Chrysler plant in Detroit, Michigan, to make four aluminum masters for the Fat Man explosive lenses, shutting down the B-29 bomber engine assembly line for three days. From the masters, molds were built in the Los Alamos machine shop using Cero-Tru non-shrink bismuth alloy. The Kellermatic, named for K. T. Keller, president of the Chrysler Corporation, was a state-of-the-art machine, using a paper-tape program and capable of making weird, very precise shapes that were impossible by any other means. The Americans did not do without CNC machines, and it was unlikely that the Pakistani effort could succeed without this technology plus some others in short supply.

making bomb-grade U-235, in Kahuta.[142] The project was disguised as the "Airport Development Workshop" and was headquartered in a nineteenth-century army barracks, complete with hot tin roofs, fans to move the hot air around, and lots of snakes.

In 1975, A.Q. Khan quit his job in the Netherlands and moved to Pakistan, ready to assume his correct place in the larger scheme of things. The Pakistan bomb program was now moving ahead with power and resolve, starting with a massive buying spree in Germany. From Germany came a high-compression rolling mill from Düsseldorf, intended for making steel artillery shells but ideal for fabricating ultracentrifuge casings. Leybold-Heraeus in Hanau supplied vacuum equipment to convert UF4 powder into UF6 (uranium hexafluoride) gas for use in centrifuging and a machine to make centrifuge rotors. An electron-beam welder was ordered to connect the aluminum pipes used to cascade uranium hexafluoride gas from one centrifuge to another. In all, Pakistan did business with eighty German firms selling equipment configured to make a uranium enrichment plant. By 1976, Pakistan had built the Baghalchor BC-1 uranium processing facility, including an ore storage mill, ball-grinding mill, a sulfuric acid plant, a solvent extraction plant, and a tunnel dryer. All this, with a great deal of help from Germany, was accomplished by a country that twenty years before had trouble spreading asphalt on a road.

Project 706 was run by Sultan Bashiruddin Mahmood. He unfortunately belonged to the Ahmadi sect, which was believed to be anti-Muslim or at least non-Muslim by many of the Sunnis working on the project. Ethnic and religious rivalries and grudges were a seething,

---

142 Highly enriched uranium (HEU) is defined as uranium with a 20% or higher content of fissile U-235, with the rest being mostly U-238. In theory, a bomb could be made with any concentration of highly enriched uranium, but a bomb of reasonable size and weight requires an enrichment of about 80%. The higher, the better. The practical limit for enrichment is 93.5%, and as the degree of desired enrichment goes up, the amount of effort, time, and the number of enrichment machine stages grows rapidly. The number of cascaded stages of enrichment grows into the thousands, as the total enrichment per stage becomes smaller and smaller with rising enrichment percentage. A critical mass (the amount of fissile material necessary to support a self-sustained fission reaction) of 93.5% enriched uranium, with an enclosing beryllium neutron reflector, is only 14.1 kilograms. A critical mass of reflected 80.0% uranium is higher, at 19.3 kilograms. The same thing in 20.0% uranium is a hefty 245 kilograms. The size and weight of a simple uranium bomb depends greatly on the ability of the enrichment facility.

grumbling background problem in Middle Eastern countries at the time, and even in an educated group it was undermining the work. Whenever something went wrong, which was usually several times in a day, Mahmood was accused of sabotage due to his unpopular religion. In April 1976, Mahmood was canned and A.Q. Khan took over as director of research and development, and this bold action did not change things very much. Khan was a muhajir, an Urdu-speaking immigrant from India, and he was working in a place loaded wall-to-wall with Punjabis. Work was tense, and Khan's expansive ego did not help, but he was intensely focused and able to ignore anyone else's opinions.

In June 1976, the first P-1 ultracentrifuges, built to A.Q. Khan's specifications, were tested at the pilot plant in Sihala. It was a modest cascade of 52 machines, eventually growing to 164. By December, Canada, suspecting the obvious, cut off all support for the KANUPP reactor, including fuel, heavy water, spare parts, and technical assistance. Six months later, on July 8, 1977, it was time for another military coup, and Prime Minister Bhutto was overthrown. What was left of relations with the United States rapidly deteriorated.

At 2:00 A.M. on June 6, 1978, a detectable amount of U-235 was separated by a single ultracentrifuge running at Sihala. Around that time, a shepherd was tending to his flock outside the walls of the separator facility when his dog started going crazy over a rock on the ground. He barked, growled, and scratched at it, trying to intimidate it and make it get up and run. Curious, the shepherd went over and turned the rock over. It was an electronic device, collecting data of some sort and storing it for retrieval later. It might as well have had "USA" painted on it in red, white, and blue letters.

In case anyone would prefer that he be in charge, Zulfikar Ali Bhutto was hanged in Rawalpindi on April 4, 1979, and two days later, the United States' President James E. Carter, outraged by the execution, instituted formal sanctions against Pakistan. A few months later, Belgonucleaire of Belgium constructed a large heavy-water plant in Multan. By 1981, the outer ring of cascading ultracentrifuges was completed at Kahuta. President General Zia-ul-Haq visited the plant to see it in operation, admire the rows and rows of sleek vertical cylinders, and hear the shrill whine of hundreds of rotors spinning at 90,000 RPM. Thrilled with the progress brought by A.Q. Khan, he renamed the engineering research lab the Khan Research Laboratories. The project was progressing nicely.

One night in September 1981, the day crew at the Kahuta centrifuge plant was bounced out of bed. The floor seemed to be rolling back and forth. A magnitude 6.1 earthquake was rocking Islamabad, which is only 30 kilometers to the southeast. Over the ever-present whine of the centrifuges, they could hear the explosions and the crashing sounds getting louder and starting to run together as the entire plant went up in a cloud of escaping uranium hexafluoride gas and bits and pieces of aluminum tubing, clattering off walls and ceilings.

The Zippe-type centrifuges, the P-1s and P-2s, were tall columns of polished, precisely machined maraging steel, spinning at rates that broke the sound barrier at the outer surface of the cylinders. At the bottom, each rested on a conical journal sitting in a static, inverse-conical bearing surface. The two metal cones never touched, with the column supported by a thin layer of constantly replenished oil. The top of the cylindrical column was kept running true by a permanent ring magnet, also never touching any metal part of the spinning mechanism. The assembly was spun as if it were the rotor in a synchronous AC motor, with alternating current vibrating at a computer-controlled rate through a coil assembly outside the bottom of the rotor.

Uranium fluoride gas, which is very dangerous to breathe, was introduced through a tube in the bottom of the centrifuge cylinder. Caught in the rapidly spinning cylinder, the very slightly heavier gas containing U-238 would tend to hug the wall, while the very slightly lighter gas containing U-235 would be pushed to the center, where it was sucked out of the centrifuge through a tube at the top and sent to the next stage in the cascade.[143] The cylinder was encased in an outer

143 This is a simplified explanation of the Zippe-type centrifuge (known in Mother Russia as the Kamenev centrifuge). This design also uses the thermal-column principle, as used in the steam-driven S-60 process at Oak Ridge in World War II, in tandem with centrifugal force to separate the isotopes. The bottom of the rotating maraging steel cylinder is induction heated, causing the lighter U-235 fluoride to float to the top and the heavier U-238 fluoride to sink to the bottom. There is one input pipe, an output pipe for lighter gas at the top, and an output pipe for heavier gas at the bottom. The cylinder spins in a vacuum to reduce the aerodynamic friction, and the aluminum cover acts as both a shrapnel catcher and a vacuum chamber. Each P-1 or P-2 ultracentrifuge has ninety-six component parts.

tube made of aluminum, not touching the spinning part but designed to save nearby centrifuges if this one happened to explode.

Things were running well as long as the floor, to which the centrifuges were bolted, did not move. If the floor slipped sideways, the spinning cylinder, experiencing an intense gyroscopic effect, would prefer to remain in the same place. It would jump out of the bearing, hit the inner wall of the aluminum case, and disintegrate into a thousand pieces trying to dissipate the kinetic energy of revolving at 90,000 RPM. The result was utter disaster, and the entire Pakistani centrifuge project went literally to pieces.

Faced with having to start again from scratch, President General Zia sent his best man, Lieutenant General Naqvi, to China to collect a debt. He returned in triumph, with 50 kilograms of highly enriched uranium (not all in the same container!), 10 tons of UF6 (uranium hexafluoride) gas ready for enrichment, 5 tons of UF6 pre-enriched to 3%, and the complete plans to build a CHIC-4 nuclear device, the perfect A-bomb for a pariah regime on a strict budget. With their 50 kilograms of 93.5% U-235, Pakistan could put together two CHIC-4s.

What exactly was a CHIC-4? That information seems unavailable, and the original sales organization for it has been disbanded. There is, however, enough information to back-engineer a possible configuration.

The CHIC-4 was a solid-core, uranium bomb yielding 12 kilotons of explosive force. It was dropped on the test range by a ballistic missile the size of a large power pole, so it had to be physically small and as light as possible, given the early stage of Chinese missile design.

Highly enriched uranium is very expensive to manufacture, requiring a large, technically sophisticated plant of a sort that could not be built in 1945. Still, highly enriched uranium is three to four times cheaper to produce than plutonium. An ultracentrifuge uranium-separation plant can be hidden in a chicken farm, while plutonium production begins with a large, above-ground nuclear reactor with an obvious radioactive signature. Process control computers, once the impossible component of the centrifuge plant, can now be bought at Best Buy. A plutonium bomb demands a terribly complicated implosion design. Uranium can be used in an implosion bomb, but that isn't necessary. A uranium-based atomic bomb, like Little Boy (L-11) used on Hiroshima in 1945, is nothing but two subcritical pieces of uranium assembled into the correct configuration by being thrown together very quickly. The design is so absolutely foolproof, the first A-bomb dropped on Japan was not the tested, plutonium-based Fat Man implosion bomb. It was the Little Boy uranium "gun-assembly" weapon. It had never been tested. There was no need to. It could not do anything but explode.[144]

Little Boy was a "single-gun" assembly weapon, consisting of a 6.5-inch smooth-bore cannon barrel, 6 feet long, poised to fire a subcritical mass of 80% enriched U-235 into a similar mass of U-235 held stationary in a block of forged steel, 28 inches in diameter. The entire bomb assembly was 10 feet long and weighed a hefty 9,700 pounds, but it was lighter and smaller than the Fat Man implosion bomb dropped on Nagasaki, and there had been no particular effort to design it to be smaller.

The criticality coefficient of an assembled mass of fissile material is "k," where k equals 1.0 denotes perfect criticality, in which exactly the same number of neutrons produced in fission equals the number of neutrons lost by all means of neutron loss. In this state, the mass releases energy at a steady rate.[145] At k less than 1.0, the mass

---

144 It was so easy to make a uranium assembly weapon explode, it remains one of the most dangerous bombs ever deployed. It has just one detonation point, and a spark, a fire, or electrical glitch could activate it. (A B-29 bomber catching fire was not a rare occurrence.) The implosion bombs, on the other hand, have as many as ninety detonation points, and if one fails to go off at exactly the right time, the bomb is a dud. An assembly weapon can go off without any detonation at all, just from being dropped on the nose. And, it must be kept dry. If water were to seep inside, it would enter the cannon through air-pressure relief slits at the front of the barrel and act as a neutron moderator, causing the touchy projectile to go critical. Safeguards for the Little Boy assembly weapon included not installing the detonation components until it was on the way to Japan, installing three copper pins in the smooth-bore barrel of the assembly gun to keep the uranium projectile from sliding forward to a critical configuration, and including a timer to turn on the radar altimeter late in the drop, to keep it from interpreting the bottom of the airplane as the ground.

145 To be precise, the coefficient is "k-effective." It is a divisor, located under the eigenvalue multiplying the fission-source term in the neutron diffusion equation, and as a solution to the equation, it is used to predict the degree of criticality of a nuclear assembly. The "equation" is actually a set of twenty-three (typically) partial differential equations to be solved simultaneously, expressing the addition of all sources and the subtraction of all losses of neutrons in the assembly, set to zero indicating perfect criticality and rendering the equations solvable. The k-effective term forces the collection of terms to add to zero by artificially lowering or raising the eigenvalue that multiplies the neutron source term representing nuclear fission. To solve this equation in three, two, or even one physical dimension is impossible without electronic digital computation, a capability that did

is subcritical, more neutrons are lost than are produced, and the power level drops exponentially. The k of the subcritical projectile in the Little Boy cannon was 0.89. At a k greater than 1.0, the neutrons lost are less than the neutrons produced, the power level rises exponentially, and the mass is supercritical. If k is greater than or equal to 2.0, the power level rises explosively, and the mass is hypercritical.[146] Little Boy had k equals 2.4 the instant that the movable U-235 mass slammed into the stationary one, driven by rapidly burning nitrocellulose propellant in the cannon breech. Only 1.4 percent of the uranium actually fissioned. The rest was wasted as the bomb instantly became a large ball of fire.

By September 8, 1957, twelve years after the Little Boy was designed, the scientists and engineers at the Los Alamos National Laboratory had reduced the U-235 assembly weapon down to a thing that was only 8 inches in diameter, 22 inches long, and weighing 243 pounds. Explosive yield was variable, up to 12 kilotons, and adjustable by the amount of U-235 in the two subcritical masses. The W33, as it was designated, was contained in an artillery shell made of titanium and could be fired from a cannon. A thousand of these things were deployed in Europe for decades in such loosely coupled locations as Belgium and Greece, and a couple of them may have been misplaced. The purpose of these weapons was to be used against approaching tank battalions of the Soviet army, launching a surprise takeover of Western Europe, and as such they were at the very least an effective threat. Its small size, weight, cross-sectional area, uranium fuel, and gun assembly would make the design extremely attractive for countries in which large bomb-delivery vehicles werent available and explosive lens system design was overwhelming.

---

not exist when the Little Boy was designed. The best that could be done was to solve the equation in zero dimensions, mathematically treating the bomb core as a single point and tolerating the inaccuracy thereof. The exact subcritical masses of the projectile and target components were thus found experimentally and not mathematically.

146 Under the hypercritical condition, it takes 0.560 microseconds to go from zero power to the power of 20 kilotons of TNT going off all at once. That's what we mean by "power level rises explosively." In that fleeting length of time, a little more than half of a millionth of a second, there are fifty-six generations of neutrons, in which neutrons are "born" from fissioning uranium and are then lost as they crash into adjacent uranium nuclei and cause further fissions.

Should the Soviet Union have acquired one or two W33s, they would have been delighted to find how the Americans had miniaturized it. First, it used 93.5% U-235, which became possible to make in quantity just after the war, when the K-25 gaseous diffusion uranium-enrichment facility at Oak Ridge became fully operational. With the original 80% enriched uranium as used in Little Boy, it takes 19.3 kilograms of uranium to make a reflected criticality. With the 93.5% U-235 as used in the W33, it takes only 14.1 kilograms to do the same thing.[147]

The second feature was the use of a double-gun assembly system. Instead of firing a projectile against a stationary target, two guns were pointed at each other, simultaneously firing two subcritical uranium projectiles, smashing together to make a hypercritical mass. Two elongated bullets melted on contact and formed into a surface-area-optimized ball at the midpoint, surrounded by a hollow, spherical beryllium reflector acting as the casting form. Cutting the speed of the projectiles by half meant that the pressure in the gun barrel was cut to one-quarter, and the weight and size of the gun were reduced to one eighth that of the single-gun configuration. It was ingenious.

After deployment of the W33 in 1955, the Soviet Union began development of uranium-powered nuclear artillery in 1956. They had by then fully implemented state-of-the-art Zippe-type uranium-enrichment facilities capable of turning out 93.5% U-235. By 1965, they were using the RFYAC-VNIITF nuclear artillery shell and had mercifully shortened its designation to ZBV3.[148] Being shamed by the People's Republic of China for not having given their kindred-spirited neighbor just one little bomb design, they probably gifted China with the ZBV3 design plus plans for the SS-3 Shyster missile. The nuclear

---

147 Or, it could have been any of the other miniaturized, gun-assembly nuclear warheads developed in the United States, including the W7, W9, W19, or the really tiny one, the W48. I have centered speculation on the W33 because it was the only nuclear artillery round that was made in 1,800 units and deployed widely in Europe. It was the warhead most likely to be found in the illicit novelty-weapons market.

148 There is a rumor, only a rumor, that the Soviets improved the W33 design by making it a "complex gun" assembly. In this design, the cylindrical neutron reflector is actually parked a couple of centimeters away from the impact point of the two uranium projectiles. This allows a further increase in the assembled k-effective (hypercriticality) by adding the reflective reactivity only after (by microseconds) the core has been assembled. This trick is accomplished using a two-dimensional shaped explosion that does not require three-dimensional lens design or manufacture.

artillery shell mechanism was repackaged and modified to make the rocket-deliverable CHIC-4 with an explosive yield of 12 kilotons.

By 1983, a large fraction of the Kahuta ultracentrifuge plant had been repaired, put back into business, and destroyed again by another earthquake, but in January 1984 in a reckless moment, A.Q. Khan claimed to journalists that Pakistan was now able to enrich uranium. In 1985, Linde, A.G., in Germany was very helpful, supplying Pakistan with a heavy-water detritator using cryogenic distillation. Its purpose was to separate tritium gas out of the coolant used in the KANUPP power reactor. The Germans didn't ask why they needed it.[149]

The eager engineers in Pakistan were already building CHIC-4s, modified to fit under the wing of an American F-16 fighter plane or a Mirage 5 from France. The Pakistani Air Force was of modest size, but it was equipped with some American and French planes, acquired in friendlier times. To test these bombs without setting off a nuclear explosion, they were "dry-fired," or exploded with every part functioning except the uranium core segments. Substituting depleted uranium (mostly inert U-238) for the 93.5% U-235, these devices could be proof-tested by detecting a neutron burst upon detonation, caused by crushing the bomb's explosion initiator between the two correctly fired projectiles.[150] If the bomb passed this test, then

---

149 They needed it to boost the CHIC-4. The original W33 artillery shell was rated at 10 kilotons yield unless the fission was boosted by a vial of a few grams of tritium in the center. Tritium, under the stress of being compressed and heated by a collocated nuclear explosion, fuses, releasing a large burst of neutrons, which hasten the fission in the U-235 and amplify the explosion. With the tritium booster, the W33 was rated at 12 kilotons. The Chinese insisted that the CHIC-4 and the Pakistani CHIC-4 derivative bombs were not boosted.

150 The initiator, or "modulated neutron source," used in nuclear weapons is technically unnecessary in a gun-assembly weapon. At least one free neutron is necessary to start a chain reaction in U-235. A neutron is always available from U-238 and U-234 impurities, which fission spontaneously and release neutrons at random. In the Little Boy device with 80% U-235, there were 12.8 kilograms of U-238/234 in the 64 kilograms of uranium, producing a background count of one fission per 14 milliseconds. For a precise detonation time, such as is necessary for air-dropped bombs, waiting for a random neutron to start the reaction does not work, and a neutron source that can be turned on instantly is necessary. This is accomplished by smashing together and mixing a sample of polonium-210 (an alpha particle source) with a sample of beryllium-9, which blasts apart under alpha bombardment and releases neutrons. At least 3 curies of polonium-210 are needed, resulting in 1 neutron per 100 nanoseconds. Polonium-210 for this "po-be" neutron source is made by bombarding bismuth with neutrons in a nuclear reactor, such as KANUPP.

there was no reason to waste highly enriched uranium on a full-scale explosion, unless there was need to rattle the seismometers in India and cause worldwide concern. There was nothing to keep a correctly fired gun-assembly weapon from exploding if loaded with the right concentration of U-235.

By January 1987, A.Q. Khan was sufficiently impressed by his own superiority and his indispensable value to the Pakistani atomic bomb project. He felt that he could get away with anything, and for the next thirteen years, he did just that. That month he spilled everything about the project to a Pakistani journalist, boasting that not only could Pakistan now make bomb-grade uranium, but it now possessed nuclear weapons. Later that year, he personally sold the drawings, specifications, descriptions, calculations for a 2,000-unit isotope separation plant, vacuum and electrical equipment, uranium casting and conversion capabilities, and a disassembled P-1 ultracentrifuge to Iran. At that point he was not only in a top position in the Pakistan government, he was wealthy.

Three years later, the Pakistani atom bomb project was plowing new ground. The Pakistanis had developed a lightweight, slim-profile fission weapon using implosion assembly.[151] Such a weapon was far more complicated than the double-gun, and it would have to be tested. They didn't want to let India and the rest of the world onto just how sophisticated they had become, so they couldn't test it in Pakistan. No problem. The Chinese set them up with a test at Lop Nur, Area C, on May 26, 1990. It was an off-the-books test, simply listed as "event

151 What exactly was this device? It appears to have been a "linear implosion," which is a lot simpler than a three-dimensional, spherical implosion, plus it makes a thin bomb that can fit under a fighter plane. The barely subcritical U-235 mass is shaped like a football with a hollow center. To become hypercritical, it must be mashed into a sphere by being pushed simultaneously at both ends of the football, and the void in the center must be crushed down to nothing, improving the density and the surface-area ratio of the mass. This is accomplished by embedding the mass in a cylinder of high explosive, with a detonator on each end, set off precisely at the same time. (The twin-detonator technology had been worked out for the double-gun design of the original CHIC-4.) An inert wave-shaper is cast in on both ends, between the ignition points and the two ends of the oblong mass. It's not an efficient use of uranium, but it gives you the bragging rights of having developed an implosion weapon. This technology, the W54, was developed by Los Alamos National Laboratory in 1956, and was famously used in the Davy Crockett recoilless gun and the "back-pack" weapons designed to be carried by one person.

35." The "CHIC-4 derivative" went off underground at 8:00 U.T. The Pakistani observers were pleased.

The Iranians had been annoyed by their first purchase from A.Q. Khan. The P-1 centrifuge he had sold them was junk, put together from broken or worn-out parts, but it whetted their appetite for nuclear technology, and they wanted more of it. In 1994, the Iranians bought five hundred used P-1 machines, which by now were obsolete in Pakistan, plus the designs for the enhanced P-2 machines. Khan, headquartered at his Khan Research Laboratories, by now had a network of technical representatives making thirteen trips to Iran in the next five years, helping the Iranian technicians and scientists with the gritty details of uranium enrichment by ultracentrifuge.

In 1995, the Khan network expanded the sales perimeter to the ultimate pariah government, giving twelve old, worn-out centrifuges to North Korea, along with drawings, sketches, technical data, some depleted uranium hexafluoride gas to play with, a catalog of stuff for sale, and $210 million cash. For that he got a truckload of Rodong-1 ballistic missiles and the plans for making them with a North Korean–level economy.[152]

Khan began sales to Libya in 1997, with twenty complete L-1 ultracentrifuges, customized for the Libyan taste, components to be assembled for another two hundred units, two sample L-2 premium centrifuges, uranium hexafluoride from North Korea, a machine shop, the CHIC-4 design, and instructions for everything. It was suggested that Libya hide its uranium-enrichment activity by camouflaging the installations as goat or camel farms. Count on prying eyes, looking straight down.

On May 11, 1998, India demonstrated their level of superiority by conducting three nuclear weapons tests 70 miles from the Pakistani border at the Pokhran Test Range. These peaceful devices were the Shakti 1-1, a 43-kiloton two-stage thermonuclear weapon for missile

---

152 The Rodong-1 is basically a copy of the old Soviet SS-1 and is usually called the "Scud missile." There are, by now, several variations, including the Pakistani Ghauri and the Iranian Shahab-3. North Korea probably got the design from Egypt (Scud-B) or from China (Scud-C) and repackaged it as a longer distance bomb hauler. Its range is about 900 kilometers. As happened with the CHIC-4 bomb design, the SS-1 design got around.

deployment, Shakti 1-2, a 12-kiloton tactical warhead, and Shakti 1-3, a 200-ton dud, rumored to have been made with reactor-grade plutonium. Oddly, these three bombs were set off simultaneously in underground shafts, probably to shake the ground as much as possible.[153] Two days later, Shakti 2-1 and Shakti 2-2 were fired, just to make sure that the point was made.[154]

That was quite enough for Pakistan's Prime Minister Mian Muhammad Nawaz Sharif. On May 16, 1998, he gave the order shoot off everything at once in the Kharan Tunnel in the Ras Koh Hills as soon as could be arranged. Five weapons, all named Chagai, were assembled in five "zero-rooms," branching off the kilometer-long tunnel. Ten days later, sealing of the tunnel was completed using six thousand bags of cement and twelve thousand bags of sand. They gave it two days to harden.

The prime minister gave President Bill Clinton a heads-up call to tell him that they were about to set off some nuclear weapons on the morning of May 28. At 10:16 U.T. (3:16 in the afternoon, local time), Chief Scientific Officer Muhammad Arshad, who had designed the triggering mechanisms, shouted "All praise be to Allah" as he hit the firing button. The earth shook noticeably as a small rock slide started on the Ras Koh Hills, and Chagai 1-1 through Chagai 1-5

---

153 The problem with setting off three bombs at the same time is that they tend to influence one another, skewing test results with neutron cross-contamination. One kiloton of fission yields 3×10^{24} free neutrons. That is a neutron fluence of 1.5×10^{10} neutrons per square centimeter at a distance of 500 meters. Most arrive at a distance in a 3-millisecond pulse, with thermal (slowed down) neutrons arriving much later. Delayed neutrons continue for 1 minute. Delayed neutrons are only 1 percent of the total, but that is still 2.5×10^{6} neutrons per second per square centimeter for a 1-kiloton explosion half a kilometer away. Solid rock does not stop neutrons. The fact that three weapons were exploded simultaneously indicates that there was more showmanship than science involved with this test. The neutron fluence from weapon detonation is especially important for battlefield use, in which a warhead blast a kilometer away can send an entire stack of safe W33s into supercriticality, instantly heating the barely subcritical cores and "cooking off" the stockpile all at once.

154 It is interesting to note that Shakti 2-2 was made using U-233, a derivation of the Indian thorium-cycle power-reactor program. It yielded 200 tons, which was probably enough to prove the concept. India has 846,000 metric tons of thorium-232 reserve in the ground, and using it to make U-233 for energy production will make them independent of any other country for power resources.

disintegrated.[155] Using the format established by India earlier that month, they set off another one, Chagai-2, two days later.

In February 2000, the latest military takeover had installed Zafarullah Jamali as the new prime minister, and on his watch, A.Q. Khan was caught doing three bad things. First, he requested a chartered flight from Turkey and was unable to explain why he refueled at Zahedan, Iran. Second, a group of North Korean scientists visited the Khan Research Laboratories without telling the president. Third, a chartered flight heading to North Korea was reported to be carrying centrifuge components. In this case, Khan was tipped off, the parts were unloaded before inspection, and the question became, Why is a chartered plane flying to North Korea with nothing on board?

Dr. A.Q. Khan was a touchy problem for the new military regime. On one hand, he was the father of the Pakistani atomic bomb program, but on the other hand, he was sowing a lot of wild oats in other places, and this could only be an embarrassment for Pakistan. On April 1, 2001, he was removed from any access to the Khan Research Laboratories and was appointed "Scientific Adviser" to the government. Eyes would be following him. By this time, he had moved most of his network operation to his Dubai office under the corporate name SMB Computers. His agent, B.S.A. Tahir from Sri Lanka, kept up with frequent shipments to Libya and Iran. Dubai was fortuitously in the Jebel Ali Free Zone.

By that summer, Sultan Bashiruddin Mahmood and Abdul Majeed, formerly employed by the Khan Research Laboratories, went rogue and had a talk directly with Osama bin Laden, head of al-Qaeda, about nuclear weapons. They had hoped to plow ahead of Khan and sell him a centrifuge plant, but as the discussion spiraled out of control into the technical intricacies of nuclear weapons, they actually managed to talk him out of using A-bombs as a mayhem vector. Even with that backhanded accomplishment, the two men were arrested on October 23, and they vanished from the nuclear scene.

---

155 That, at least, is the Pakistani story. They claimed to have set off a total of 36 kilotons of nuclear explosive in five tritium-boosted fission devices. Western observations put it closer to one device yielding 6 to 12 kilotons.

Although Pakistan and India had signed a localized nuclear test ban treaty, the Lahore Declaration, in 1999, on December 15, 2001, India's Cabinet Committee on Security decided to mobilize for war against Pakistan. By May 31, 2002, the two countries come to within a uranium nucleus of firing on each other.

Khan's network began to unravel in 2003, when the International Atomic Energy Agency in Vienna, Austria, confirmed that Iran had a centrifuge plant at Natanz, and it was all A.Q. Khan's fault. Shortly thereafter, a German ship, the BBC *China*, was caught in the Mediterranean carrying nuclear equipment, labeled "used machinery," to Libya. It had all the marks of an A.Q. Khan transaction. Libya had ordered ten thousand L-2 centrifuges and a uranium hexafluoride piping system. It was lost in delivery. That March, Saif al-Islam Gaddafi, son of Colonel Gaddafi, Brotherly Leader and Guide of the Revolution of Libya, was secretly negotiating with the United States to lift the sanctions imposed for the Pan Am Flight 103 bombing over Lockerbie, Scotland, in 1988. He disclosed A.Q. Khan's dealings with Libya and agreed to a total dismantlement of the A-bomb program. There was now irrefutable evidence of what Khan was doing, and in November, Pakistan was prodded to begin an official investigation.

To the apparent shock of the Pakistani government, Dr. A.Q. Khan had sold ring magnets, aluminum, maraging steels, flow-forming and balancing equipment, vacuum pumps, noncorrosive pipes and plates, end caps, centrifuge baffles, and the entire inventory of P-1 machines to the most notorious pariah states in the world, managing and negotiating brisk sales from Scomi Precision Engineering in Malaysia, ETI Elektroteknik and EKA in Turkey, Trad Fin and Kirsch Engineering in South Africa, Bikar Metalle in Singapore, and Hebando Balance, Inc. in South Korea. And, he was too rich to kill.

Khan was fired as Scientific Adviser in 2004 and charged with operating a nuclear proliferation network while in the employ of the Pakistani government. He pled guilty and was immediately pardoned by President General Pervez Musharraf. The man who was seen by many as a national hero to Pakistan was placed under "protection," or house arrest. The last we heard of him was in 2012, when he announced an intention to form a new political party, the Movement to Protect Pakistan.

Pakistan can now produce 100 kilograms of highly enriched uranium per year using P-3 and P-4 ultracentrifuges. The 50-megawatt-thermal plutonium and tritium production reactor has been running since 1998, and plutonium warheads are on the production line. Three additional heavy-water reactors are under simultaneous construction. A fuel reprocessing facility has been constructed at Chashma. Solid and liquid-fuel ballistic missile technology has been transferred from North Korea and China. Examples of the American Tomahawk cruise missile have landed on the ground in Afghanistan, have been transferred to Pakistan, and have been reverse-engineered, copied, and are now equipped with a miniaturized, 12-inch plutonium implosion bomb of Pakistani design. In the farms, cities, mosques, and government, and inside the gates of the nuclear laboratories, reactor plants, and manufactories, religious extremism is increasing in Pakistan on a daily basis.

The United States is diplomatically trying to coax the genie back in the bottle in Iran, preventing them from sharing the thrill of owning the Islamic bomb with Pakistan. It may be too late for that. Remember, it all started on April Fool's Day, 1936.

CHAPTER 7

Japan's Atomic Bomb Project

"Strive for the fire,
seek the fire,
so thou wilt find the fire,
put fire to fire,
boil fire in fire,
throw body, soul, and spirit into fire,
so thou has dead and living fire."

—from *Codex Rosae Crucis: the D. O. M. A. Document*, 1775

PAKISTAN, A POLITICALLY SHAKY ISLAMIC republic in which ten regional languages are spoken, has armed itself with an impressive array of state-of-the-art, homebuilt nuclear weapons. This notable accomplishment occurred, to quote Pakistani nuclear physicist Dr. Noor Muhammad Butt, "in an ocean of ignorance in a country that possessed lame high technology." If Pakistan could do this in the 1980s, could Japan in its technical and economic configuration of World War II have developed an atomic bomb?

Japan in 1939 was hardly analogous to Pakistan. The Japanese navy was the third largest in the world, and it was being used as part of an extended military operation to annex the entire Far East. There were ten Japanese aircraft carriers. The United States had seven, and only three in the Pacific Ocean. The Japanese had the biggest, the fastest, and the smallest submarines in the world in a dazzling range of styles, from human-guided torpedoes to a submersible aircraft carrier. The navy was equipped with radar, fast and nimble fighter planes, and an unquestioned willingness to die for the Emperor.[156] It is true that the small island-country of Japan lacked the vast energy resources and the wealth of raw materials that graced the United States, but still, they built the biggest battleships in the world. There was plenty of manufacturing activity on the home islands, including among the industrial giants Mitsui, Mitsubishi, Sumitomo, and Yasuda, but the largest, war-winning industry was not in Japan. It was across the Sea of Japan in a big plot of land that had been won in the Russo-Japanese war of 1905. It was in the Japanese-owned northern section of Korea, out of sight and bomber range. The Japanese conquerors ignored the Korean name Hungnam and called the area Konan.

The build-up of this largest industrial complex in Asia began in 1873, with the birth of Shitagau (Jun) Noguchi. Noguchi was small

156 When the Second World War started, Japan did have a frighteningly good navy, but the Americans were confident of their superiority to Japan and to the other Axis powers when it came to electronics and communications. After the war, detailed studies confirmed that the Japanese radio equipment was pathetically behind almost everything the Americans used in the Pacific Theater. The Japanese didn't even have decent batteries to power their radios, components were not protected against humidity and fungus, and radio coils were wound on cores made of dried mud. There were, however, two stunning developments in Japanese electronics. In 1926, the very advanced Yagi radio antenna had been invented at the Tohoku Imperial University, and Japanese voice transmitters from World War II were equipped with a new type of microphone: the electret. Today, every telephone in the world, including your smartphone, uses an electret microphone, and if you have ever seen a television antenna on top of a house, then you have seen a Yagi. Mounted on the airframe of both the Little Boy and the Fat Man atomic bombs were four, three-element half-Yagi UHF antennae, used by an array of four AN/APS-13 "Archie" aircraft tail-warning radar units. The Archies were used as radar altimeters to ensure a precise detonation altitude, and they would not have worked without the Japanese-invented Yagis.

and of slight build, even by Japanese measure, but he possessed an unquenchable drive to succeed. He graduated from Tokyo University in 1896 with a degree in electrical engineering and immediately found employment at the Japanese branch of Siemens, the company that had electrified Europe.

He saved his earnings, started forming his own companies in 1906, and in 1908 he merged two of his companies into the Japan Nitrogen Fertilizer Company. He built his first factory in 1909 at Minamata on Kyushu Island, across the bay from Nagasaki. A year later, he built another factory in Osaka. Four years later, another one in Kumamoto. In 1919, he built his first 50 Hz, Siemens-equipped electrical power plant and bought the rights to the German "Haber process" for making ammonia by electrolysis. The next year, he built another power plant plus the Japan Mining Company. By 1923, he owned a large chemical factory at Nobeoka, and he turned his sights to Korea—a vast, vacant lot.

The Yalu River runs down the mountains of North Korea, branching into the Fusen and the Chosin Rivers. In 1926, Noguchi started an impressive hydroelectric project by damming these two Yalu tributaries. In a few years, he was generating 600 megawatts of electricity with the Chosin Reservoir.[157] By 1939, his industrial complex strung out in a roughly northwest line, 200 miles long, from the coast of Korea almost to the Russian border, with the largest cluster of factories stretching 68 miles. Korean labor was cheap, there was an abundance of electrical power, and Noguchi's industries made graphite, chemicals, synthetic oil and gasoline, gunpowder, dynamite, nitroglycerine, magnesium, and, as a sideline to the vast Korean Nitrogen Fertilizer Company (Nitchitsu), heavy water. He also owned a uranium mine in Chuul, a settlement just north of Eian and 125 miles up the coast from Konan. Its product was used by the tableware industry to make a pretty yellow color in ceramics.

---

157 During the war, estimates of the industrial capacity of Japan were used to dismiss any speculation that they were working on an atomic bomb. The extremely large uranium enrichment effort at Oak Ridge, Tennessee, using electromagnetic calutrons (Y-12), gaseous diffusion (K-25), and thermal diffusion (S-50) required 250 megawatts of electricity as supplied by a string of TVA hydro plants on the Tennessee River. The entire electrical power inventory in Japan was 3,000 megawatts. Adding a 250-megawatt load to the power grid that was already inadequate to keep the lights on in Japan was not an option, but these estimates did not take the Chosin Reservoir into account.

Japan of 1939 thus had a large and sophisticated industrial base, capable of building whatever was needed to make war in the twentieth century, both copying technology from the greater world and hatching new ideas. To develop the technology for a nuclear weapon, however, much more was needed. At the beginning of the Second World War, a few key participants—Great Britain, France, Germany, and the United States—all started atomic bomb programs from scratch. At the time, there was no Chinese bomb design being passed around, no plans to steal from anybody, and no Atoms for Peace program. Everyone was starting from zero. The most vital beginning resources were nuclear physicists, and by pure chance, the United States had sopped up most of the refugee scientists who had been bounced out of Nazi-governed Germany.

Japan had no such windfall, a phenomenon that both enriched America and stunted Germany's efforts, but there was serious nuclear physics work being conducted by Japanese scientists at the time, and a healthy scientific infrastructure did exist. There were nuclear physicists, theorists, neutron specialists, cosmic-ray authorities, chemists, geologists, future Nobel Prize winners, experimentalists, electronics experts, artillery men, particle accelerator jockeys, and at least one unrepentant Marxist who understood fluid dynamics. There were scientists available in Japan for radical weapons work who had known Albert Einstein, had lived in Niels Bohr's house, had worked with Lord Ernest Rutherford at the Cavendish Laboratory, and had worked with Ernest O. Lawrence on his new cyclotron at Berkeley. Many of them had published papers in the prestigious American journal *Physical Review*. There was no critical lack of expertise in Japan at the start of the war.

There was one large, expensive piece of important scientific equipment necessary to make the measurements for atomic bomb development: a cyclotron. It is a machine that will take individual charged particles, such as protons, from an atomic nucleus and accelerate them from rest to extremely high speed in a hard vacuum. There are other ways to do this task, such as the Cockcroft-Walton linear accelerator, but nothing at the time was as powerful, accelerating the most particles to the highest speed, as a cyclotron, and one needed as much accelerator action as possible to make the measurements that were needed.

The cyclotron is compact, spinning particles around in a spiral as they gather speed before exiting the machine in a blue glow that looks like the deadly end of a ray gun. A huge electromagnet, weighing several tons, establishes a vertical magnetic flux in a gap at the machine's center, constraining the charged particles to run around in circles as they are pulled back and forth by an oscillating electrical field.

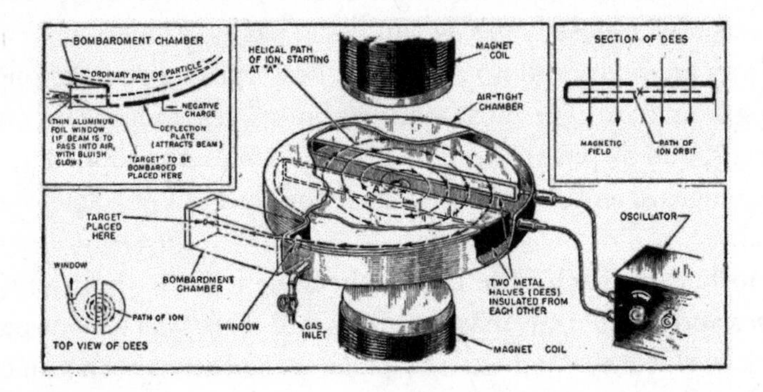

This diagram from Radio-Craft *in 1947 shows very clearly how a cyclotron works. The vacuum chamber is set in the gap of a huge magnet, and the vertically running magnetic field in the gap constrains charged particles (deuterons) to run in a circle as they are pulled back and forth by the oscillating electrical field introduced on the right. As the particles gain speed, they spiral out and finally escape out the window.*

Great Britain had a small cyclotron at the Cavendish. They gave up the atomic bomb hunt early and joined the American effort. The French had one in Paris. It was commandeered by the Germans upon invasion, which brought the French A-bomb work to a stop, but it wasn't working and it was too big to move to Germany. The Germans had two, but one was privately owned by Manfred von Ardenne, the fellow who saved the Hermann Göring Aircraft Research Institute from Ronald Richter. It spent most of the war apart on the floor of von Ardenne's basement. The other one was at Heidelberg University. It was intended purely for cancer treatment, and construction was not completed until the war was almost over in 1945. The Soviets had plans to build the biggest one in the world, of course, at Bolshaya Kaluzhskaya Ulitsa in Moscow (50 MeV) and another, smaller unit at the Physical-Technical Institute in Leningrad.

Ernest O. Lawrence, the professor at the University of California at Berkeley who had built the first cyclotron in 1931, had two of them in operating condition, a 37-inch unit making 8 million electron-volt (MeV) deuterons, a 60-incher doing 16 MeV, and a 184-inch monster that wasn't completed until after the war.[158] There was a cyclotron at Harvard, but it was moved to Los Alamos in 1944. In 1941, a new one was being built at the University of Pittsburgh, and General Electric at Schenectady was buying one.

In Japan, they had five cyclotrons, and one was a replica of Lawrence's 37-incher that they bought from the University of California in 1939, right before things started getting ugly in the Pacific.[159]

The critical job for a cyclotron in the early days of fission research, which started right at the beginning of World War II in Europe, was to make neutrons. The fission of uranium-235 by bombardment with neutrons was found in 1938 without the use of a powerful cyclotron, but the effects were barely detectable. Neutron sources of the era consisted of a mixture of radium powder and beryllium powder. Alpha particles emitted by the radium would crash into the beryllium and cause free neutrons to release from the beryllium nuclei, making a simple but weak neutron source. Radium was exceedingly rare, and no one lab had much of it. Using a cyclotron, protons from hydrogen or deuterons from heavy hydrogen can be accelerated and crashed into a beryllium target, simulating the alpha particles on a massive scale and making a dense, continuous source of neutrons, heading off

---

158 The inch-size of a cyclotron refers to the diameter of the flat, circular vacuum chamber in which the charged particles are accelerated. The bigger the chamber is, the more times a particle can go around the spiral before it exits, and therefore the faster it becomes. Also, the diameter of the vacuum chamber determines the size of the electromagnet gap that must be maintained, and this determines the thickness of the concrete floor under the cyclotron plus the amount of electricity it requires to run. Lawrence's 184-inch cyclotron spun particles up beyond 100 MeV, and it was designed to simulate cosmic rays, confirming the existence of meson particles.

159 At the end of the war, when the Japanese scientists were instructed to identify their five cyclotrons, they couldn't put their hands on the fifth unit. It was a 39-incher, built by the Eighth Army technical labs in the shops at the Ischikawajima dockyards. It was probably never finished and was lost in the confusion at war's end. The week before the search for cyclotron number five was conducted, the U.S. Army had begun the systematic destruction of all cyclotrons in Japan.

in all directions. Using this enhanced neutron source, the interaction properties of neutrons hitting any material can be quickly measured and plotted. At that time, there was no Barns Book or Knolls Nuclide Chart compiling fission properties, absorption cross sections, or scattering coefficients. Cyclotron-generated neutrons were used for everything from finding the number of neutrons emitted in fast fission to checking the efficiency of uranium enrichment by inducing fission.[160] Just about everything needed to work out a bomb design had to be found out using a cyclotron, and in Japan, scientists were as well equipped for these tasks as anywhere else in the world.

Japan was also equipped with excellent research and development facilities for advanced concepts, including the University of Tokyo, Tokyo Engineering College, Kyoto University, Nagoya University, Tohoku University, Chiba University, Osaka University, the Eighth Army Technical Laboratory, and the Navy Technical Research Laboratory at Yokosuka. Japan also had the Institute of Physical and Chemical Research (or RIKEN).[161] For atomic-bomb-making purposes, it would have been the Japanese equivalent of the Clinton Engineer Works at Oak Ridge, Tennessee, the headquarters of the Manhattan District engineers. Oak Ridge was built in a wartime frenzy in the middle of nowhere, out of sight and out of connection to any obvious population center. It originally had more than fifty buildings and included everything from ore refineries to machine shops. RIKEN was located in the Segoku section of the Tokyo

---

160 Dr. Lawrence at Berkeley also used the cyclotron-generated fast neutron stream to convert uranium to plutonium, making the first measurable quantity of Pu-239, and he used the intense magnetic field at the cyclotron gap to demonstrate electromagnetic isotope separation. This exercise made a measureable quantity of pure U-235, and it led to the alpha and beta calutron clusters built at Oak Ridge at location Y-12. The Little Boy atomic bomb dropped on Hiroshima was made of uranium enriched using these cyclotron-inspired devices. There is no record of any Japanese use of cyclotrons for plutonium production or experimental electromagnetic separation during the war.

161 RIKEN is an acronym, made from the first syllables of the words Rikagaku Kenky sho (the name of the institute in the original Japanese), sounded out in English. RIKEN is still in business, with three thousand scientists and an annual budget of about $760,000. It relocated to Wako, Saitama Prefecture, just outside Tokyo, in 1963. Scientists at RIKEN investigate everything from terahertz radiation to supercomputer research and development.

suburbs, and it had more than fifty buildings, including everything from ore refineries to machine shops.

There was, however, one noticeable difference between the RIKEN and Oak Ridge. At RIKEN, when walking between buildings one was surrounded by beautifully landscaped Japanese gardens with koi ponds and benches where scientists could sit and contemplate the universe or eat lunch and converse with colleagues. At Oak Ridge, as millions fought the World War on two fronts, between buildings you were walking in mud and gravel and strengthening your resolve to get this job done and get the hell out of here.

The war with Japan was officially kick-started on December 7, 1941, with the Japanese navy's well-played attempt to wipe out the US Pacific fleet and its support facilities at Pearl Harbor, Hawaii. The military intelligence services in the United States noticed something immediately: The Japanese navy knew a great deal about Pearl Harbor, a place that most Americans had never heard of. In fact, they seemed familiar with every building, fuel dump, landing strip, and military installation on the Big Island. Some large group of agents had to have been feeding them information for months or even years. There was a large presence of Japanese citizens and Americans of Japanese heritage in Hawaii, and it was only logical to assume that a percentage of them had been working for Japan. Japan was now, by definition, an enemy of the United States.

There were also a lot of Japanese Americans living on the West Coast, which seemed wide open for an attack or even an invasion. On February 19, 1942, President Franklin D. Roosevelt authorized the relocation of such citizens with Executive Order 9066. About 110,000 Japanese Americans were thus forced to spend the war in camps located outside the exclusion zone of the West Coast.

An American and British cryptanalysis project named "Magic" had been decoding Japanese diplomatic and navy codes since 1932. The Allies were thus looking at a continuous stream of secret messages sent to and from Japan, but even decoded and translated into English, Japanese messages could be difficult to interpret. With the Pearl Harbor revelation, some of the message traffic now became clear. Japan had been collecting military data from the United States for a long time, and with a new top secret project in place to exploit nuclear fission, the Japanese government was hungry for information about this new endeavor.[162] All the agents were now incarcerated or staying very quiet, and a worker of Japanese character was not allowed anywhere near nuclear physics, regardless of credentials. A new intelligence channel back to Japan would have to be devised.

The obvious source of new spies for Japan was Spain. Spain was neutral in the war, and Spanish citizens could, in theory, go anywhere they pleased. They could even blend in with the citizens of Mexico, and posing as day workers, they could stream through the border crossing in Juarez/El Paso or Tijuana, infiltrating everything from aircraft factories to the crown jewel of espionage, the new top secret laboratory complex at Los Alamos, New Mexico. American boys of employment age were off to fight the war, and there was a desperate need for construction workers to scratch-build new industries in a frantic rush. Mexicans and Native Americans were put to use by the thousands. Workers who could claim not to speak English were prized. If they were to hear something of a secret nature, then they wouldn't even know what was going on. In fact, most of the security guards employed at Los Alamos were Spanish-Americans, and few of them admitted to understanding English.

Baron Hiroshi Ōshima, the Japanese ambassador to Germany, came up with the idea of using Spaniards to gather technical intelligence in America. He named the operation "TO," harking back to the word *door* in Japanese. The government of Spain, sensitive about America's failure to provide any support in the recent revolution and still hurting from the Spanish American War of 1898, jumped at the suggestion.

Japan had to abandon the Japanese Embassy in Washington immediately following the declaration of war, and Spain offered to take it over, relaying diplomatic messages, should any occur. As promised, the new occupants found $500,000 dollars cash in the safe. It was to get the ball rolling for spy work, and there would be more coming,

---

162 Technically, the United States began a government investigation into the military use of uranium fission on October 21, 1939, with the first meeting of the Advisory Committee on Uranium, or the "Briggs Uranium Committee." A budget of $6,000 was grudgingly approved. The organizational meeting of the S-1 Project, which was fully dedicated to the development of nuclear weapons, was held on December 18, 1941, eleven days after Pearl Harbor was bombed. It was eventually named the Manhattan Project.

deposited in the Berlin branch of the Yokohama Specie Bank. On the payroll were a scientist in Chicago, a major in the Pentagon office of the Army Air Corps, a supervisor of floating piers in New York, and a Jewish officer somewhere. Eight Spaniards were imported, each managing a group of operatives.[163]

Much remains secret, and most of our knowledge of the Spanish espionage of World War II comes from a single source, the loosely coupled memory of a flamboyant, talkative fellow named Ángel Alcázar de Velasco. He, a former bullfighter, Falangist, calculating anti-Semite, and dependent astrologist, believing in Satan, a coming Aryan messiah, and the power of black magic, was recommended to Ambassador Oshima by the Spanish foreign minister, Ramón Serrano Súñer, who happened to be the brother-in-law of Francisco Franco, the *Caudillo* of Spain.

Velasco drew pay from SIM, the Spanish intelligence service, and in the fall of 1940, he was dispatched to London to be a spy for the Nazis.[164] Spain had signed a very secret agreement to help Germany against Great Britain. Súñer was impressed by what Velasco was accomplishing in London, and he was first on his list for agents to be sent to America for the Japanese.

Velasco was born in Mondéjar, Guadalajara, in 1900, and at the age of seven did his first secret agent job. His village was in a dispute with another village, and there was a message that would have to be sent to their leader, who was holed up in the mountains. Young Velasco volunteered. His horse died of exhaustion in the effort, but the message was relayed successfully, making him a hero in his village. He was rewarded with notable respect plus 1.5 pesetas. He never forgot the thrill.

By age twelve, he was an amateur bullfighter, going pro in 1928 under the name "Our Young Gypsy of Madrid." He was gored badly

163 There was at least one double agent working for the United States out of the American Embassy in Madrid. The FBI and the OSS (Office of Strategic Services) were acutely aware of the Spanish spy rings, and many of the participants were tailed for the entire war, making them nervous and ineffective as spies. The OSS formally warned the Joint Intelligence Committee of Spanish espionage on March 10, 1942. There are thousands of pages of documentation in the FBI files and what is left of the OSS files. They are difficult to get to. The FBI code name for TO was Span-Nip.

164 Some sources call the Spanish intelligence agency SECED, but this is Velasco's story, and he called it SIM.

in 1932 and gave up the sport. He next attended the University of Salamanca and was persuaded to leave in exchange for a degree in philosophy. Gaining interest in the occult, he wrote several books on the subject. In 1934, he met Wilhelm Oberbeil, a director of the Hitler Youth organization, an Abwehr official, a fellow occultist, and a member of Adolf Hitler's inner circle, who persuaded him to attend the Abwehr Intelligence School in Berlin. He returned to Spain in 1934 with a German code name: Guillermo. When Franco seized power in 1936, Velasco was thrown in jail and sentenced for execution because of a political disagreement, but he managed to gain Franco's admiration and a job at SIM by slamming and locking a gate during a mass escape at the prison. From there, life was downhill.

Most agents were usually sent to Central America on Spanish ships, then slipped over the Mexican border at night into California at Tijuana, but Velasco was special. He was taken by German U-boat through the Caribbean to the east coast of Mexico, then overland to Mazatlán, a port on the Pacific coast. A headquarters was there, manned by technically savvy Japanese, passing themselves off as simple fishermen. Information from agents was forwarded by shortwave either to Spanish ships off the coast of the United States or down to Chile, where the Japanese used the relay service of Juan Perón, future proponent of fusion power in Argentina. Velasco's encryption code was his Madrid phone number, 29824. To his Japanese handlers, he was agent X27Z. They often found his intelligence reports and his expense statements difficult to believe, but they sent them on to Tokyo anyway.[165]

Velasco, in a book he wrote on the subject, claims that his agents penetrated the A-bomb project in New Mexico in 1943, just as the new

165 Velasco was capable of some taller-than-life tales. He, for example, claimed to have taken Martin Bormann, Hitler's personal secretary, to Argentina, escaping Soviet-held Berlin on May 1, 1945. Artur Axmann, a Hitler Youth leader, remembered it differently, having escaped from Hitler's underground bunker with Bormann at 11:00 A.M. that day. Axmann was sure that Bormann died near the railroad bridge at Lehrter Station, just outside Berlin. Bormann or his remains were nowhere to be found, not even in Argentina, until December 7, 1972, when some construction workers accidentally dug him up near the tracks at Lehrter. Identification of the remains was eventually confirmed by dental records, facial reconstruction, and definitively by DNA matching with his living children.

Los Alamos laboratory complex was opening but still being built. The first report, collected by a Spanish doctor turned spy, seemed technically distorted, describing a new type of detonator.

By sometime in the summer of 1944, the FBI had seen enough of what Velasco was doing, and they captured him. He was threatened with death unless he would agree to become a double agent for the United States. He refused on professional grounds and was released, with the FBI thinking that he was so inept, it was better for the Allied cause to keep him working for the Japanese. Immediately, on July 4, 1944, Velasco fled to Germany and never set foot in America again.

Velasco chronicled many exploits in America, but his most interesting account came not from his postwar book but from an interview in 1980. He was told by his Japanese handlers to recruit a "couple of Chicano boys" and send them to a certain spot in New Mexico. There was going to be a big explosion nearby. After the explosion had gone off, they were to bring back soil samples in their pockets, just by scooping up a little bit of dirt off the ground around them. That was all they had to do.

The boys made it back, and Velasco took them to the Majestic Hotel in Mexico City, where two Japanese agents had a room. The agents did not want to touch the soil samples or get any on their hands, but they instructed the boys to empty their pockets into two glass bottles, which were then sealed and moved to a far corner of the room. Velasco thought it odd. It was as if the agents were afraid of catching typhoid by touching the dirt. Had the Americans tested a plague-spreading bomb? They then gave the boys a most thorough medical exam, right there in the room, examining their fingernails, their skin, and taking blood samples from the fingers, ears, and arms. A test solution was swabbed on the skin, wiped off, and the wipes were saved as samples.

It was the most elegant piece of spy craft in World War II, and probably the most heartless. The two Mexican boys were used as self-propelled, auto-positioning radiation dosimeters to monitor a test firing of the new atomic weapon, as predicted by an interpretation of the garbled reports coming out of Velasco's southwestern intelligence quadrant. "Human dosimetry" has been used many times to find the radiation exposure received by individuals caught in criticality accidents or isotope spills. By counting the number of white blood cells per volume of blood, the recent dose of ionizing radiation can be estimated with precision as good as any electronic or solid-state dosimeter. Exposure to neutron flux is determined by counting gamma rays per second coming from the metastable sodium-24 in the blood. This time, it was used to instrument a weapon test. If caught, the boys had no idea what they were doing or why. They had no hardware, written instructions, or embedded instrumentation on their persons. Samples of their blood, plus topsoil where pulverized radioactive isotopes would have landed, were sent for analysis to the laboratory in Mazatlán. It was brilliant.

Near as the Japanese could tell, the Americans were working on an atomic weapon at Los Alamos. Scientists back in Japan assured the intelligence organization that there was no practical way for the Americans to build an atomic bomb. While it was theoretically possible, it simply took too much effort to isolate the rare U-235 isotope out of natural uranium to make an explosive fission. By an estimate of the amount of electricity going into Los Alamos, there just was not enough industrial activity to accomplish isotope separation. Instead, the Americans had obviously built a nuclear reactor running on natural, out-of-the-ground uranium, and they were using it to make dangerously radioactive fission products, such as strontium-90. Rather than risk a massed beach landing, they planned to poison Japan and kill everyone there by dropping bombs full of radioactive by-product.

The soil samples showed no radiation contamination at all, and the two test subjects were completely unharmed. This was a better outcome than could have been imagined. The American radiation bomb was a complete failure! No matter how the forces of the United States sowed fear and leaflets as warnings, the Japanese people would be calmed and told they had nothing to worry about from these new "weapons."

Unfortunately for the mission, they had instrumented the wrong explosion. On May 7, 1945, a pile of 108 tons of Composition B explosive was set off 800 yards away from the ground-zero tower being configured for the plutonium bomb test, which was scheduled for July 16. It was, at that time, the largest purposeful explosion ever detonated, and everyone within 30 miles of the Trinity Test Site had been warned to stay out of the area. Its purpose was to calibrate the instrumentation for the later atomic blast. The Spanish agents could not help but notice the warnings, and it may have been interpreted as the atomic weapon test.[166] [167]

The Allies had excellent signals intelligence (SIGINT) against Germany and Japan, but lacked the third-party information-gathering from scientific research that the Axis side enjoyed. Our tracking of the German atomic bomb project came in the form of dire rumors, at best, but we had nothing from Japan.

We had no clue as to what was going on until Germany, suffering from armies coming in from all sides, surrendered to the Allies on May 8, 1945. Admiral Karl Dönitz, the newly appointed president of Germany, radioed his beloved underwater fleet, commanding all U-boats to surface, raise a black flag, and proceed to the nearest Allied port for surrender.

One of these boats was U-234, a Type XB mine-laying submarine, refitted to carry cargo. The captain, Johann-Heinrich Fehler, had been given a mission to deliver 240 tons of critical war material, technology,

---

166 There is a problem with this account. Velasco did not mention this incident in any of his postwar books or articles, and it just came out when he was specifically questioned about atomic bomb espionage. If he was bounced out of North America on July 4, 1944, then he was not around to manage spy craft in New Mexico when any large-scale explosions were happening. The data-collection method makes perfect sense, and it seems that this misreading of an explosion test is in agreement with Japan's initial response to the Hiroshima bombing on August 6, 1945. The military was so certain that the Americans did not have an atomic bomb, it was hard to come to grips with the fact that Hiroshima had just been destroyed by one. Velasco may have acquired the story by swapping war stories with one of his fellow Spanish spies long after the war. The other guy was too reluctant to claim it in writing, but Velasco had no such qualms. No other Spanish intelligence accounts are as easy to get to as are Velasco's. Or, maybe it never happened.

167 This test shot on May 7, 1945, actually contained a radioactive sample. A spent fuel slug from the Hanford B-reactor was dissolved in acid and placed in a sealed glass container at the center of the explosive pile. It was 1,000 curies of beta activity and 400 curies of gamma activity from a mixed bag of fission products, uranium-238, and what remained of the small percentage of uranium-235. The purpose of this sample was to test radiation instruments in place at the Trinity site, and technically, it was the first "dirty bomb" ever exploded. The radiation sample was dangerous to handle in its concentrated condition, but when the explosive pile went up, it was spread evenly in an aerosol across square miles of desert. In this modified concentration, it was indistinguishable from background radiation, and it did not show up as harmful radiation contamination on the two test subjects.

and expert personnel to Japan, Germany's last remaining ally. On March 25, 1945, U-234 began its journey, first stopping at Kristiansand, Norway, to pick up its load. The cargo included one complete Me-262 jet fighter plane (some assembly required), a Henschel Hs-293 radio-controlled glide bomb, a high-altitude pressure cabin, two Lorentz 7H2 bombsights, two airborne radar sets (B/3 and FUG 10), 25 bomb fuses, examples of the latest electric torpedoes, tons of technical drawings, plus one ton of mail to Germans in Japan, and in silo-location 38 on the forward deck, labeled ST1270/1-10, were ten gold-lined, welded aluminum cans containing 1,235 pounds of refined uranium oxide powder, colored a bright fluorescent yellow. Glued to each can was a brown paper label. On the labels was "U235," in Japanese script.[168]

On May 10, the surrender order from Dönitz was received by radio, and the course was changed, heading west toward the United States.[169] The boat was intercepted by the destroyer USS *Sutton* two days later and was taken under guard to the Portsmouth Naval Shipyard in Maine. The unloaded cargo was a technical bonanza for the United States, but of greatest interest was the uranium bound for Japan. The U-boat crew had no clue as to what it was for, and the two Japanese passengers, Lieutenant Commander Hideo Tomonaga and Lieutenant Commander Shoji Genzo, had ended their lives with lethal doses of

---

168 So says the account by Lietenant Junior Grade Karl Pfaff, who had directed the loading of U-234 at the port of Kiel and was taken prisoner upon surrender. There is a conflicting report by Wolfgang Hirschfield, the radio operator, who claimed that the uranium oxide was contained in lead cubes. The "gold-lined" cans version is supported by the personnel who X-rayed the containers before opening them with an acetylene torch, but the gold-lining story does not make sense. The Germans were technical show-offs, but not to that extent. The cans were probably treated inside with an iridite-chromate process to prevent the uranium oxide from reacting with the aluminum. The treatment looks just like dull gold plating. (The Islamic shrine Dome of the Rock in Jerusalem was roofed with iridite-chromate-treated aluminum in 1963. It was replaced with 24K gold leaf in 1993, and both treatments look about the same.) The "U235" labels did not mean that the uranium was the pure, fissile isotope. It was natural uranium, with no U-235 enrichment. The X-rays revealed what looked like a stick of TNT in one of the cans, and a booby trap was feared, but Pfaff assured the Americans that it was only a wooden stick used to tamp down the uranium oxide powder and was somehow left in the can.

169 On May 8, 1945, another radio transmission was received reporting that the Togo government in Japan had severed all relations with Germany and was now arresting all German citizens. The boat was turned around and was prudently headed back to Bergen, Norway.

luminal after learning that the destination had been modified. What was the implication of this cargo? Was Japan building a bomb?

Meanwhile, General Leslie Groves's Alsos Mission was plundering its way across occupied Germany, mopping up nuclear physicists, documents, and anything that could have been used in the Third Reich's obviously unsuccessful attempt to develop nuclear technology. Alsos was sending daily reports back to the Manhattan Project. From what the Alsos team could tell by translating truckloads of reports, memos, and material requisitions, the German research and development seemed inadequately organized and lacking certain expertise.[170] In retrospect, there had been nothing to worry about from the German nuclear physics community. There did pop up, however, an interesting report from Japan. It said that a scientist addressed the Japanese House of Peers to announce that "he is succeeding in his research for a thing so powerful that it would destroy an enemy fleet within a few moments." Was this a progress report on nuclear weapons development?

Operation Olympic, the invasion of Kyushu, the southernmost island of Japan, was scheduled to begin on November 1, 1945. It was to be the largest amphibious landing in history, but it never happened. Japan unconditionally surrendered on August 14, and the event was celebrated throughout the Allied world as V-J Day. A mission was immediately assembled to answer hanging questions about Japanese atomic bombs. It would be similar to Alsos, the expedition that had found a surprising lack of progress in the German nuclear weapon effort. It was named Group Three, the third in a cluster of scientific investigations. The first two were to visit Hiroshima and Nagasaki to study the effects of atomic bombings, and the third, the Tokyo Group, was to find out what Japan had been doing in nuclear research. Major Robert Furman, who had worked in the Alsos Mission, was in charge. The Tokyo Group was further divided into three sections. Brigadier General T. F. Farrell, headquartered in Tinian Island, hunted for military scientists involved in

---

170 The ultimate prize of the Alsos Mission was the German heavy-water reactor hidden from Allied bombing raids in a cave under the cathedral in Haigerloch, Swabian Alb, Germany. A first impression of the reactor setup was that the Germans were lucky that the thing was incapable of working. It was inadequately instrumented and had no controls of any kind. If it had somehow gone critical, it would have quickly killed everything in the cave, down to the bacteria on the floor.

nuclear research; Dr. Karl T. Compton, president of MIT and a physicist, investigated the academic scientists; and Robert D. Nininger, an expert in geology, looked for uranium mines.

The group arrived in Tokyo on September 7, 1945, and the project was given a less cryptic name: Atomic Bomb Mission. The investigators spread out. Farrell decided to start in Korea, as it was quickly being blocked off by the invading Soviet army, and the timing was becoming crucial. He was making his way there when the last battle of World War II occurred near Konan. A ragged brigade of un-surrendered Japanese holdouts tried one last banzai charge against the Soviets to save their industrial enclave. The group wrapped up the investigation and submitted their summary report, "Atomic Bomb Mission, Investigation into Japanese Activity to Develop Atomic Power," on September 30, 1945, after only twenty-three days of looking around. There would turn out to be a lot more going on than the Atomic Bomb Mission found in this too brief investigation.

David Snell, a journalist-turned-soldier from Atlanta, Georgia, added greatly to the excitement almost exactly a year later. For most of the war Snell worked as a newspaper reporter for the *Atlanta Constitution*. Wanting to get closer to the action, he volunteered for the U.S. Army on July 30, 1944, and reported to Camp Chaffee, Arkansas, for basic training. Given his specialty as a writer, Private Snell was assigned to write the history of the army, but as the war ended in 1945, he was reassigned to the 24th Military Police Criminal Investigation Detachment (CID) and landed in Seoul, Korea, as a Tech 5. There were Japanese war criminals to be gathered, indexed, and documented.

There was a mind-numbing array of war-weary Japanese individuals that Snell had to interview. Most of the officers had nothing of value to report, and they just wanted to go home. One fellow, however, dropped a bombshell upon questioning in May 1946. The army had given him a pseudonym to protect him from reprisals: Captain Tsetusuo Wakabayashi, in charge of Japanese atomic development in Korea.

Wakabayashi rolled out his war story as he and Snell sat in the Chōsen Jingū Shinto shrine, repurposed as a refugee center and overlooking Seoul on Nam Mountain. He claimed to be one of only a few involved in the Japanese atomic bomb development project that had escaped the invading Soviet army as the war abruptly ended. After midnight on August 10, 1945, four days after Hiroshima was obliterated and five days before the surrender, he had accompanied a convoy of Japanese trucks from Konan carrying a new type of destructive device down to the coast near an islet in the Sea of Japan. All day on the 10th and all that night, badly shot-up naval ships, old fishing boats, and anything of size that would float were anchored near the islet, in a configuration to simulate an American invasion flotilla preparing to send forces ashore. Just before dawn on August 12, a radio-controlled boat carrying the special bomb was sent through the tangle of anchored boats and beached itself on the shore of the islet.

Observers were posted 20 miles away, and all were issued welders glasses, the black goggles meant to protect the eyes from the intense, white-hot electric arc used to join metal by melting it. At the moment of sunrise, the bomb was detonated, making a blinding light and a multicolored fireball 1,000 feet in diameter. The vertical column of dust hit the upper atmosphere and spread out in a mushroom shape. On the water, those ships that had not completely vanished burned brightly and quickly sank.

Unfortunately, the successful test of an atomic bomb was too late. It was built to wipe out the American fleet with a single kamikaze attack, but the Soviets were now making significant headway, moving south from Manchuria. When they returned to Konan from the test, the scientists and engineers had to destroy everything and dynamite the caves to keep the Russians from capturing the technology they had developed. Most of the scientists did not evade capture. They were shipped to Moscow for further interrogation, along with entire factories and laboratories.

The story was difficult to believe, but other stragglers were giving similar accounts of the offshore A-bomb test.[171] There were also stories

171 There is an eerily similar story of a German A-bomb test, within days of the German surrender, near Rügen Island, off the coast of Pomerania in the Baltic Sea. Again, the Soviets were right there soon after, and everybody had to demolish the equipment and flee. One soldier escaped and lived to tell the tale. These stories seem fabrications by the vanquished, claiming that they had all the technical ability of the Americans as well as the determination, but they were morally superior to the Americans and therefore did not use these terrible weapons against human beings. This story is not backed up by any German accomplishment. During the war, they could not make a simple, self-sustained chain reaction, much less an explosive one.

of a second, possibly less successful test the same day, supported by the Kwantung army technical detachment, Unit 731, on the eastern edge of the Gobi Desert. In his G-2 intelligence summary for Konan, May 1 through 15, 1946, Colonel Cecil W. Nist concluded, "It is felt that a good deal of credence should be attached to these reports." This information was classified secret.

Snell was discharged from the army on September 11, 1946, and he immediately got his job back at the *Constitution*. He filed his story about the Japanese atomic bomb just twenty-one days later, and it was a screamer headline on the front page, October 3, 1946: JAPAN DEVELOPED ATOM BOMB; RUSSIA GRABBED SCIENTISTS.

The story was a sensation, and it was picked up by papers in the United States from San Mateo, California, to Troy, New York. The United States Army was not pleased, nor was the Soviet Union, which controlled Korea north of the 38th latitude parallel, including Konan. Snell never revealed the real name of the Japanese officer who had given him the story.[172]

The Atomic Bomb Mission of 1945 had found absolutely no nuclear work during the war. The scientists they interviewed would shrug and say that they had tried to keep from being drafted and spent most of their effort growing potatoes, which they would cook on Bunsen burners and eat for lunch. There wasn't enough electrical power to fire up a cyclotron. There wasn't enough of anything. Not even chalk. One group admitted to having worked long hours on the "death ray," a fantasy weapon longed for by all participating countries in the war.

---

172 Without naming a credible witness who could be further interviewed, Snell's sensational story could not be confirmed, and it was generally considered to be a hoax. The army didn't step forward with any information to support it. In January 2014, Dwight R. Rider, a senior intelligence associate with Intelligence Decision Partners of Virginia, published his impressively researched paper, *Tsetusuo Wakabayashi Revealed*. In it he makes an excellent case for identifying Lieutenant Colonel Tatsusaburo Suzuki, physicist, Imperial Japanese Army, as Snell's source, working under the pseudonym of Tsetusuo Wakabayashi. Suzuki graduated from Tokyo Imperial University in physics. Although he was in the Imperial Japanese Army, he was attached to RIKEN, a non-government institute, after graduation. There he worked with Yoshio Nishina, the most honored physicist in Japan, and he was tasked with writing a plan for the development of atomic power in April 1940. Tracing his whereabouts as the war ended, Rider makes a convincing argument that Suzuki was the man who told Snell about the A-bomb test. Suzuki's obvious participation in the beginning of the Japanese nuclear weapons effort will be detailed shortly.

They used focused microwaves, and after years of research, they were able to put the hurt on a small monkey at a distance as great as 1 meter, but only if he would stand still.

So, at one end of the spectrum we have a successful test of a fission weapon, and at the other end we have "What's a fission weapon?" The truth is somewhere between these widely separated poles.

Japan began its quest for an atomic bomb about the same time everyone else did. In April 1940, Major General Takeo Yasuda, director of Japan's Army Air Force Technical Research Institute, was aware of the nuclear energy release caused by uranium fission, the discovery of which was announced by Otto Hahn of Germany in January 1939. It had become a rapidly developing scientific topic, and there were two obvious military applications. Marine propulsion was one, and an extremely powerful explosive was another. To his judgment, Japan should work on controlling fission to produce an alternative power source.

Japan had a big navy and the largest warships in the world, but they didn't have enough reserve fuel to move that big navy from one end of the Pacific Ocean to the other, once. If they were to fulfill their destiny and annex the western end of the Pacific Rim, they would need a better way to move floating iron. Yasuda graduated from the Imperial Japanese Army Academy in 1909 and went on to earn a degree in electrical engineering at Tokyo Imperial University in 1916. He knew a young staff officer, Tatsusaburo Suzuki, from his stint as an artillery specialist. Suzuki was born in Nagoya, the center of Japanese porcelain and the only place where uranium was in every factory and shop. It was used to make dishes yellow. He had a degree in physics from Tokyo Imperial, and he was twenty-nine years old. Yasuda assigned him an important task. He was to study the fission of uranium and come up with a written plan for military exploitation. For this task, he would be headquartered at RIKEN, where he could take advantage of its large population of physicists.

At almost exactly the same moment, on April 10, 1940, the first meeting of the MAUD Committee met in Great Britain to plan how they were going to use nuclear fission for military purposes.[173] On April

---

173 Although the committee met on April 10, 1940, the name MAUD was not adopted until June of 1940. It was not meant to mean anything, as is appropriate for the name of a top-secret operation, but it later came to mean "Military Application of Uranium Detonation."

27, the Briggs Uranium Committee in America met to discuss preliminary lab experiments before proceeding with fast fission research. The worldwide physics community started to notice something. American physicists stopped publishing nuclear work in *Physical Review*. There was no doubt who was working on a bomb.

Suzuki submitted his twenty-page report in October 1940, concluding that a nuclear weapon could be built. He estimated that a runaway fission of 1 kilogram of pure uranium-235 would make the same hole as setting off 18,000 metric tons of gunpowder. From a military standpoint, the concept was irresistible.

Suzuki reasoned that it would be logical to first build a controlled chain-reaction demonstration, but it was not absolutely necessary for the task of developing a nuclear weapon. There was no doubt that the fission would increase exponentially in the right circumstances. A controlled reaction would be used to power ships, but that might take a decade of development. That should be left for postwar work. Although a fission engine would be an excellent neutron source, all of the cross-section measurement work could be handled by the cyclotrons. So, skip the controlled fission experiment. This war was going to be one quick, massive takeover of the Pacific, and it would be over before they could make significant headway in nuclear power research.[174]

There was one sticky point. Uranium occurs in most minerals in the earth's crust, but rarely in large concentrations and usually as mere traces. Moreover, the particular uranium isotope that is needed is less than 1 percent of the naturally occurring uranium, and it is

174 The Japanese bomb project was unlike the American effort but similar to the German work, in that it was scattered and divided, with different factions competing for the same limited resources and having divergent goals. In the United States, we had one military commander, Leslie Groves, in charge of everything, and the entire mammoth undertaking was focused on one outcome. A Japanese project did pursue a nuclear reactor design, but the only thing that was eventually built was a cardboard scale model, made by Professor Eizo Tajima, a neutron specialist. A problem was the availability of heavy water, thought to be necessary for the moderator material in a natural-uranium reactor. Heavy water was produced at the fertilizer factory in Konan, similar to the German source in Vemork, Norway, but the output was in 20-gram batches. For a reactor, they needed to start with 5 tons of it. The heavy water produced in Korea was used as a source of deuterium, the projectiles that were cyclotron-accelerated to make neutrons by striking a beryllium target. There is no record of criticality experiments in Japan using heavy-water moderation.

a difficult job to separate it out. To determine the most practicable isotope separation scheme, Suzuki studied the several methods invented during the 1930s and recommended thermal gas diffusion as having the lowest number of moving parts and the least amount of needed development. He and his team would require hundreds of tons of uranium ore to process down into that 1 kilogram of U-235 for a bomb.

It was an excellent start, but the government dragged it down in decision-making. Some military leaders feared that the Americans were working on an atomic bomb, and they would have to have one, too. Some thought it a waste of resources, which were all in short supply, and if they didn't knock the Americans out of the Pacific Ocean soon, things were going to get heavy. Finally, in July 1941, RIKEN was given a contract and a budget to develop nuclear weapons.[175] The 6th Army Technical Lab would be the military liaison and provide technical assistance. In charge would be Dr. Yoshio Nishina, the most famous nuclear physicist in Japan and their equivalent to the Manhattan Project's J. Robert Oppenheimer.

Nishina, described by his fellow scientists as a chubby, round fellow with a notoriously quick temper, had earned his bachelor's degree in electrical engineering at Tokyo University in 1918. In 1921, he studied in England at Cambridge under Ernest Rutherford. From there he went to Copenhagen, working with Niels Bohr at the Institute of Theoretical Physics and, with Sweden's Oskar Klein, wrote the Klein-Nishina formula for predicting the scattering of cosmic rays. In 1928, he moved back to Japan, and by 1931, he had his own laboratory at RIKEN.

His office was on the second floor of Building 29, with a chalkboard, a napping couch under the east window, and on the wall a picture of him in full uniform with the brother of Hirohito smiling with his arm around him. In 1940 and 1941, right before the Pearl

175 It is difficult to pin down the amount of money spent by the Japanese government on the atomic bomb project at RIKEN. The best estimate is ¥20 million, or $35 million in 1940s dollars. In wartime Japan, that was a lot of money, but in America the Manhattan Project spent $2 billion (1940s dollars) in the same length of time. To put that large number in perspective, $3 billion was spent to develop and test the B-29 strategic bomber that dropped the atomic bombs.

Harbor attack, he wrote four papers on nuclear fission for *Physical Review* and *Nature*. He is credited with discovering U-237, and on May 3, 1940, his paper, "Induced (beta) Activity of Uranium by Fast Neutrons," was published in *Physical Review*. That year, Japanese nuclear science was running parallel to effort in the United States, based on Nishina's work alone. He was an expert experimentalist as well as a theorist, and he made extensive use of the two cyclotrons at RIKEN.

Things were off to a slow, characteristically academic start, and immediately upon the awarding of the government contract, Japan started to slip behind the manically enthusiastic effort in America. A month later, in August 1941, Enrico Fermi and his team at Columbia University were already assembling a subcritical pile consisting of 30 tons of synthetic graphite and 8 tons of uranium oxide, achieving a k-effective of 0.83. At RIKEN, they didn't start measuring neutron interaction cross sections, work upon which all other work depended, until December 1941.

On May 3, 1942, the Imperial Japanese Navy, having enjoyed total victories in the war so far, began Operation MO, which was to invade Port Moresby in Papua New Guinea. In what would be remembered as the Battle of the Coral Sea, the United States Navy stopped them in their tracks, and a month later they suffered an even more punishing defeat at the Battle of Midway. This was a shocking change of fortune, and on top of that, espionage sources were bringing in bits of information about a major push in the United States to develop some kind of atomic device. Changes in priorities were necessary.

In August 1942, very little progress had been made, and Nishina was told that his project was being taken over by the Army Air Force headquarters. Asked what he was lacking, he mentioned money, scientists, and technical assistance. He was given ¥50,000 ($125,000), priority assistance from the 8th Army Technical Lab, and his choice of ten young men from universities, each to be given a draft deferment.[176]

---

176 At that time, Enrico Fermi was spending $2.7 million to build CP-1, the world's first nuclear reactor, in Chicago. Nishina should have mentioned that he also needed a few tons of metallic uranium.

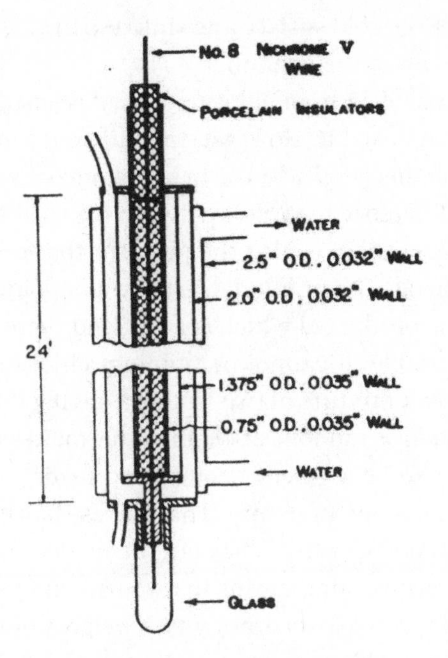

A 24-foot-tall thermal gas diffusion column, as described in Physical Review *in 1940. The isotope-separation device used in Project NI was based on this design.*

The critical path was the isotope separation process that would give them purified uranium-235, which was necessary for either a bomb or a ship-sized nuclear power plant. Suzuki's idea to use thermal diffusion seemed reasonable. The process had been invented by a German team, Clusius and Dickel, and had been successfully used to separate chorine gas isotopes in 1939. The apparatus consisted of a tall vertical column made of nested metal tubes. The innermost metal tube was electrically heated by a nichrome wire running through the middle. That tube was surrounded by another tube, and a gaseous form of the element that was to have its isotopes separated ran through the gap between the two tubes, introduced at the bottom of the column. An outermost tube with a gap conducted chilled water from bottom to top, giving the gas passage a cold side and a hot side. The lighter U-235 gas would tend to hug the hot surface and rise, while the heavier U-238 was expected to prefer the cold surface and sink. Purified U-235 gas would be drawn off the top of the column.[177]

Separating chlorine-35 from chlorine-37 had seemed almost simple when Klaus Clusius tried it. However, the difference in the masses of the two isotopic atoms is 5.7% of the lighter atom, whereas in the case of uranium, the difference in masses between the U-238 and the U-235 atoms is only 1.3%. There is also the problem that uranium is not a gas. It is a solid metal. The uranium has to be compounded with some other element, the product of which is a gas, and there is a restriction on which gas is usable. It cannot be uranium chloride, for example, because chlorine is a mixture of two isotopes having different masses, and this would add a random error into the mass of the resulting compound. It has to be a gaseous compound made with an element that has only one isotope in nature. That leaves fluorine, and the gas is uranium hexafluoride, called "hex" in the nuclear business.

Having to use a compound means that you are not really separating U-238 from U-235. You are separating two weights of uranium hexafluoride, at 349 and 352 grams per mole, and the difference is only 0.8%. The task of uranium isotope separation using a thermal column is therefore anything but simple.

By October 1942, the research group had decided on uranium hexafluoride as the separation gas, and Kunihiko Kigoshi was assigned the task of making the compound. He was too embarrassed to walk over to the Poison Gas Lab and ask how to do it, so he and his assistant, Takehiko Ishiwatari, consulted their college textbooks to find out where to get fluorine. They found that fluorine could be made by electrolyzing potassium fluoride. Rather than risk the embarrassment of ordering a few kilograms of potassium fluoride, they used some soldering flux on hand.

For days, running into weeks, they tried to derive fluoride from soldering flux by running electricity through it. Nothing seemed to

---

177 Instead of picking one of the possible isotope separation methods, the American bomb project tried them all at once. The first method to work was the thermal diffusion column, but the American version used liquid diffusion instead of gas diffusion, and the innermost tube was heated by steam and not electricity. The thermal diffusion separation plant at Oak Ridge, S-50, was a 21-stage cascade, and when working perfectly could improve the concentration of U-235 in natural uranium (mostly U-238) from 0.720% to 2.0%. It was better than nothing, and it was used to feed the gaseous diffusion plant (K-25), which improved the separation product up to 20.0% U-235. The output from K-25 was then sent to the electromagnetic (calutron) separator plant (Y-12) where in two stages, alpha and beta, the separation was again improved up to 84.0% U-235.

work. Finally, they had to humiliate themselves by going to Sendai and asking two professors at Tohoku University, Fusao Ishikawa and Eizo Kanda, the only two people in Japan who had ever made fluorine gas. It turned out that they had simply not given the reaction enough time. Turning off the power at 5:00 and going home was not the way to make fluorine. It has to run overnight.

Now, they needed some uranium, and they humbly asked the secretary, Sumi Yokoyama, to please order some. Eventually, 100 pounds of uranium nitrate arrived, packed into small glass bottles. On a roll now, they tried exposing the uranium compound to some fluorine gas. It exploded, reducing the expensive glassware to an embarrassing spray of shrapnel. They decided to first make some uranium metal out of the uranium nitrate and try again.

Consulting the chemistry book, the two researchers found that they needed to heat the nitrate to 1,500 degrees Celsius in an oven to make a dry, crystalline oxide and then combine it with pure carbon in the same oven. They had no pure carbon on hand, so they decided to make some by burning sugar.[178] By heating the uranium oxide powder combined with powdered carbon they could make detectable quantities of metallic uranium, exhausting carbon dioxide. Then, by heating the metal in a quartz vessel with fluoride salt, they made the first sample of uranium hexafluoride, condensing it in a flask as a single, green crystal. It had taken them twelve months. The American team doing the same thing took thirty-six hours.[179]

---

178 Unfortunately, sugar was an expensive, strategic resource, and it had to be flown in from Formosa. As they perfected the carbon reduction, batches of sugar would disappear overnight from the lab. People were sneaking small quantities to take home. Kigoshi found that common starch could be used instead of sugar to derive carbon, and the supply problem was solved.

179 While researching this incident, I noticed an important mistake in this procedure as described in army interrogations of the researchers. The Japanese researchers had not, in fact, produced uranium hexafluoride gas. They had, instead, made uranium tetrafluoride, which is called "green salt" in the nuclear business. Green salt is a solid at room temperature, and it must be vaporized with heat to make a gas. Uranium hexafluoride (hex) is a gas at room temperature, and it is made by heating green salt to 320 degrees Celsius in fluorine gas. For the purposes of thermal diffusion separation of isotopes, there is a big difference between the two substances. The green salt vapor will condense on the cold pipe surface in the separation tower instead of exiting out the top port. This explains all the trouble that the project was in for trying to separate U-235. I don't think that this problem had been noticed before, at least in print.

Although it had taken a long time to come up with a diffusion gas, the slow progress had not slowed down anything. The different methods of isotope separation had been debated, argued, and studied, sometimes at Nishina's house after hours, weighing the advantages and disadvantages of centrifuges and electromagnetic separators, and finally on January 4, 1943, Nishina and Masa Takeuchi decided to stick with the original concept: thermal diffusion. On March 15, 1943, it was made official in writing, and design got underway. A month later, Building 49 was erected at RIKEN to house the uranium enrichment equipment, next to Building 29. It was two floors, with 700 square feet per floor. At that same time, the ground for the K-25 gas diffusion enrichment plant at Oak Ridge was being surveyed. Its floor space would be 2,000,000 square feet.

Hidehiko Tamaki, short and fat with very bushy eyebrows, was made head of the theoretical design group at the start of the project in 1941. He was thirty-two years old, and his position was the equivalent of Hans Bethe in the Manhattan Project at Los Alamos. His degree was from Tokyo Imperial, and he had worked with Nishina on cosmic rays.

Tamaki's report outlining a hypercritical, explosive uranium mass was forwarded to General Yasuda in April 1943. He had too little experimental data to pin down a critical mass, but he predicted that a U-235 enrichment level of 10 to 20% would be necessary.[180] Word came back directly from General of the Imperial Army and Prime Minister of Japan Hideki Tojo. Intelligence from America indicated great activity on an atomic bomb project, and Japan was ordered to keep pace. Nishina's bomb project was given a new designation: NI, meaning "Nishina, large bomb." NI was given top priority for materials and manpower and another ¥700,000, and seven young army officers

---

180 Was it possible to build an explosively assembled atomic bomb with only 20% U-235? The L-11 "Little Boy" bomb that destroyed Hiroshima was built using 80% enriched uranium metal, explosively assembled into a cylinder 6.5 inches in diameter weighing 64 kilograms, making nearly three critical masses. A single, fast-neutron critical mass of 80% enriched uranium, beryllium reflected, is 19.3 kilograms, and the bomb required 3.3 times as much material. A critical mass using 20% enriched uranium (beryllium reflected) is 245 kilograms, or an explosive, hypercritical mass weighs 809 kilograms. That is 1,784 pounds, just for the bomb core. Explosive assembly of the Little Boy required a 6.5-inch ack-ack gun barrel. Assembly of Tamaki's weapon would require a 15-inch naval gun weighing many tons. In World War II, there wasn't an airplane or torpedo boat in the world that could carry such a bomb.

were assigned to the project to offer technical assistance. Back in the United States, 20,000 construction workers were building the uranium enrichment facilities at Oak Ridge; 1,000 researchers were working at Columbia University and Kellex to develop a gaseous diffusion barrier; 5,000 men were building the cooling systems for plutonium production reactors at Hanford, Washington; and 4,800 men and women were hired to work at the Y-12 electromagnetic separators. Project NI was to be administered directly through the air force headquarters. Nashina issued one plea back to the administrators: "Get more uranium."

On July 24, 1943, Toranosuke Kawashima, the Leslie Groves of Japan and now military head of the project, radioed Ambassador Baron Hiroshi Ōshima in Berlin, asking for the Third Reich to please send a boatload of high-grade pitchblende uranium ore, preferably from occupied Czechoslovakia. There were other pressing matters in Europe, and the request was not at the top of the pile. Another request was made on August 24, 1943, this time with the diplomatic language made brief. The Germans had a question: "What's it for?"

"It's for use as a catalyst to make butanol," Kawashima lied.

The Germans were skeptical. It would be terribly inefficient to send uranium ore, but they agreed to send two tons of refined uranium oxide by cargo-carrying U-boats. They would require a report on how well that "catalyst" was working.

By October 1943, all the parts had been manufactured for one thermal diffusion column. It was 16 feet high, and they had to take out a section of the floor on the second story of Building 49 so that it would fit. A spiral staircase was built for access to the entire height of the device. After a lot of debate, the design team had decided to make it out of copper so that the highly corrosive uranium hexafluoride gas would not corrode it, and the gap between the hot and cold walls, where the gas would separate into light and heavy molecules, was only 2.0 +/-0.1mm.[181] Construction was farmed out to Furukawa

181 The design was almost a direct copy of a thermal diffusion column described in *Physical Review*, volume 57, from 1940, but that column was 24 feet high, and the gap between the hot and cold walls was 16mm. The team in Japan reasoned that they could get better isotope separation in one stage by making the gap very narrow, and their building was not tall enough for a 24-foot column. The American version, using a liquid solution instead of gas, was made of pure nickel.

Engineering Company in Nikko, which was a four-hour drive from RIKEN. Unfortunately, it was a rough section of road, and the long, delicate copper tubes were bent by the time they arrived at RIKEN. It took a month to straighten them out, and 2mm copper pegs had to be silver-soldered to the hot-leg tube to maintain the strict 2mm spacing.

Materials, such as asbestos and gum for sealing and insulation of the column, steel for holding up the copper tubes, and a thermostat to regulate the heating wire, were needed, and Project NI had to overcome a natural timidity and use its priority status to order things. In general, the war, going on longer than was expected, was shortening supplies of just about everything. To deal with this growing problem, the government decided to complicate the act of ordering, thus decreasing the demand. One could not simply order a pump for thermal-column cooling water. A pump was broken down into component parts, and each part had to be ordered separately and then sent to a central location to be assembled. To order one part, thirty-two forms had to be filled out. It could take months. Getting desperate, Masa Takeuchi, still in charge of isotope separation, appealed to Colonel Kenji Koyama to break up the logjam. Koyama was their military liaison to the air force, and he tried to keep on top of things, roaming around in Building 49 daily.

Materials deliveries improved, and by November 1943, the separator column was ready to be tested. Nothing worked right. The inner pipe was too thick, and the heat wasn't uniform. The thermostat was erratic, and the cooling water system leaked. The uranium hexafluoride induction system would have to be changed. It took a month just to get a thin-walled pipe.

By March 12, 1944, the thermal diffusion column was ready to test. To start with, they tried a simple problem: separate argon-36 from argon-40, and confirm it using the only mass spectrometer in Japan, over at Osaka University. Curiously, they had no success, and by May 27, the team had given up and was going to try a load of uranium hexafluoride. On July 14, 1944, a 179-gram crystal of uranium hexafluoride was heated in a flask, and the resulting fumes were piped into the bottom of the 16-foot column. For some reason, nothing worked. The pressure in the gap between the pipes dropped to zero, and nothing came out of the top port. Puzzled, they tried beating on it with a mallet. Bang! The top of the column split open, and noxious green vapor spewed into the top floor. It took a month to get it back together and working.

They were ready to try again by August 1944, but first Nishina held a meeting to evaluate the progress. Asao Sugimoto from the cyclotron group was quite vocal. He thought they should shut the project down and do something useful instead. The young researchers started to look very nervous. "If we stop the project, will we have to go fight the Americans?" one asked. Nishina challenged everyone to work harder and ended the meeting.

Takeuchi, still in charge of isotope separation, was under growing pressure. He knew a guy, Mitsuo Taketani, who was a physics genius and knew a lot about fluid dynamics. Takeuchi had a theory about the turbulence in the gas channel, and he wanted to run it by Taketani. Unfortunately, Taketani was in jail for Marxist tendencies. Back in 1938, he had been put in prison by the "thought police" for being a communist but had been released to work for Sin-Itiro Tomonaga, a famous scientist.[182] In 1944, with the Soviet Menace on the move, he was thrown back in prison. It didn't help that he was married to a Russian. Takeuchi used the priority status of Project NI to get Taketani set up in a private office in the prison, complete with research materials and a courier to pick up his results. He theorized that the transfer of energy between colliding particles containing uranium might not be as unimpeded as one would expect.

By November 10, 1944, all the thermal column problems were not quite solved, and someone mentioned that Eiichi Takeda had been using a cascade of three thermal column separators for years over at Osaka University to separate inert gas isotopes. Why not just go over there and use his? They were 61 feet high, using four stories of his research building. It pained Nishina to have to go outside his fiefdom, but the pressure was on him to build a bomb that could wipe out the American fleet in one swat, and he had no choice but to act on the suggestion.

---

182 Sin-Itiro Tomonaga shared the Nobel Prize in physics with Richard Feynman and Julian Schwinger in 1965 for having developed quantum electrodynamics. He worked at RIKEN with Nishina after graduate school, and during the war he worked on magnetron vacuum tubes (for radar transmitters), meson theory, and his "super-many-time" theory, which sounds like something Feynman would have done.

Working quickly, Suzuki decided they would need a five-stage cascade of thermal columns, built 61 feet high like Takeda's. His assistants knocked out floors two through four in Takeda's building at Osaka, and construction was contracted to Sumitomo in Amagasaki. Suzuki, striving to improve on the Osaka design, decided to line the innermost copper tube with steel, to ensure even heat distribution, and he decreased the gas-channel spacing to 1.6mm. A window was added, so they could watch the gas swirling around in convection currents.[183]

With the coming of the New Year in 1945, Project NI now had a five-column isotope-separator cascade. It was set up temporarily in a hangar-like building at Sumitomo's Amagasaki plant, near the outskirts of Osaka. The Americans captured Iwo Jima in February. Their heavy bombers were now only three hours away from Tokyo, and they were bombing everything in sight. On March 13, as the team was getting the building ready for installation of the new system, a bombing raid near Osaka University knocked out all electrical power and water. The delicate separator system, sensitive to injury, would have to be moved somewhere else, out of bomber range.

The uranium enrichment team, now under the strict supervision of Sumitomo's scientist, Dr. Toshio Ikejima, ran isotope-separation tests on nitrogen and carbon dioxide. The system worked perfectly, and the differing temperature expansion coefficients of copper and steel in the heated tube wasn't giving them a problem. Feeling confident, the team ran a "hexafluoride" batch through the new system. They took the results over to RIKEN, where a cyclotron-driven neutron source setup was used to evaluate the enrichment percentage. Hitting a crystal of non-enriched uranium compound with neutrons caused fission in the U-235 portion of the sample. By counting the resulting gamma-ray rate, a benchmark was established. They then substituted an enriched crystal of the same weight, turned on the neutrons, and gamma-counted it. The two samples showed identical response to

183 Having a window to watch the gas was a bit odd. Gas is invisible. What they were watching was uranium tetrafluoride vapor in which molecules were bound together into visible droplets, like the "steam" from a teakettle. In this clumped state, molecules of different weights cannot be separated by diffusion. They had put the cold and hot walls so close together, the cold wall was above room temperature and the vapor did not condense in the column, so it appeared to be working.

neutron bombardment. There had been zero isotope separation. A larger percentage of U-235 in the sample should have resulted in a proportionately larger gamma-ray count. Nishina was devastated. "Run it again!" he demanded. No joy was forthcoming.

As if that were not enough, on the night of March 9 and into March 10, 1945, the United States Air Force, under Colonel Curtis "Bombs Away" LeMay, ran the biggest fire-bombing raid in the entire war against Tokyo, killing more than 220,000 people and making five million homeless. RIKEN was reportedly in flames.

The NI team raced back to RIKEN, and found that Building 49 was surrounded on all sides by burning buildings. Everybody—scientists, engineers, assistants, and maintenance workers—fought the fire any way they could, keeping the flying embers away from their project base. Completely exhausted and convinced that the fires were out, they went home to find sleep. Sometime before daybreak, a spark landed on the building, and by the next morning, it had burned to the ground. Their initial separator experiment, all their documents, and the entire stock of uranium compound were gone.

In the Pacific Ocean, the Tinian Island air base was now being specially configured for the 509th Composite Group, a unit of the U.S. Air Force's 58th Wing that would drop A-bombs on Japan. Separate, air-conditioned shops were being erected for final assembly of the Little Boy uranium bomb and the Fat Man plutonium bomb.

Japan was now under a "state of siege," according to the high command. All industries that could not be moved were to be relocated underground. Those that could move were to move to Korea. Nishina suggested that they move the new separator cascade to Kanazawa, over on the west coast and resume debugging the equipment.

At that point, the story goes dark. Kanazawa was not out of range of the B-29s, and an attempt was probably made to ship the entire project to Korea. There simply wasn't time to set up the system in Konan and do any useful work before the Soviet army poured in over the border in early August. Whatever and whoever made it to Korea in the chaos wound up in Russia after the war.

There was, however, another atomic bomb project. It was started by the Japanese Imperial Navy and was destined to compete with Project NI for the limited resources strained by a full-scale war that was not going as planned. In early 1941, a navy captain with a science education, Yoji Ito, was assigned the task of compiling a report that evaluated the concept of developing a nuclear weapon. It took him over a year, consulting with every expert he knew of in the field. On July 8, 1942, his efforts were evaluated by a distinguished committee including Yoshio Nishina, Hantaro Nagaoka, Ryokichi Sagane, Juichi Hino, Tsunesaburo Asada, and Seishi Kikuchi. Nishina, who was made chairman of the committee, did not emphasize that he had his own bomb project. Nagaoka was the father of Japanese physics. He had invented the short-lived "saturnian" model of the atom in 1904, and had inspired Ernest Rutherford to come up with a better model. Sagane had an international reputation, to the extent that American scientists later in the war had a B-29 drop a letter on him appealing for the surrender of Japan.[184] Asada had been pushing the navy to do nuclear development since 1937, and he became a major proponent of the death-ray weapon. Kikuchi was Ito's colleague at Osaka University who worked on radar and a related form of the death ray.

Ito's report began with what he thought to be the most practical application of uranium fission, the synthesis of radioactive isotopes for use in luminous paint. The committee wanted to talk about atomic bombs. The group, now named the Physics Colloquium, decided to meet once a month to discuss a development project. Some were opposed to a bomb-development effort. Nishina pushed for it, but not for this war. It was going to be over too fast, with the American navy collapsing in the middle of the ocean, and there would be no need for such an advanced piece of technology. A conclusion was that the United States would not be able to come up with anything of a nuclear nature, so there was no rush. Secretly, Nishina did not want a competing project going after the limited uranium supplies. The committee adjourned the last time at the May meeting in 1943, without having put anything in motion.

The ordnance branch of the navy ignored the Physics Colloquium's findings and funded a top-secret project to develop an atomic bomb,

184 It was an interesting idea, to avoid massive destruction and loss of life by convincing a renowned scientist that resistance was futile, but despite Sagane's worldwide fame in science circles, Japan was not his to surrender.

headed by the distinguished physicist Bunsaku Arakatsu. He was born in 1890 and graduated from Kyoto University in 1918 with the appropriate fire in his belly. When he moved to the University of Berlin to study physics under Albert Einstein, the two ambitious scientists became friends. After a year of soaking up Einstein, he moved to the Technische Hochschule in Zurich, Switzerland, and from there he bounced to the Cavendish Laboratory at Cambridge, where, like Nishina, he studied under the master of experimental nuclear physics, Ernest Rutherford. By 1934, he was back in Japan, scratch-building a Cockcroft-Walton linear particle accelerator, a cloud chamber to display the paths of particles as miniature jet contrails, and home-brewed heavy water. There wasn't a person in Japan better qualified to implement an atomic bomb plan. He was given ¥6,000 to get started, with another ¥600,000 in the mail. The Colloquium, including Nishina of Project NI, was not advised of the existence of this new project.

By this time, April 1943, the navy bomb project was way behind what the Americans were doing. In a magnificent bonding of industry and science, the United States had already discovered plutonium, built and studied a working nuclear reactor, built the atomic cities of Oak Ridge, Hanford, and Los Alamos, and had a unique idea: achieve hypercriticality by crushing a ball of fissile material down to an unnaturally dense nodule using a surrounding ball of high explosives. It was the single idea that made the plutonium-powered bomb a deadly reality. Plutonium would fission under neutron bombardment like U-235, and it could be made in vast quantities using otherwise worthless U-238 in the new "converter reactors" at Hanford, Washington. There was no tedious isotope separation necessary to use plutonium in a bomb, as was required for uranium, but it tended to pre-detonate in the simple, gun-assembly bomb design. The crush-explosion idea solved that problem. Using plutonium, there was virtually no limit to how many atomic bombs the United States would be able to manufacture.

Legend has it, the Japanese navy project was named "F-go," where the F stood for fission or fluoride, depending on whose memory is plumbed. Some say it was named *Nichi*, Japanese for "sun," and others say it was "F Research." Whatever it was named, the director of the project, Arakatsu, homed in immediately on the need to enrich natural uranium. Nothing else mattered or could be proven until they had made an enhanced concentration of the active ingredient, U-235. He decided that the method of isotope separation with the best chance of succeeding was the gas ultracentrifuge, running at 150,000 RPM, the speed of an aircraft turbocharger rotor at full throttle.

Sakae Shimizu, a physicist at Kyoto University, was put in charge of the centrifuge development, and he and Arakatsu threw everything into the project. It would be a large, complicated machine, requiring extreme precision and balancing to prevent it from vibrating and flying asunder at full speed. An air turbine on the perimeter of the rotor would spin it up, and in the working diagrams, it resembled a General Electric axial-flow jet engine standing on its nose. The main rotor, containing the uranium hexafluoride, would spin at extreme speed, causing the heavy U-238 to hug the walls and displace the lighter U-235 to the center axis, physically separating the two uranium isotopes. It looked big and fat, running on air-bearings and ringed with turbo-fins. A static magnetic field at the top was supposed to stabilize it and dampen vibrations.[185] Component designs were farmed out to Hokushine Electric, where navy gyrocompasses were built, and to Tokyo Keiki, a large electronics firm. Construction and final assembly went to Sumitomo.

While that was going on, the project needed uranium so that they would have isotopes to separate. First, the ceramics shops in Kyoto were raided, and the team was triumphant when they had scrounged 100 pounds of uranium oxide. This was less than they needed, which was closer to 1.5 tons. In September 1944, Arakatsu, feeling the heat, went on a scouting expedition to Korea, looking for uranium mines. There was some modest mining, but the ores were all secondary sources, such as fergusonite, euxenite, and monazite. What they

185 This was an interesting design, different from what the Germans were working on during the war and what would become the standard ultracentrifuge configuration of the 1990s, as is now used in certain Middle Eastern countries. The flattened, large-radius rotor would probably have self-destructed at the very high angular velocity. (Newfound documents indicate that the rotor was to be made of super duraluminum, a fairly exotic alloy at the time.) It has since been found that the only way to make this device is to use tall, thin rotors instead of short, fat ones. Instead of hoping for complete separation in one stage that breaks the sound barrier at the rotor rim, it is better to have a cascade of many rotors running at a less dangerous speed. Using air-bearings and a stabilizing magnetic field, however, was definitely a correct way to do it.

needed was a big mine full of pitchblende, but what they had was a hole in the ground in Ishikawa, on the west coast of Japan, worked by schoolchildren and producing only 750 kilograms of samarskite ore. An old gold mine in Central Korea, the Kikune, was reopened and sifted for a pitiful 4% fergusonite, which is only 8% uranium. A message from Manchuria claimed that they had found a uranium mine with 3 to 5 tons of unrefined ore ready to ship. Communications went dead, and nothing came of it. The army in Shanghai had confiscated 800 kilograms of uranium oxide, but they were in the process of evacuating and they could not bring it with them. The supply of uranium was becoming a weak link in the F-go project.

The theoretical work in F-go was performing well, with Hideki Yukawa, a future winner of the Nobel Prize in physics, in charge.[186] Minoru Kobayashi was one of his best theorists. Grinding numbers for two days using a hand-cranked calculator, he estimated that it would take a sphere of pure U-235 that was 20 centimeters in diameter to get an explosion. The estimate was optimistic. The isotope-separation team was hoping for a U-235 enrichment as great as 10%.

By this time, the American bombing campaign was in full swing, and Japan's task of turning theories into reality was becoming difficult. The death ray that was intended to knock bombers out of the sky was bogged down in development. The Hokushine electric factory where the centrifuge was being built took a direct hit. Records are confused at this point, but the F-go centrifuge components may have been shipped to Sumitomo before the explosion took out the assembly building. According to Commander Kitagawa, the navy liaison officer with F-go, the completed machine was on its way to Kyoto, stopped at a train siding, when it was wiped out by another aerial bomb.

On July 21, 1945, the navy convened a meeting of the F-go project to discuss progress, held at scenic Lake Biwa just north of Kyoto. Everyone was there. Arakatsu and his theoreticians Yukawa and

---

186 Hideki Yukawa won the Nobel Prize in physics for having predicted the existence of the meson particle, explaining the relationship between the proton and the neutron. In 1943 the Japanese government, as the world was beginning to come down around it, awarded him the Decoration of Cultural Merit. In 1977, he was named the Grand Cordon of the Order of the Rising Sun. He died of pneumonia at his home in Kyoto in 1981 at the age of 74.

Kobayashi, centrifuge designer Shimizu, and the radiochemists, Sasaki and Okada, were there representing the project. The navy sent a commander, two rear admirals, and a captain. Several scientists from Osaka University attended, as well as Shoichi Sakata from Nagoya University. Toshikazu Kimura, a neutron specialist from the NI project at RIKEN, may have sneaked in.

Arakatsu opened the proceedings with a summary of everything that they had achieved so far, followed by Rear Admiral Nitta explaining the centrifuge action, pointing to a blackboard drawing using a bamboo stick. The meeting fell quiet, as the crushing load of what had not been accomplished started to sink in. Someone spoke up. "Couldn't we do something with thorium?"[187]

A fine meal was served to the attendees, the F-go project was formally dissolved, and everyone went home.

On August 6, 1945, at 8:14:00 in the morning, Captain Matao Mitsui was at his desk at the Kure naval shipyard, working on the problem of synthesizing jet fuel. Downtown Hiroshima was 12 miles northwest of his office, and from his office window, he could see the atmosphere shimmering from the heat rising from the awakening city.

Almost directly over Mitsui's office, three B-29 bombers were flying at 31,600 feet, looking like the usual weather-gathering expedition. It did not resemble a bombing raid, as had happened in June and July, so the alert siren did not even start wailing.

The lead plane, *Enola Gay*, held L-11, the Little Boy atomic weapon, in the bomb bay. The plane was specially equipped with two AN/ART-13 high-frequency radio transmitters, each tuned to a different frequency, warmed up, and ready to send a 100-watt radio signal. At 8:09 A.M., Colonel Paul Tibbets, the pilot, flipped the switch that transferred directional control of the airplane to the Norden bombsight. Five minutes later, at 8:14:17 A.M., one of the AN/ART-13s came on, transmitting a continuous audio tone. The tone, indicating that the Hiroshima bombing

---

187 There was a thorium factory in Niihama on the Japanese island of Shikoku, making lantern mantles. The one thorium isotope occurring in nature, Th-232, is not fissile. However, it can be neutron-activated into a uranium isotope, U-233, which is fissile and can be made into a bomb. No isotope separation would be necessary. If they had a decade to work on it with no American bombing missions, they could have done something with thorium.

run had commenced, was heard on a radio receiver at the Tinian air base, 1,600 miles away. It was also picked up by one of the other B-29s, *The Great Artiste*. Sixty seconds later, the tone stopped, and bomb away. At Tinian, the received signal indicated that the Little Boy was on its way down, and from *The Great Artiste*, two radio-equipped instrument pods were released on parachutes to monitor the explosion. On the *Necessary Evil*, the third plane, the high-speed motion-picture cameras, aimed at the ground, started rolling. The three B-29s made hard left-banking turns, losing altitude fast and accelerating to maximum speed.

At 8:15:27 A.M., a flash of white light lit up Mitsui's desktop, and he instinctively turned to the window. Had an electrical substation blown up? The floor shook with a rolling explosion. He rushed to the window, but he could see no burning substation. The sound seemed to have come from behind the hills to the northwest, in the direction of Hiroshima. He ran out of the office, into a gathering crowd of curious onlookers, and went up the hill where he could get a better look. A huge, gray cloud was forming up where Hiroshima should have been. Did an entire trainload of munitions explode?

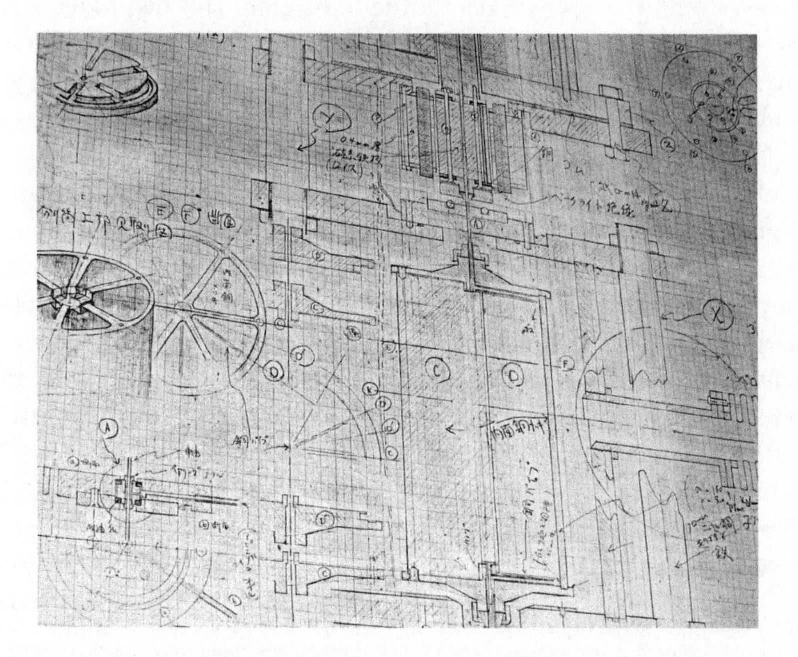

Plans for the centrifuge, discovered recently in the Kyoto University library.

At 8:30:00 A.M., Captain William S. "Deak" Parsons, the atomic bomb specialist aboard the *Enola Gay*, instructed radioman Richard Nelson to send a coded message to Brigadier General Thomas F. Farrell, aide to General Leslie R. Groves, director of the Manhattan Project, at Tinian, using the other AN/ART-13 transmitter. The message, sent in Morse code, consisted of: A1296. The cypher was known only to Parsons and Ferrell, and they had written down a list of phases with corresponding single-digit numbers two days before. The "A" indicated the start of the four-number message. The "1" meant "Clear-cut, successful in all respects," describing the general outcome of the mission. The "2" meant "Visual effects greater than Trinity." The "6" was "Hiroshima primary target," and "9" was "Conditions normal in airplane, proceeding to regular base."[188]

Mitsui saw something odd about the cloud rising over Hiroshima. It was streaked with crimson. He knew enough nuclear physics to realize exactly what he was looking at. That particular shade of red was made by ionizing a sample of strontium in a Bunsen burner flame. He was looking at the results of a sudden, short-lived uranium-235 fission event, from which one of the most dominant by-products is the strontium-90 isotope. Nothing else makes that particular glow. The world had just entered the Atomic Age, and all that Japan had contributed was the target.

The 6:00 evening news on Japanese radio mentioned that a few B-29s had fled in fear after dropping some incendiary bombs on Hiroshima. Damage was being surveyed. At 7:30 A.M. the next morning, Mitsui led the Kure naval base survey team into Hiroshima, where they met up with the Imperial Japanese Navy survey team in front of where the historic Hiroshima Castle had proudly stood the day before. On August 8, they submitted a discouraging report that 80 percent of the city had been wiped off the Earth along with an estimated 100,000 people.

In North Korea, Japanese Unit 731, purported to have managed an atomic bomb test in the Gobi Desert, was aware of the rapidly

---

188 I have never found the handwritten cypher cooked up by Parsons and Ferrell, but one can imagine that the message "A0000" would mean something like "Unable to reach primary or secondary target; all four engines on fire."

advancing Soviet army. The unit began destruction of its extensive facilities at Pin Fan on August 9, 1945, using dynamite and all the excess aerial bombs they could find. For three days, they tried to erase their presence in Korea and then fled for the coast. The facility remains amazingly intact to this day, and is now a museum dedicated as a reminder of the dark days of Japanese occupation.

⚛

On the morning of November 22, 1945, the United States military occupation forces, operating on an order from on high in Washington, began a campaign to rid Japan of any ability to develop nuclear technology. Each cyclotron was located, broken to pieces, loaded onto a boat, and dumped overboard into Tokyo Harbor. Credit for this order has never been claimed.

Did the Japanese test a nuclear weapon off the coast of Korea in the last hours of World War II? I do not see how. It would have been a third atomic bomb program, completely hidden at Konan in North Korea. There were no cyclotrons in Korea. Nobody noticed any nuclear physicists missing after the war, all of them having been spirited off to Moscow for questioning.[189] In general, the Japanese projects suffered from the same lack of enthusiasm that plagued the German efforts. They were never able to "throw body, soul, and spirit into fire," as prescribed by the Rosicrucian alchemic text quoted at the beginning of this chapter. Both the Germans and the Japanese knew that World War II would have to be over quickly, and strategic success depended on prompt surrenders by the Allied powers. Neither empire had access to enough liquid petroleum fuel to power ships, airplanes, trucks, and tanks for a long fight. Any realistic evaluation of the development of a new, previously unimagined technology from scratch indicated that it would take more time than was available before the end of the war.

Yoshio Nishina, head of Project NI, died of liver cancer on January 10, 1951. A crater on the Moon is named for him.

---

189 The chances of a nuclear physicist of any nationality surviving the global conflict of World War II was about 100 percent. It was not until a few days after V-J Day that we started losing them to criticality accidents.

CHAPTER 8

The Criminal Use of Nuclear Disintegration

> "It's a good thing we won the war. If we hadn't, I'd be hanged as a war criminal."
>
> —General Curtis "Bombs Away" LeMay

THE HAGUE CONVENTIONS OF 1899 and 1907 were one of the first statements of the law of wars of civilized nations and a formal definition of war crimes. Many techniques of warfare were strictly prohibited, from the discharge of projectiles and explosives from balloons to the laying of automatic submarine contact mines. Convention IV specifically prohibited the use of poison or poisoned weapons in warfare. This specification was ratified by all major powers except the United States.

Although always prepared to do so until 1972, the United States never used poison in a war. A letter sent to Enrico Fermi from J. Robert Oppenheimer on May 25, 1943, seems to indicate that the use of a nuclear-derived poison was at least considered during World War II. Fermi was a nuclear physicist, imported from Italy, who conducted the first nuclear reactor experiment at the University of Chicago in 1942, and Oppenheimer was the technical head of the Manhattan Project, developing nuclear weapons at the Los Alamos laboratory in New Mexico.

The letter, which was not declassified until 1984, begins:

> Dear Fermi:
>
> I wanted to report to you on the question of the radioactively poisoned foods, both because there are some steps that I have taken, and because Edward Teller has told me of the difficulties into which you have run.

The document goes on to suggest strontium-90 as the isotope of choice and to read "I think that we should not attempt a plan unless we can poison food sufficient to kill a half million men . . ." This letter implies that there was a plan to use strontium-90, an otherwise worthless by-product of the plutonium production at the Hanford site, to poison the enemy. To do so would have been a war crime, as defined by Convention IV, ratified by a representative of the Empire of Japan. There is no available evidence that the concept of radioactively poisoned food went any further.[190]

If somehow the Japanese navy had managed to destroy the United States invasion fleet with a single atomic bomb and had turned

190 This document seems odd and out of place in the steady stream of important letters bouncing around the country during the hectic days of the Manhattan Project. It is vague and wordy, but there is obvious effort to mention several people, including theoretical physicist Edward Teller, Harvard president James Conant, General George C. Marshall, Nobel Prize winner Arthur Compton, medical physicist Joseph Hamilton, director of the Metallurgical Laboratory Samuel Allison, and director of the chemistry division of the met lab, James Franck. No documented response or follow-up to this letter seems available. It is almost as if it were a dummy proposal, meant as a test to find any security leakage. This was definitely a tactic used during the war, to drop a false letter-bomb into the stream of communications and listen for an immediate response from Japan or Germany in encrypted messages. In this case, it may well indicate a security leak. The Japanese military was never convinced that the United States had an atomic bomb, but they were braced for an attack using gravity bombs and artillery shells filled with radioactive poisons.

victorious in the war in the Pacific, General Curtis LeMay would probably have been near the top of the list of military personnel to be hunted down and prosecuted for war crimes. Not just for orchestrating the atomic bombing of Hiroshima and Nagasaki would he be found guilty, but for his even more destructive cremation of Tokyo, the largest and most important city in Japan.

Since November 1944, when airfields were firmly established on Saipan and Tinian Islands, Tokyo was harassed on a weekly basis by B-29 airplanes in LeMay's XXI Bomber Command. The most destructive mission of the entire war, flown on the night of March 9 into March 10, 1945, was Operation Meetinghouse. Loaded with 1,665 tons of incendiary bombs, 334 B-29s took off and flew to Tokyo. Of the 334 planes, 282 actually made it to the aiming point, and 226 of them were able to unload over the target, releasing 500-pound E-46 cluster bombs. Each E-46 carried 34 incendiary bomblets, loaded with napalm and impact detonators. As the bomb reached an altitude 2,000 feet above ground, a barometric sensor tripped a catch, and all 34 bomblets were set free to fly on their own. Each ignited about 4 seconds after touchdown, hosing down everything in reach with flaming gobs of jellied gasoline. The densely populated, highly flammable residential neighborhoods, unprotected by the most basic firefighting equipment, burned to the ground as the wind gusted at up to 28 miles per hour. More than 100,000 people died that night in Tokyo, about a million people were injured, 286,358 buildings were destroyed, millions were left homeless, and a quarter of the industrial production in Tokyo was lost in that one operation.

It took Japan a long time to get over it, and by 2007, the Japanese people were even starting to blame their own government for the highly destructive bombing raids. By November 1944, it should have been obvious that they were not going to win the war, it was reasoned, and they should have surrendered before Tokyo was wiped out. The wisdom of having attacked Pearl Harbor back in 1941 was even reexamined, and blanket apologies were issued to South Korea for uncivilized behaviors during the war. However, in 2013, in his second term as prime minister of Japan, Shinzō Abe turned it around. The bombing raids were "incompatible with humanitarianism, which is one of the foundations of international law." He had effectively defined the

American bombing raids as war crimes, but was too late to do anything about it. Technically, Curtis LeMay may not be guilty of having committed a crime using nuclear disintegration. He was simply directing warfare operations on a scale that had never before been attempted.

There are many other incidents involving electromagnetic radiation, often of nuclear origin, plus attacks with alpha and beta particles, and at least one mischief involving neutrons. All instances are definitely criminal acts. There have been murders, attempted murders, attempted suicides, and at least one attempted illegal fetal abortion. Many nuclear crimes have involved the theft of valuable or seemingly valuable radiation sources without an inkling of the danger involved with handling highly radioactive objects. When working a crime on the ragged, outer edge of technology, a lack of knowledge can be deadly.

Scheming toward ways of using radiation to maim or kill probably began shortly after Nikola Tesla published the first mention of painful irritation and open, oozing wounds caused by exposure to X-rays in *Electrical Review* on December 1, 1896. An ability to kill someone while standing off at a distance, not having to actually touch or be touched by the victim, had appealed to the bad side of humanity since the invention of the spear, sometime in prehistory. Using a firearm, as became popular in the fourteenth century, was better than swinging a blade, gouging eyes, beating with a rock, or choking with the hands, but the ultimate murder instrument would accomplish an absolutely clean kill, without spraying blood all over everything, tearing loose body fragments, and sending projectiles to bounce off walls and cause collateral damage to the furniture. This perfect weapon would leave no trace of itself, no bullet, cartridge case, or propellant residue. A singed spot on the shirt and the smell of burnt hair would be acceptable, or perhaps even a neatly cauterized hole in the abdomen, but nothing that bleeds.

Inventors, scientists, engineers, and technologists of every stripe and heritage have worked on this problem or at least thought about it. It has been given many names: the death ray, the death beam, the heat ray, the directed-energy weapon, the laser weapon system, the teleforce, the Ku-Go, and the peace ray. It even became a common device used in science fiction, in which every hero from Flash Gordon in *The Flight of the Hawkmen* to Han Solo in *Star Wars Episode VII: The Force Awakens* carries one and uses it with abandon. This weapon class has become such a familiar, expected technology that sometimes we wonder why it isn't yet available at the local gun shop.

Claims for successful development of a ray gun started in 1923, when Edwin R. Scott made the announcement from his laboratory in San Francisco, California. He hoped, of course, to sell ray-gun units to the United States Army by the tens of thousands, and his device was advertised as capable not only of cutting down enemy soldiers—it could melt an airplane in midair. Scott had worked for nine years in Yonkers, New York, under Charles "Forger of Thunderbolts" Steinmetz, the famous electrical engineer/socialist who pioneered everything from magnetic hysteresis to artificial lightning, and he had picked up a trick or two. Unfortunately, his ray gun did not work as advertised. A similar disappointment was created by inventor Harry Grindell Matthews a year later in the United Kingdom. Proposed demonstrations to the British Air Ministry of bringing down airplanes flying overhead were unsuccessful.

More credible was Nikola Tesla's public announcement of his new design for a death beam, or "teleforce," in the *New York Times* on July 11, 1934. Tesla had been a widely publicized electrical engineer since 1888, when he licensed his polyphase induction motor design to George Westinghouse and helped introduce the United States to alternating current. His list of inventions was impressive, including a radio-controlled boat demonstrated in 1898, a vaneless turbine using aerodynamic friction in 1906, and his famous high-voltage transformer operating at radio frequencies, producing millions of volts of electromotive force, the "Tesla coil." He had been somewhat idle ever since 1917, when his Wardenclyffe power plant project on Long Island, New York, was abandoned due to lack of development funds. He had hoped to use an electrical resonance of Earth to send electrical power wirelessly, to be used freely all over the world. The concept of giving electrical power away for free did not appeal to those who had enough money to fund the endeavor.

By 1937, Tesla claimed to have a working model of his "charged particle beam weapon," and that it was capable of dropping armies dead in their tracks and bringing down a fleet of ten thousand airplanes at a distance of 200 miles. It sounded large. He described the design in a paper written the following year as an open-ended vacuum tube used to accelerate charged particles using high-voltage electrical attraction. It is impossible to accelerate charged particles in air. It must be accomplished in a vacuum, but the particles have to move through air to reach the target, and the end of the evacuated tube will always discourage particles from venturing out into the atmosphere. Tesla solved this problem by making a virtual tube-end, consisting of a high-speed air jet blowing across the open end of the tube, provided by a Tesla turbine used as a blower. It was ingenious, in principle.

His "charged particle" electrically accelerated projectiles were not what we think of as charged particles, which are usually electrons, protons, or ionized atoms. They were small pellets of tungsten, possibly the size of talcum powder dust but still macroscopic and enormous compared to subatomic particles. Tesla had been working on the idea of an electrical weapon ever since 1900, when he noticed that it really stings to be hit by pieces of a metal anode flying off an overstressed, overheated vacuum tube. No demonstration was forthcoming, and on Tesla's death by natural causes in 1943, nothing found in his effects indicated the location of his working model of the teleforce, the "weapon to end war."

During World War II, there were two death-ray projects in Germany and one, the Ku-Go, in Japan using a high-powered magnetron vacuum tube to produce a tight beam of microwaves. The plan was to exploit a resonance in water molecules. Water tends to vibrate in resonance with radio waves oscillating at around a gigahertz. The energy of vibration expresses as heat, and the water can quickly boil if enough power is applied. A live human being is made mostly of water, so it would seem that a good way to kill somebody at a distance is to hit him with a beam of microwaves and boil him inside out. Research proved inconclusive.[191]

---

191 The only useful thing related to these projects was the microwave oven. Although the formal military projects remained classified, navy personnel could not help but notice that the way to clear birds off a radar antenna, which blasts out hundreds of watts of pulsed microwave power, is to turn it on, which makes them instantly flop over dead. The dead birds falling on the deck appeared cooked and ready to eat. The microwave oven was patented by Percy L. Spencer of Raytheon on October 8, 1945, when he noticed that a candy bar melted in his pocket when he got too close to a radar transmitter being tested on a bench. Microwave ovens were used to make food in cramped submarine galleys long before the Radarange became a popular accessory in American kitchens.

In 1960, light amplification by stimulated emission of radiation, the laser, was invented. The earliest laser demonstrations could pop a blue balloon several feet away by making a brief, tight beam of red light using a rod made of synthetic ruby, stimulated by a xenon flashtube. Making it into a death ray was only a matter of scale, and it was quickly predicted that in a few years we would be hunting deer with a laser rifle connected to a backpack full of batteries. This application did not materialize, but the laser principle has since been applied to everything from playing video disks to cutting patterns in stainless steel.

There were several technical problems to scaling a laser up to a destructive weapon. For one thing, the laser beam was scattered by fog, clouds, dust, or smoke, ruining the energy concentration. A laser weapon was susceptible to active armor, in which the protective covering on a target would flash off when struck by the beam, diffusing its ability to do damage. Just burning off the paint on a tank could break up the hit, and the weapon gave its position away every time it flashed. However, a laser weapon could pop that blue balloon at a distance of several miles.

In the 1980s, the Strategic Defense Initiative proposed several death-ray methods of knocking out incoming ballistic missiles, and physical experiments included everything from an X-ray laser powered by a hydrogen bomb explosion to a visible-light laser carried in a Boeing 747. Research continues to this day on particle beam, laser, and microwave weapons for military purposes. The main problems with using any existing ray-gun technology for committing crimes are the monstrous size, weight, and cost of the weapon. It is nearly impossible to rob a liquor store with a piece that has to roll in on a tracked vehicle dragging a diesel generator behind it. The scale of any electronic destructive device, from the teleforce to the Excalibur Plus X-ray laser, is simply out of range for normal criminal activity.

An exception to this too-big rule may be an obscure weapon, developed in the Soviet Union in the 1980s, when the Cold War between the USSR and the western world was running at its highest temperature. Nuclear annihilation was not impossible in those tense times, and the United States and its Soviet counterpart were still locked in competition on all fronts, including the conquest of outer space. At that time, the next frontier in space travel was to build a "permanent" space station, orbiting Earth and eventually acting as a way station on trips to the Moon or Mars. The Soviet Union was the first up, assembling a modular habitat in low orbit, beginning on February 19, 1986, when the base block, module DOS-7, was lofted into place using a Proton-6 launch system. The United States was far behind but had developed, at great cost, a reusable spaceship, the Space Shuttle. The Space Shuttle *Columbia*, or STS-5, first gained orbit with a human crew on November 5, 1982, and was successfully recovered in fairly good condition.

It was a time of great secrecy, distrust, and irrational fear verging on paranoia between the two great powers. High officials in the Soviet space program were honestly concerned that the United States, jealous of their successes in space and secretly coveting their sparkling new Mir space station, would come alongside in a space shuttle, throw grappling hooks, forcibly board the peaceful habitat, and claim it as captured territory. To the hard-core Soviet mind-set, the American government was an unstable combination of cowboys and gangsters, unpredictable and capable of any insane action. The cosmonauts would have to be armed against outrageous aggression.

It turns out that all cosmonauts were sent into space armed. The purpose of carrying a concealed weapon was protection from wolves at the landing point, which was usually in the Siberian wilderness, or from peasants with pitchforks having mistaken you for a Martian. The first man into orbit on April 12, 1961, Colonel Yuri Alekseyevich Gagarin of the Soviet air force, packed a 9mm Makarov PM semiautomatic pistol. By the time the Mir space station was orbiting, cosmonauts were each issued a TP-82 triple-barrel survival gun, with two 12.5 by 70mm (40-gauge) shotgun barrels on top and a 5.45 by 39mm (0.22 caliber) rifle underneath. The removable buttstock doubled as a machete.

One thing you could never do with a TP-82 was shoot it in outer space, even if the guy who was locked up in the enclosure with you was driving you insane by incessantly popping his chewing gum. Inside the space station, no matter where you aimed, it would blow a hole in the airtight hull, completely ruining the mission, and outside, floating around in space, setting off one of the cartridges could blow you out of orbit from the recoil. Protection from astronauts/pirates would have to be accomplished with a zero-recoil device that didn't shoot hard projectiles: a ray gun.

An ingenious laser pistol was developed by the staff of the Military Academy of the Strategic Missile Forces in Moscow, directed by the Honored Scientist of the Russian Soviet Federative Socialist Republic, Doctor of Technical Sciences, Professor, Major-General Victor Samsonovich Sulakvelidze. The goal was a high-power, mode-locked mono-pulse laser capable of burning a hole in an enemy's space suit, blinding any optical instrument including human eyes, and doubling as a medical, self-cauterizing laser scalpel. These requirements were met using a flexible fiber-optic cable as the laser resonator. The optical fiber was durable, reusable, driven by a flash of light, and capable of an energy pulse that could melt through steel, but it had to be several meters long. The ray gun had to be the size of a small military sidearm.

This mismatch of lengths was solved by winding the optical fiber on a spool, making it a short solenoid with the output end of the fiber bent so as to direct the beam out the "barrel" of the pistol. The fiber was driven to lase by a 10-millisecond flash of light striking the entire length of the laser medium. A small, disposable, one-shot flash lamp, the size and shape of a 10mm bullet cartridge, fit neatly into the cavity running down the center of the solenoid, and eight lamps were loaded into a magazine that slid and locked into the pistol butt. Each lamp was stuffed with a crumpled piece of thin zirconium foil and back-filled with pure oxygen. A tungsten-rhenium filament, coated with a dried pyrotechnic paste, ignited the zirconium by connection to a piezoelectric crystal struck with a cocked hammer, similar to a butane lighter ignition.[192]

To fire the weapon, you cock and load by pulling the bolt on top. Aim at the center of mass of the approaching enemy and pull the trigger without staring at the aiming point. It makes a slight bang as the flash lamp goes off and the air expands suddenly around the fiber

192 The effect is very similar to an old-fashioned flashbulb, once used as a light source for indoor photography, but flashbulbs used crumpled magnesium wire or, in the most powerful bulbs, magnesium foil. The zirconium foil gives a threefold increase in the specific light energy over magnesium, with a 5 to 10 millisecond burn at 5,000 degrees Celsius. Certain metal salts are added to the zirconium foil to skew the light spectrum toward the absorption resonance of the laser medium.

solenoid. Push the catch on the left side of the gun with your thumb and the barrel tilts up. The spent lamp pops out on a spring. Push the barrel back into locked position and it snaps into place. Repeat until enemy is neutralized and police your empty rounds.

A working prototype weapon was built and was in testing when the Soviet Union started falling apart in late 1989, before it was put into production and given a military designation. The project was abandoned, American astronauts were invited to stay at the Mir while their bigger space station was being constructed, and the military academy was renamed for Peter the Great.[193] The great Cold War in which no armies clashed, filled with prioritized and imaginative creativity in the destructive weapons category, simply spun down and closed shop. The production of ray guns would have to wait for another stimulus.

There is no manufactured ray gun or even a radiation-throwing assault rifle to fill the needs of antisocial behaviorists, but there is something even better. It can be small enough to fit in a pocket. It is completely solid state, and it has no moving parts. It makes no sound, and it has no odor, but if you lose it, it has a built-in locator beacon. If used correctly, it is 100 percent lethal, and the victim is unaware that he or she is being killed until it is too late. Death can be slow and agonizing. The simple threat of its implementation is strongly terrifying. It is the industrial radioactive source.

The use and proliferation of industrial sources of radiation have increased since shortly after the discovery of radium-226 in 1898 by Marie and Pierre Curie. Radium-226 emits alpha rays, which are heavy, highly energetic subnuclear particles emitted as a radium nucleus decays down into radon gas. It has a half-life of 1,600 years. It is only a rare contaminant in uranium ore, but its use as a cancer treatment made it, gram for gram, the most valuable substance on Earth in the

---

193 This was not the only laser sidearm in the works. There was also a smaller revolver version, having six flash lamps in a cylinder that would rotate into position and activate a ruby-rod laser. The flash was ignited by a percussion cap instead of an electrical pulse from a piezoelectric crystal. There may have also been a single-shot derringer version that a cosmonaut could keep hidden in a pouch on his right boot. It was rough out there in space. These devices are now on display in the Peter the Great Military Academy museum in Moscow.

early twentieth century. The alpha particles destroy what they slam into, but they have an extremely short range. In air, they travel about 1 centimeter before they are stopped by crashing into air molecules. Total shielding from the dangerous alpha particles can consist of a sheet of typing paper. Oncologists found that inserting a needle coated with radium into a cancer tumor would destroy whatever was touching the needle without causing damage to anything else in the body of the patient. It was an excellent therapy for an otherwise untreatable disease. One needle could be reused indefinitely, or until it stuck to a bandage and was lost. The radium needle was even self-sterilizing, as no bacteria could exist on a radium-226 alpha-ray source.

The use of solid-state radiation sources plus electrically generated X-rays became an important division of medicine with many applications. Although radium is now rarely used medically, radiation sources are still used to treat specific cancer tumors, both with implanted "seeds" and with externally directed radiation beams. In the technically advanced medical world, beams of tightly controlled radiation for disease treatment are now made artificially, using electrically driven particle accelerators, but similar therapy can be used in less developed situations where there isn't even any electrical service, using metallic gamma-ray sources, such as cobalt-60 or cesium-137. These constantly radiating sources are controlled and kept safe by confinement in heavy metal shielding structures.

Radioactive sources are also used for medical diagnosis, from tracing the motion of blood flow using technetium-99m to observing metabolic localizations using fluorine-18. Both procedures involve injecting a radioactive isotope into the patient, and computer-assisted tomography (CAT scanning) is used to construct readable images, mapping the locations of radiation concentrations in the patient's body. Dangerously large concentrations of gamma rays, controlled by aggressive shielding techniques, are also used to sterilize a wide range of medical devices, including surgical instruments. Bacteria can be wiped out on the surfaces of delicate plastic tools that could not withstand the high temperature of autoclaving by exposing them to high-energy gamma rays. The same technique is used to pasteurize dairy products and to keep fruit looking good and edible way past its expiration date.

Manufacturing and heavy industries also use radioactive sources for a wide range of tasks. Alpha emitters, such as americium-241, are used to ionize the air with alpha rays in smoke detectors, or to neutralize the electric charge that builds up on the product in a soft-drink can factory as it rolls through the fabrication process. Penetrating gamma radiation sources, such as iridium-192, cobalt-60, and cesium-137, are used to examine the quality of welds in applications ranging from shipbuilding to the plumbing in a nuclear power plant. The use of a simple, shielded metallic radiation source is much more practical in the field than using a fragile electronic X-ray machine that requires a power cord. Lighted buoys in the ocean, remote weather stations, every space probe sent out beyond the Moon, every lander on another planet, asteroid, or moon, and every Apollo Moon mission uses a chunk of very active plutonium-238 to generate electricity and make heat to prevent freezing. There are radioactive sources all over the Earth and the solar system, working twenty-four hours a day, never turning off, and covered with DANGER signs.

Small radioactive sources can do no harm, as long as the user is trained to be careful, always knows where the source is and where the open end is pointed, and, when the business dries up, does not abandon the thing. A new and appropriate home must be found for it. Owning a radioactive source is like owning a trained attack dog. It is useful to have when you need it, but when the job is done, you cannot just set it free.

The most common crime involving nuclear material is the petty theft of an abandoned industrial radiation source, leading to indirect death or injury due to ignorance of the danger involved. No matter how well a source is locked up when it is no longer needed, a scrap-metal thief will find a way to get his hands on it. Metal scroungers will take down light poles and industrial air-conditioner evaporators, which is a labor-intensive feat, just to get the copper for resale. A radiation source is particularly attractive because it is locked away securely, implying great value in a small, easily carried piece of metal.

A classic example of this type of crime occurred between February 27 and March 5, 1996, in Houston, Texas. A company named Larpen of Texas owned two gamma-cameras, and they provided radiographic services to a steel manufacturer on its 37-acre site in east Houston, near highway I-10. These cameras were heavy, shiny, and potentially dangerous. The larger of the two devices weighed 1,665 pounds and held 35.3 curies of cobalt-60. The small one weighed only 631 pounds and had 8.6 curies of cobalt-60. Most of the weight was very dense gamma-ray shielding, which in this case was depleted uranium. Uranium is much heavier than a like volume of lead, and its density makes it an excellent radiation shield.[194] It takes a lot of shielding to protect yourself from 35.3 curies of a gamma-ray source. A naked source that size glows blue-green from Cherenkov radiation when under water, and it generates its own warmth.

Larpen of Texas went bankrupt in 1992, and by October the Texas Bureau of Radiation Control ordered the sources to be impounded and locked away. Proper disposal of the sources was mandated by 1994. It is difficult and very expensive to dispose of 43.9 curies of cobalt-60. You cannot just put it on a truck and send it to another state. Instead, the bureau had the door on the storage facility welded shut. The steel company went out of business, and all the structures surrounding the storage facility were torn down in 1995 until 1996, making the welded door look even more abandoned than it actually was.

An industrial radiographic camera of this type is a solid block of uranium-238 with a narrow, S-shaped passage through it, running back to front. A stainless steel covering protects the metallic uranium from corrosion, and there is a lifting lug on top. The cobalt-60 is a round or cylindrical pellet, as small as 1.5 millimeters in diameter, welded into a stainless steel capsule. The capsule is connected to a short piece of flexible steel cable called the pigtail. The cable is run into the entrance hole on the back of the uranium block until the source at the end is in the middle of the S-shaped passage. Gamma rays can only travel in a straight line, so in the S-shaped passage they cannot escape the shield, even though it is open at the front and back of the uranium block. In this safe condition, gamma rays from the powerful cobalt-60 source are not detectable.

---

194 Larpen of Texas actually had three radiography "cameras," used to expose X-ray film to find hidden flaws in steel. The radioactive sources were shielded so as to direct gamma rays out a small opening in the front of the device, and a shutter would close when it was not in use, protecting people from the directed rays. The third camera used iridium-192, but it was old, and its radiation source had decayed down to the point where it was no longer useful or even terribly dangerous.

To operate the camera, a long, flexible cable is connected to the pigtail and a hollow, flexible tube is connected to the front of the shield. The end of the tube is placed where a radiographic image is desired, usually on one side of a steel weld, and an X-ray film is affixed to the other side of the weld. Standing far away from the camera, the operator turns a crank on the end of his cable, the source capsule snakes out of the S-shaped passage and through the hollow tube, until it touches the test subject, and the weld is penetrated by the gamma rays, exposing the film. When the exposure is complete, the operator winds the source back into its place in the S-shaped passage.

On February 27, 1996, three thieves breached the door with vigorous use of hand tools and stole the radiographic cameras to sell to the highest bidder in the metal recycling trade. Fortunately for the men, the extremely active cobalt-60 sources were enclosed in the heavy shields, but they peeled off all the yellow-and-purple labels reading DANGER RADIOACTIVITY to make them easier to sell. Marketing the devices as stainless steel, they sold both to the Lockwood scrap yard for $200 cash.

Lockwood then resold them to A-1 Metals the same day. A-1 quickly determined that they were not junked stainless steel, as assumed. They mixed the two radiographic-cameras into a big load of scrap, and sent the lot to Gulf Materials Recycling Company, hoping that nobody would notice the hot potato that A-1 was trying to palm off.

Just as the truck from A-1 Metals entered the gate at Gulf Materials, the radiation alarm went off. Any big recycling yard has radiation detection equipment with a preset alarm point, specifically to prevent what was happening. Although the cobalt-60 radiation sources were deeply buried in the devices and could not give off radiation, the comparatively slight radiation from the uranium shields was enough to set off alarms. Gulf Materials segregated the cameras out of the scrap heap and sent them back to A-1 with a hearty "No, thanks."

On the afternoon of February 29, A-1 Metals tried to return the cameras to the Lockwood scrap yard. Unfortunately, as it was being forklifted onto the truck, the pigtail in the larger camera was torn loose, and the source capsule was now outside the uranium shield. The truck driver showed up at Lockwood, but the yard was closed for the day. A-1 Metals got the Lockwood yard operator, Jesse Santana, on the phone to inform him that the stainless steel scrap he had sold them contained radioactive sources, and they were returning it. On March 1, the truck from A-1 returned to Lockwood. After the unwanted objects had been unloaded, the driver noticed a small piece of steel cable lying on the bed of his truck. The pigtail had fallen through the wooden pallet when the larger camera was off-loaded. He picked it up by the source capsule end and tossed it on the ground in the Lockwood yard, not realizing that it was a naked, 35.3-curie gamma-ray source. An employee noticed it on the ground, and wishing to neaten the yard, he kicked it under the corner of the office building.

Not knowing that the source was lost out of the big camera and was under his office, Santana sold it to another scrap yard. He sold the small camera, knowing full well that both devices were dangerously radioactive, to yet another scrap yard. If he had simply turned the devices over to the authorities, he would not have been paid for them. The yard that bought the small camera, realizing what it was, then palmed it off to another scrap yard. In three days, the radiographic cameras, which were supposed to be tightly controlled, made a tour of five scrap yards in Houston, Texas. Meanwhile, Jesse Santana and everyone at the Lockwood yard who came to his office was being hosed down with the 35.3-curie gamma source immediately under the wooden floor.

The right hand of the A-1 truck driver sustained a nasty radiation burn, and doses were sustained by everyone working at the Lockwood yard, the yard owner's two children, five investigating police officers, and two Radiation Bureau personnel. When tracking down a naked 35.3-curie cobalt-60 source, "Show me where it's at" is the incorrect way to proceed.

The two cobalt-60 sources were recovered by the Radiation Bureau on March 5, 1996, and six men were arrested for stealing and for knowingly receiving stolen property. The truck driver's thumb received 3,000 rem of radiation. The scrap yard owner, 1.8 rem; the yard manager, 53 rem; his wife, 55 rem; his children, 39 rem; workers at the yard, 15 rem; and customers at the scrap yard, 0.16 rem. In perspective, the Nuclear Regulatory Commission sets a maximum permissible radiation dose to the public of 0.1 rem per year. A nuclear industry worker is permitted a dose of 5 rem per year.

Similar shenanigans have occurred in Russia but in greater number and with a larger death count. In 1990, when the Soviet Union caved in, it was a season of unprecedented plenty for metal thieves. The radionuclide power sources and heaters, abandoned in careless quantities all over the vast Union of Soviet Socialist Republics, were free for the taking once the government was no longer interested in owning and managing them. As was the case everywhere else, having them locked away only gave them an apparent high value, and the average person did not have a clue as to the danger of touching highly radioactive objects or even being in the vicinity of them.

The record time for dying by stealing an industrial radioactive source was set by a male resident of the town of Urus-Martan in the Chechen Republic, Russia. He was one of a gang of six Islamic Chechen guerillas attempting the theft of cobalt-60 from a chemical factory in Grozny on September 13, 1999. Anticipating an invasion of Grozny by the Russian army in what would become the Second Chechen War, the gang was probably planning to build a weapon using the stolen radioactive material. Grozny is the capital of the Chechen Republic. The Russian siege, attempting to regain central control over Chechnya, began a month after the attempted theft, on October 25, 1999.

The metallic cobalt-60 was formed into rods, each 12 centimeters long. There were nine rods per container, and twenty-eight containers were stored in a shielded underground vault in the basement of the building. Each rod had an impressive radioactivity of 27,000 curies, and without shielding, it was dangerous to be in the same building with it. The thieves managed to break open the vault and pry off the lid on a container. One of them grabbed a cobalt-60 rod and immediately felt unwell. He was dead in 30 minutes. Two of the others who were next to him died later, and the remaining three made it to a hospital in Rostov. Specialists from the Chechen Ministry of Emergency Situations worked to decontaminate the site as the city came down around them in a Russian artillery barrage.

That same year, three scrap-metal thieves broke into an unmanned lighthouse near St. Petersburg, Russia, and stole the radio-thermal generator, used to power the light. They managed to remove the strontium-90 radioactive core and drag it 50 kilometers to a bus station in the town of Kingisepp, before falling over dead from the 1,000 rad-per-hour of beta and gamma radiation at the surface of it.

Two years later, in May 2001, a similar thing happened in the Kandalaksha Nature Reserve near Murmansk, Russia, when four scroungers stole the radio-thermal generator out of a lighthouse and tried to remove the three strontium-90 radiation sources, each producing 35,000 curies of radiation.[195] All four men were hospitalized with radiation poisoning.

The most completely analyzed case of radioactive-source theft was the Tammiku incident of 1994. It is unique in that it resulted in the death of both one person and one pet dog.

In 1963, a management facility for low- and medium-level radioactive waste was established at Tammiku, Estonia, in the Union of Soviet Socialist Republics. A concrete vault was buried in the ground, according to criteria established in Moscow in the 1950s, in a remote, wooded area about 12 kilometers south of the town of Tallinn. A barbwire fence surrounds the vault at a distance of 500 meters, and entrance is through a gate with a guardhouse. There were two metal signs on the gate, painted yellow with the international symbol for radiation in purple and the warning, in Estonian: DANGER RADIATION.[196] When the central Soviet government dissolved in the early 1990s and Estonia became an independent country, no guard was left in the guardhouse.

After more than thirty years of operation as an active repository, 97 tons of radioactive waste were left in the vault, producing a combined radioactivity of 200 trillion becquerels, or 5,400 curies. By January 1993, the radiation had decayed down to 76 trillion becquerels, or 2,900 curies. That was still a great deal of radiation. Of those 2,900

---

195 There were actually eight different types of radioisotope thermal generators used in lighthouses in the Soviet Union, ranging from the Beta-M, generating 10 watts of electricity and 230 watts of heat, up to the IEU-1M, making 120 watts of electricity and 2200 watts of heat. The Beta-M weighed only 560 kilograms, and it was the easiest one to steal. The IEU-1M weighed about 2,100 kilograms, and it was just too bulky to make off with. The electricity from a radioisotope thermal generator is made by heating up one side of a thermocouple array using a strontium-90 heat source, made into a hard, ceramic block, and cooling the other side with air moving by convection over a circle of fins. There are no moving parts. A shield made of depleted uranium is supposed to protect handlers from the strontium-90 radiation. The 10 watts provided by a Beta-M was almost enough to run a small fluorescent light, warning ships in the dark several meters away.

196 Signs cautioning about the dangerous presence of radiation are officially available in most languages, including Klingon.

curies, 24% of it was sealed cesium-137 radiation sources, probably retired from use in a sterilizing irradiator for medical instruments. The rest of it was a mixed bag of strontium-90, plutonium-239, radium-226, and americium-241.

There was also a separate, underground liquid-waste storage facility, consisting of a 200-cubic-meter cylindrical tank lined with stainless steel. The entrance door was secured with a padlock.

After midnight on October 21, 1994, three brothers, known only by RiH, RaH, and IH, sought to enrich themselves by stealing valuable metal from the Tammiku waste disposal facility and selling it to a scrap dealer in Tallinn. They easily scaled the 6-foot fence and cut the padlock on the steel door, bypassing the alarm system. The above-ground portal looked like a lean-to, with the doors on the slanted side. The doors creaked as they manhandled them open.

RiH climbed down into the open vault, and he grabbed the first thing on the top of a stack of fancy, machined metal containers. He passed it up to his brothers, and in transit, a metal cylinder fell out of the metal tube that was nestled in the container. It was 18 centimeters long and 1.5 centimeters in diameter. It was a dummy spacer, used in the tube to prevent a cesium-137 source from rattling. RaH picked it up and tossed it back into the pit. RiH noticed that another cylinder had fallen out at his feet. It was only 3 centimeters long. This was a 190-curie (7 TBq) cesium-137 radioactive source, producing a dose rate of 250,000 rem per hour at a distance of 1 centimeter from the cylindrical source.[197] Bear in mind that a nuclear power worker has a maximum allowed dose of 5 rem per year. RiH picked it up and slipped it into his coat pocket. RiH had an odd feeling. Before long, the odd sensation evolved into a very sick feeling.

Wishing to explore further, the brothers defeated the lock on the liquid waste facility and entered, finding it stacked with nice-looking, shiny aluminum drums. They emptied several of them onto the floor and carried them out. Feeling woozy, RiH let a drum slip out of his hands, and it slammed one of his legs against the concrete wall of the entranceway. It hurt, and he was bleeding, but not much.

---

197 "TBq" is short for terabecquerels, or trillions of becquerels. One becquerel is 1 nuclear disintegration per second.

The brothers dragged the metal container and the drums through the woods 50 meters to the road, dropped them there, walked to their car, drove back to the drop point, and loaded the metal container into the trunk. They would figure out what to do with the drums later.

IH drove, and first he stopped at RaH's house to let him out, and then to RiH's house in Kiisa. By now, RiH was vomiting, over and over again. There was nothing left in his stomach to bring up, but still he vomited. The other occupants of the house were his stepson, RT; his wife, BK; his wife's grandmother, AS; and the family dog. RiH hung his coat, containing the cesium-137 source, in the hallway and went to bed. Curious, the stepson, RT, went through the coat pockets, found the little metal cylinder, examined it up close, turning it over and over in his hand, and transferred it to the tool-drawer in the kitchen. IH drove on to Tallinn. He wanted to be at the recycling center when it opened and collect money for the metal in the empty container.

RiH was not getting any better, and a few days later, on October 25, 1994, he was checked into the hospital. He claimed that he had injured his leg working in the forest, and didn't mention that a drum had fallen on him as he tried to steal metal objects at the Tammiku radiation waste facility. He was diagnosed with a crush injury, was medicated accordingly, and died on November 2, 1994. He was twenty-five years old. There was no reason to think that the death was related to radiation, except for the classic radiation sickness symptoms.

Six days later, on November 8, there was a need to dispose of some radioactive waste at the Tammiku facility. The workers immediately noticed that the padlocks had been cut off the doors to the vault and to the liquid-waste facility. Always monitoring their activity with radiation instruments, they noticed that the dose rate at the entrance had gone down by a factor of 100 since the last time they were here. They didn't think to tell anybody about the broken locks or the decreased activity.

On November 9, RT was working on his bicycle, and he rummaged in the drawer in the kitchen, looking for a tool. He noticed the small metal cylinder that he had found in the coat pocket. He picked it up, examined it for a second, and put it back in the drawer. RT was thirteen years old. The dog, only four months old and living in the kitchen, was vomiting and urinating blood.

On November 16, the dog died. The next day, RT was admitted to the hospital with severe burns on his hands, and he had started vomiting. With questioning, it became clear that the injuries were caused by radiation. The police were summoned.

The Estonian Rescue Board dispatched a team to the house in Kiisa, arriving at 11:30 on the night of November 18. Their instruments picked up radiation before they even entered the house. All neighbors within 200 meters of the house, living in fifteen houses, were quickly evacuated. The rescue team cautiously opened the front door and stepped in. In the front hall they were measuring 5 rem per hour, meaning that if they stood there for an hour, they would receive their entire dose allowed for the next year. The team found the source to be a drawer in the kitchen, close to which their instruments registered 120 rem per hour. They left quickly and ordered up a lead shield box.

It took 2 minutes and 15 seconds to retrieve the cesium-137 source and drop it in the shield box. Two team members were outfitted with lead aprons and rubber gloves, but unfortunately, they lacked handling tongs. One of the team picked up the source with his index finger and thumb and dropped it into the shield box. The 2-second exposure to his hand was significant. With the lid screwed down, the dose-rate outside the lead box was 10 rem per hour.

By 2:30 P.M. the next day, November 19, the site was declared safe, and the neighbors were allowed to return. The source, never leaving its new lead shield box, was taken back to the waste facility at Tammiku. Six members of one family had been exposed to dangerous gamma radioactivity for twenty-seven days. The worst cumulated dose was to RiH, who received 183,000 rem to his thigh, the location of the pocket on his coat. Seven other individuals who had been in the house were also exposed.

It had been a sobering incident, instilling caution bordering on fear, and it pointed to problems in the control of radioactive sources in Estonia. The team, still spooked by the incident, kept their radiation detection gear turned on at all times. On January 14, 1995, they were on the road between Tallinn and Narva, on a routine inspection trip, when the dose set-point on the Geiger counter was exceeded and the instrument made a beeping noise. It took a while, digging through the snow, but eventually they found the cause. It was another cesium-137 source, loose, naked, and lying by the side of the road. They never found out how it got there. The metal container that was taken to the recycling yard was never recovered.

Premeditated murder is possible using radioactivity. Over the past sixty years, at least fourteen people have been purposefully killed and ninety-three injured using industrial or medical radiation sources, and at least one person was injured using an X-ray machine. Eight crimes were committed using cesium-137, five using cobalt-60, four with iridium-192, two with strontium-90, two with phosphorus-32, and one each with polonium-210 and californium-252, which wins the prize for the most exotic nuclide used as a weapon.[198]

Crimes by nuclear means are unusual and often purely evil. Many are difficult to document and verify, as criminal case records involving radioactivity are often sealed and unavailable. The only leakage may be through research papers or presentations in health physics symposia. An example is the case of a woman who attempted to self-induce an abortion using a medical X-ray machine. It happened in Pennsylvania, sometime between 1965 and 1968. That is all that the scientific community knows about the incident. In Suffolk County, New York, in 1996 two disgruntled citizens planned to commit murder using five canisters of radium. The intended victims were two county officials, and the radium was to be hidden in their food, homes, and cars to ensure adequate contamination. The only source of this crime report is the U.S. Nuclear Regulatory Commission Preliminary Notice of Event or Unusual Occurrence PNO-I-96-043, June 14, 1996. It would be interesting to know where the conspirators got enough radium to be dangerous.

One of the most horrifying cases of assault by industrial radioactive source involved a petroleum engineer and his eleven-year-old son. It happened in Harris County, Texas, in 1974, and interesting details of the crime are available.

---

198 The case of californium-252 used for an attempted murder in Riga, Latvia, on August 18, 1988, is insufficiently documented for inclusion in this chapter, but the successful use of polonium-210 as a poison on November 1, 2006—the murder of Alexander Litvinenko—has been publicized to the point where there is nothing I can add. I recommend the book *The Litvinenko File: The True Story of a Death Foretold* by Martin Sixsmith (Macmillan, 2007). I agree with the author's analysis and his conclusions.

A petroleum engineer in Texas named Kerry Andrus Crocker married the love of his life, Barbara, in December 1955. Their first son, Kirk, was born in 1961, and a few years later another son, Patrick, was born. By May 1970, that loving feeling was lost, and the couple divorced. Finding that they couldn't live without each other and it was all just a misunderstanding, they remarried two months later. Further discovering that they honestly could not stand each other, they re-divorced seven months later, in February 1971. Crocker had visitation rights with the two boys at his place two weekends each month, normally on the first and third weekends, and during a month each summer.

Kirk, the older boy, immediately got the impression that his father wanted him dead, possibly as a once-removed retribution against his mother. First, there was the waterskiing incident in which he almost drowned, crying for help as he bobbed to the surface ten times while his father watched with studied detachment, and then there was the potentially fatal explosion and fire in the camping trailer where Crocker had insisted that Kirk stay by himself.

Eight months after the second divorce, on November 18, 1971, the oil well logging and perforating service involving Crocker was granted a license to have 1 curie of cesium-137 for wireline well logging.[199] Thirty-seven billion atoms in 1 curie of cesium-137 decay each second. One curie is considered to be a great deal of radiation. The dangerous product of each disintegration is a 662 KeV gamma, capable of running clean through human body parts, leaving an ionized path of

---

199 For gas and oil-well drilling, a nuclear wireline log is a record of the gamma-ray backscatter intensity from a fixed radiation source into a Geiger counter probe slowly lowered into a drilled well on a steel wire. The amount of backscatter indicates the level of porosity or the density of the rock in the hole along its depth, usually recorded on a strip-chart or stored as a digital recording on a portable computer. It is particularly useful for mapping the shale beds in a drill hole. A layer of shale makes a tight cap over an underground oil or gas reservoir, and its presence indicates that you are drilling in the right place. The shale is unusually solid and impermeable to oil, and it gives an enhanced gamma backscatter over sand or broken rocks. A nuclear log has an advantage over other methods of shale-finding, in that the gamma-rays from cesium-137 can penetrate the steel pipe that lines a cased well and feel for the quality of the rocks on the other side of the tube wall. Crocker was an independent consultant at the time, and he was employing Sidney Morrison's well-logging company for oil-drilling jobs. Without his knowledge, Crocker made a copy of Morrison's radioactive source license and used it to justify access to the controlled cesium-137 pellets on demand.

cellular-level micro-injuries behind it. Damage adds up as the highly radioactive source remains in place, shooting gamma rays in random directions.

In very small, sub-microcurie doses, the danger from gamma rays is easily controlled by heavy-metal shielding, distance from the source, and limited exposure times. Severe, life-threatening gamma-ray destruction from large radioactive sources, fractions of a whole curie, is undetectable as the radiation invades the body, until later when the dead and injured tissues begin to break open and ooze. The 1 curie of cesium-137 owned by the drilling company was divided into small pellets, each a tiny, metallic cylinder about the size of a pencil eraser.

The original license was amended twice, to allow the company to have two sealed canisters, each containing enough cesium-137 pellets for a disintegration rate of 2 curies. Crocker was named in the license document as the only person to use and be responsible for these radioactive sources, all 4 curies of them.

Feeling a vague sense of danger, in April 1972, Kirk, then eleven years old, was visiting his father's apartment when he was left alone for several hours. He was told to watch television while Crocker went on some errands but to listen to the programs using the headphones. The walls were thin, and Crocker said that he did not want his neighbors to complain of the noise. Kirk could not help but notice that the earpieces were stuffed with cotton, and he started picking them apart. Out rolled two shiny silver pellets. He called his mother, expressing concern over the strangeness of it, but she told him not to worry about it. For some reason, Crocker had tried to expose his son to cross-fired gamma rays to the brain from cesium-137 radiation sources borrowed from the well-logging set. Where he had handled the pellets, Kirk's fingers turned red and painful, as if he had picked up the wrong end of a hot fire poker.

Kirk visited Crocker again on the first weekend of July 1972. This time, he was instructed to drink a glass of orange juice, and at the bottom of the glass was a partially dissolved pill, apparently a sleep-inducing pharmaceutical. He passed out on the couch in the den. He was suspicious upon waking, and he poked around in the cushions, finding a thin sock retaining two impressions of small cylinders. It looked as if the sock had recently contained two pellets, similar to the ones he had found in the headphones. On the end table, he saw some sort of unusual instrument. He was terribly nauseous. Showing classic signs of radiation sickness, he vomited until the dry heaves were exhausting. Crocker explained to him that the instrument, which was later identified as a radiation survey meter, was used to clean guns. Kirk also noticed a pair of long metal tongs nearby.

Visiting again on the third weekend of July 1972, there was a party going on in the apartment, so Crocker had him once more drink orange juice with a pill in it and go to sleep in the bedroom. He awoke the next morning, again feeling terribly nauseous and finding a medicine bottle containing three of the little cylinders rattling in his pillow. He found his father and his little brother sleeping off the party in another apartment. Back home, his mother was concerned with a red rash developing on his thighs. His right thumb, where he had handled the medicine bottle, was red, painful, and swollen.

Kirk was again obviously exposed to radiation in August 1972, when during his scheduled visit he was instructed to lie down on a couch where he could not help but notice a sock with two cylinders in it under a cushion. His father and his younger brother left for an outing, and after they returned, he noted that his father carried the sock out to his car, held at arm's length.

The last provable exposure incident occurred in a motel room during a visitation in October 1972, when again he was instructed to take a pill dissolved in orange juice. He awoke to find a sock with two cylinders in it draped over his legs and his father asleep in his car. Between April and October 1972, there were probably eight purposeful irradiations, but no connection was immediately made to the strange metal cylinders and Kirk's fits of vomiting or the ever-worsening rash on his thighs. The hair on one side of his head was coming out in clumps.

Kirk was under a doctor's care by this time, but the cause of the rash remained a mystery. He wasn't responding to any attempted treatment. The skin was dying, leaving bare sores that would not heal. A plastic surgeon, Dr. Thomas Cronin, was called in to repair the affected areas, and he immediately recognized the symptoms. His specialty was rebuilding tissues that had been severely damaged in cancer treatment by directed radiation beams. The injuries he saw on this child, necrosis ulceration, were caused by prolonged exposure to gamma rays. His right ankle was obviously affected, showing "deep indolent ulceration" and "a tendon which was dead due to the effects of radiation." His inner thighs, right at the groin, showed "extensive ulceration and scarring with telangiectasia," or "spider veins." Further testing revealed that his right testicle was no longer a functioning organ. Its internal mechanism had been replaced with inert fibrous tissue, and the left testicle was reduced to a nonfunctioning, hardened lump. In medical terms, he had been castrated. Kirk would have twenty-three reconstructive surgical procedures over the next six years, he would have to take testosterone replacement injections on a regular basis to prevent eunuchism, and he was likely to develop leukemia in the next twenty years.

On January 31, 1974, Kirk's enraged stepfather, Harrie Smith, called the Texas Department of Health Resources, Radiation Control Branch, to accuse Kerry Crocker of having willfully exposed Kirk to radiation and causing burns on his thighs, ankle, and thumb. Smith was asked to put his accusations in writing, and the department received his letter on February 4, along with a list of sixteen doctors who had examined the boy, sketches of the cylindrical pellets, the handling tongs, and the radiation survey meter. Satisfied that a felony had occurred and realizing that the laws regarding radiation handling were inadequate to deal with it, the department notified the district attorney in Houston.

A grand jury on May 2, 1974, indicted Kerry Crocker for assault with intent to murder, castration, disfigurement, assault with intent to maim, and intentionally causing injury to a minor. He was arrested later that day and was released on $10,000 bond. The trial started on March 31, 1975, in the 178th District Court of Harris County, Texas, and a six-man, six-woman jury was selected. The prosecution called thirteen witnesses, almost all of whom declined to accept a professional fee.

On April 16, 1975, the case went to the jury, and after ten hours of deliberation Kerry Crocker was found guilty of castration. He was given the maximum sentence, ten years in prison plus a $5,000 fine. He immediately posted an appeal and was set free on a $10,000 bond. He retained visitation rights to the younger child, but only under the mother's supervision. The appeal was denied in May 1978, and Crocker immediately jumped bail. It would take three years to recapture him, and he began to serve his sentence in January 1981.

A model prisoner, Kerry Crocker was released on parole in October 1986. Kirk still fears him, believing that revenge may be his only goal in life. Kirk's mother died in 1982. He works in real estate and champions the cause of children affected by marital discord.

As a method of expressing murderous intentions, using radioactive sources has been more common than one would hope. In February 1995, in Zheleznodorozhny (near Moscow), Russia, someone quietly placed a 1.3-curie cesium-137 industrial radioactive source in the left door pocket of a truck belonging to someone he did not like. The victim rode around for five months with gamma rays impinging on his left thigh, soaking it down with a cumulative dose of 6,500 rads. (A 1,000-rad exposure is considered fatal.)[200] His whole-body exposure totaled at 800 rad. On July 7, 1995, the reason his thigh was bothering him was found in the door pocket, and he immediately checked into a hospital. He was suffering from epilation of the thigh (hair falling out), moderate pancytopenia (reduced red and white blood cells), and azoospermia (reduced sperm count). After eight months of treatment, he developed myelodysplastic syndrome (bone marrow failure), which progressed into full-blown leukemia. He remained hospitalized until his death on April 27, 1997, fifteen months after having discovered the tiny pellet of cesium-137 in his truck.

Also in Moscow, on April 14, 1993, Vladimir Kaplun, director of the Kartontara packing company, was not popular among all of his

---

200 A rad is 1 erg of energy from radiation deposited in 1 gram of matter. It is considered an obsolete measure of accumulated radioactive dose, but it was in use for so long, used in so many thousands of nuclear physics documents, it remains a familiar measure. The more current, SI measure is the gray, which is equivalent to 100 rads. The rad (or the gray) is used to express the amount of radiation absorbed by anything solid. To more accurately express the amount of radiation absorbed by a human, the rem (radiation equivalent man) can be used, or the current SI unit, the sievert. To improve a rad or gray measurement into to a rem or a sievert, it is multiplied by a fudge factor that takes into account the biological damage specific to the type of radiation involved. The "quality factor" for alpha particles, for example, is 20. Neutrons have a quality factor of 10, and gamma rays, X-rays, and beta rays have a quality factor of 1. When discussing radiation damage by gamma rays, 1 rad is the same as 1 rem. Further rem improvements can involve taking into account the sensitivities and vulnerabilities of individual human organs or the age or gender of the subject, but for postexposure estimates of whole-body exposures to radiation, the quality factor based on radiation type is as good as can be expressed.

employees. Someone slipped a cesium-137 radiation source into the seat of his chair, and he sat on it for several weeks, developing what would eventually be diagnosed as symptoms of radiation sickness. He died a month later in the hospital, after which the contamination of his seat was identified by colleagues.

A similar radiological assault occurred at the La Hague site, a nuclear fuel reprocessing plant on the Cotentin Peninsula in northern France, on May 11, 1979. The plant was originally built for producing military plutonium, but by 1969, the plutonium-239 stockpile was sufficient, and the facility was turned over to civilian power companies for power-reactor waste processing. It extracts plutonium from spent reactor fuel, which is then mixed with new uranium fuel for power production. About 1,700 metric tons of spent fuel are processed yearly, and fuel from power reactors in France, Japan, Germany, Belgium, Switzerland, Spain, and the Netherlands has passed through the La Hague plant.

An employer had incurred the wrath of an employee who tried to kill him by hiding a graphite fuel element plug, warm to the touch with an energetic mixture of fission products, under the seat of his car. By the time he figured out that something was wrong, the victim had sustained a 10- to 500-rad accumulated dose to his testicles, and 25 to 30 rads to his spinal bone marrow. The employee was terminated, fined $1,000, and confined in prison for nine months, convicted of poisoning by radiation.

A similar crime occurred on August 24, 1991, in Bratsk, an industrial town in Siberia on the Angara River. Bratsk is famous for the Gulag Angara prison labor camp, a pollution of mercury in the ground equal to half the world's total mercury production in 1992, and a record low temperature in January of 95 degrees Fahrenheit below zero. It has been declared an ecological disaster zone, and it was evacuated in 2001 due to pollution from the Bratsk aluminum plant.

Two company directors at the wood fiberboard plant in Bratsk discovered cesium-137 radiation sources that had been hidden in the seats of their chairs. One of them developed radiation sickness, but not trusting anyone who worked in their plant, they found the sources before absorbing fatal doses.

On June 8, 1960, a nineteen-year-old radiological laboratory worker in Moscow having unfavorable relationships with his family committed suicide by purposefully exposing his body to a cesium-137 radiation source. He took a pellet in a hermetically sealed aluminum capsule from the "plant gamma defectoscopy" lab, put it in his left pants pocket, and walked around with it for five hours. He then shifted it to his lower abdomen and then to his back, experiencing close contact with the source for another fifteen hours. The dose to his trunk was 3,000 rad, and his whole body received somewhere between 1,500 and 2,000 rad. He developed radiation sickness symptoms in a few hours. Bloody diarrhea became intense after thirteen days of agony, and he passed away in the hospital two days later. It was a miserable way to die.

By 1978, iridium-192 had taken over as the most used radiography source, knocking cobalt-60 out of first place. Both nuclides, cobalt-60 and iridium-192, are metastable, having two sequential decay modes, with the first decay leading to no transmutation of the element. In the case of cobalt-60, the first decay has a very short half-life (10.47 minutes), but the second (beta minus) decay has a longer half-life of 5.27 years, and it emits an impressive 1,333 KeV gamma ray. The half-life is short enough to make it vigorously radioactive but long enough that a source does not play out too quickly to be practical. The iridium-192 first decay has a half-life of 250 years, making it much less radioactive than the cobalt nuclide, and the gamma ray is less energetic, at a modest 317 KeV. It takes a physically larger iridium-192 source to do the same job that a tiny pellet of cobalt-60 or cesium-137 will accomplish, but this makes the source larger and less likely to be lost, and the gamma rays require less shielding for safety. Radiography in industry, particularly of welds, is carried out routinely in the field. The monolithic radiation source acts as an X-ray machine, exposing a large sheet of photographic film and disclosing otherwise invisible cracks and voids in a metal test subject.

In 1978, in the United Kingdom, a radiographer intentionally exposed himself to an iridium-192 radiation source, trying to commit suicide. Iridium-192 was the wrong nuclide for this, and he only got a whole body dose of 152 rads, enough to make him very sick but not to end his life.

Another incident involving iridium-192 occurred in Guangzhou, People's Republic of China, in May 2002. A Chinese nuclear medicine specialist, Guy Jiming, used forged papers to get a license so that he could buy a radiography source full of iridium-192 pellets. His business rival, whom he wanted gone, had an office in the local hospital. Jiming sneaked in and unloaded the entire barrel of iridium-192 pellets on top of the ceiling panel right over his desk. Soon, the victim was reporting a list of strange symptoms to his supervisor, including memory loss, fatigue, appetite loss, headaches, lots of vomiting, and bleeding gums. Another seventy-four hospital staff members who came in and out of his office began complaining of similar though lesser symptoms. A quick examination of the office with a radiation survey meter turned up the iridium-192 pellets, which were serial numbered. Jiming was soon arrested, convicted of attempted murder on September 29, 2003, and given a suspended death sentence. He is spending the rest of his life in a prison somewhere in China.

Phosphorus atoms are found in many organic compounds, occurring in everything from metabolic pathways to the DNA molecule. A synthetic phosphorus isotope, phosphorus-32, is made by bombarding sulfur with neutrons for use in biochemistry and molecular biology studies. It emits a beta ray that can be tracked with radiation detection instruments, and this feature makes it valuable for tracing phosphorylated molecules. The beta ray, an electron, is energetic at 1.7 MeV, and in air it has a range of about one meter. It has a very short half-life of 14.29 days, which makes it extremely radioactive and potentially harmful in anything but trace quantities.

Anything radioactive can be accused of causing cancer, but phosphorus-32 is one of only six radionuclides rated as causing cancer in humans by the International Agency for Research on Cancer (IARC). The others are iodine-131, thorium-232, radium-228, radium-226, and radium-224. Although it is automatically deposited in bone tissue, phosphorus is also incorporated into DNA, and as such it can turn up in any cell in the body.

In 1996, the U.S. Nuclear Regulatory Commission announced that in the past twenty years there were ten known deliberate poisonings in biological research laboratories in the United States using phosphorus-32. In bio-research, a disgruntled employee "going postal" with the silent, undetectable, but deadly effects of a phosphorus-32 dose is apparently more probable than having the offender show up with an assault rifle. It is also possible to get away with this crime.

The worst case may have been at the National Institute of Health (NIH) in Bethesda, Maryland, in 1995, when twenty-seven researchers, including a pregnant female scientist, were purposefully exposed to phosphorus-32. The female victim, Dr. Maryann Wenli Ma, had the largest internal dose, somewhere between 8.0 and 12.7 rem, which was significant. She had been pregnant for seventeen weeks when her contamination was discovered, and she may have been the only intended victim. She and her husband, Dr. Bill Wenling Zheng, were visiting fellows in Dr. John Weinstein's lab on the fifth floor in Building 37, and there were no radionuclides in their lab.

For reasons not divulged, late in the afternoon on June 6, 1995, Dr. Zheng turned on a Geiger counter in the lab and noticed that there was a radioactive source somewhere near it. The rate meter on the instrument, indicating the number of gamma and beta rays intersecting the detector tube per minute, showed a reading that was obviously above the normal background level. The source was found quickly. It was his wife, Dr. Ma, internally contaminated with phosphorous-32. She was taken to Holy Cross Hospital that evening.

The Nuclear Regulatory Commission (NRC) immediately sent an Augmented Inspection Team to NIH to find the source and the cause of the contamination. It was clearly not accidental. There was phosphorus-32 on the floor in front of the refrigerator in the break room, where Dr. Ma had taken out her leftover Chinese dinner to eat for lunch. Dr. Ma and Dr. Zheng were quick to point to Dr. Weinstein, their supervisor, as the culprit, saying that he wanted them to abort their child so that their research would not be interrupted and their work could be patented. The water cooler in the hallway was found to be internally contaminated with phosphorus-32. Everyone in the building was required to pee in a cup, and from these samples it was discovered that this one source had contaminated another twenty-seven NIH personnel.

Weinstein was eventually cleared of any suspicion, although Zheng and Ma accused him in every way possible, including that he delayed her trip to the hospital and he interfered with her treatment for radiation poisoning. Zheng and Ma demanded that the radiation source license for NIH be suspended, but there was nothing found wrong with the NIH radiation handling procedures or records, and the request was denied.

There were questions regarding Dr. Zheng's candor. Why, for example, did he turn on a Geiger counter in a lab that contained no radioactive sources? How did he know that it was phosphorous-32? When he and his wife had used a phosphorous nuclide, it was phosphorous-33, with a longer, 25.3-day half-life and a weaker beta ray. The only ways to tell which nuclide was making the beta radiation were analyzing a sample with a beta spectrometer or reading the label on the original container. Zheng never answered these questions posed by the NRC investigation. No one was ever accused of any wrongdoing in the NIH radiation poisoning case.

This particular type of crime is not confined to the United States. In Taiwan, the Republic of China, from October 1, 1994, to February 15, 1996, a young male graduate student at the Institute of Plant Pathology was poisoned by phosphorous-32 on his eating utensils and in his drinking cup. The phosphorus-32, stolen from an adjacent molecular biology lab, was put there on thirty occasions by a fellow student who apparently had bad feelings toward him. The victim suffered from diarrhea, abdominal pains, a poor appetite, and the loss of his moustache before the perpetrator was finally caught in the act. His adverse health effects persisted through 1999.

So, what can one do to prevent injury or death by the unnoticed presence of a deadly radioactive source? Ideally, an individual should carry a Geiger counter with the beta-filter opened and the power turned on at all times. Check your seat before you sit down! Unless you are an experimental nuclear physicist or a health physicist at a nuclear plant, this is not going to happen. However, most people in the industrialized world now carry a radiation detector with them wherever they go. It is your smartphone.

All smartphones now have at least one matrix of millions of charge-coupled photo-sensors on board, used as a camera to record visible-light images. This feature is not only useful for making self-portraits and YouTube videos, it can be used to detect gamma or X-rays. This phenomenon has been thoroughly studied and verified by Rolf-Dieter

Klein of the Helmholtz Institute in Jena, Germany, home of the rapidly advancing field of laser-induced Wakefield particle acceleration. There is now an application to exploit this effect available for Apple and Android smartphones. Look in the App Store under "radioactivity counter." Most of the Geiger-counter apps are pranks, meant to simulate a Geiger counter and cause panic when you surreptitiously turn up the radiation count. After reading this chapter that should not seem amusing.

The image plane of the built-in camera is essentially two-dimensional for this application, and it lacks the depth and mass of a scintillation crystal or even a gas-filled Geiger-Mueller tube, so most of the gamma rays pass right through it without interacting with the pixels. Its lack of sensitivity is therefore worse than a civil defense ion chamber, but it has the advantage of being connected to a powerful digital computer in the telephone. The computer can integrate counts over minutes or hours, averaging continuously and displaying a remarkably accurate gamma-ray count rate and even a radiation dose rate. The camera's response is linear over a wide range of rates, and it therefore can be used for dosimetry if calibrated using a known gamma source. Light to the camera must be blocked by covering the lens with a piece of black tape. Use it in "cloud" mode so that it shows each gamma ray hitting the camera as a flash of light on the screen, have fun, and try to stay away from recycling yards.

CHAPTER 9

The Threat of the Dirty Bomb

"I am you and you are me and what have we done to each other?"
—from *The Keener's Manual*, a book that
does not exist, by Richard Condon, 1958

THE TAMMIKU INCIDENT IN 1994 in Estonia, in which three metal-scrounging brothers stole a lethal gamma-ray source, had tragic consequences. Not only did one brother and one dog die of radiation poisoning, but several other people were exposed to the radiation over a period of twenty-seven days without a clue as to what was happening. Once it was discovered that a cesium-137 radiation source was loose in the metal scrounger's house, even the professionals tasked with retrieving it were significantly dosed with radiation.

The Estonian Rescue Board found itself in over its head. It was easy enough to determine the extreme radiation doses to the most affected individuals by noting the extent of burn damage to the flesh and modified blood compositions, but determining the doses for people who had only been in the house seemed impossible. The house had not been equipped with dosimeters stationed in every room, nor was there a single radiation survey instrument to be found there, as would be necessary in any normal facility having access to a cesium-137 source.

Experts from Sweden, Finland, and the Russian Federation were called in. I. A. Gusev and G. D. Selidovkin from the Institute of Biophysics in Moscow were able to map the radiation doses that had occurred the year before in every room of the house. This amazing feat of retrospective dosimetry begs the question: How?

As a reactor operator once told me as I admired his big calcite crystal, glowing with a pretty yellow light, "Anything will glow if you hit it hard enough with gammas." The statement was almost correct. Many naturally occurring mineral crystals will glow under radiation bombardment. An incoming gamma ray will ionize an atom, disturbing the electron orbitals and kicking it into an unnatural, excited quantum state. This hyperenergetic state will instantly decay, letting the atom's outer electrons fall back to their more comfortable positions, and this action, losing the excited energy state, manifests itself as a released photon, or a single, monochromatic blink of visible light. However, minerals in nature are messy and filled with imperfections in the crystal lattice. At these imprecise points, electrons have trouble resetting themselves, and the excited states become trapped, unable to fall back, at least at room temperature. To make the crystal resume its original condition before being irradiated, the crystal lattice has to be loosened up. Heat the crystal in an oven and the jiggling speed of the atoms increases, as well as the spacing between lattice sites. It lights up like a Christmas tree as unresolved quantum excitations fall back into place. The total number of visible light photons released is proportional to the number of gamma rays it absorbed over the dosage period.[201]

---

201 Yellow calcite, a polymorph of calcium carbonate, is unusual in that for it, room temperature is the heated condition under which it will glow, releasing the light from gamma-ray exposure. To turn off the light, put the calcite in a refrigerator. It will go dark but retain the gamma-image indefinitely under this condition. Bring it back to room temperature and it starts to glow until all the trapped quantum states have resolved. In complete darkness, you can read a newspaper with a calcite crystal that had been exposed to 200,000 curies of cobalt-60.

A mineral crystal, such as quartz, therefore retains a record of the number of gamma rays that have passed through it. Any ceramic object contains quartz, and in the house were ceramic plant pots, a vase, porcelain dishes, old-fashioned fuses in the electrical box, and light fixtures. Using these objects as opportunistic dosimeters, the Russians were able to map the radiation levels everywhere in the house, from the bathroom to the veranda.[202] This phenomenon is called thermoluminescence, and it is used to keep precise track of the radiation dosage to the body of anyone working with or near radiation. Everybody in the business, from nuclear power plant refuelers to dental X-ray technicians, wears a tiny badge on a clip, called a TLD, or Thermo Luminescent Dosimeter. It contains a crystal. After the shift is over, the crystal from each dosimeter is removed and placed in a light-tight dose reader. It is heated to a certain temperature, it glows, the light intensity is measured by a photomultiplier tube, and the resulting electrical current is displayed in terms of sieverts or rems of accumulated radiation absorption.

The same method was used to determine the gamma dose on the ground underneath the Little Boy atomic bomb dropped on Hiroshima. Under the bomb when it detonated was the Shima Surgical Clinic, which was pounded flat by the shock wave, with everything but bits of masonry vaporized in the photon blast. A piece of terracotta tile from the roof was retrieved, and from the gamma-ray image retained in its quartz content, it was determined that the gamma exposure from the bomb, 1,900 feet directly overhead, indicated a fission energy release equivalent to the explosion of 16,000 tons of TNT.

There are dozens of other ways to detect invisible radiation that is beyond the reach of our senses, from chemically developed photographic film to liquid nitrogen-cooled intrinsic germanium gamma-ray

---

202 The Russians also used electron paramagnetic resonance and chemiluminescense on samples of sugar in the kitchen, medicines in the bathroom cabinet, and mollusk shells on a decorative basket in the den as dosimeters. These methods are not as well developed as thermoluminescence, but using these exotic methods allowed the Russians to assert state-of-the-art expertise. The electron paramagnetic resonance phenomenon was first observed in the Soviet Union at the Kazan university in 1944, when it was the evacuation site for scientists fleeing Moscow during the German invasion of World War II.

spectrometers. They range in size, complexity, and capability from the Polimaster PM1208M Geiger counter watch to the ATLAS toroidal Large Hadron Collider apparatus. The PM1208M, built in Belarus, weighs 100 grams, fits on your wrist, can digitally store five hundred radiation measurements, and is most often used when visiting the Chernobyl 4 disaster site in Ukraine. The ATLAS, built in Switzerland, weighs 7,000 metric tons, fits in the 25-meter-diameter Hadron Collider tunnel, produces 1 quadrillion bytes of raw data per second, and was used to discover the Higgs boson. Somewhere down near the middle of this vast range are the ubiquitous handheld radiation survey meters. They are at this moment in use all over the world, protecting workers from exposure to radiation. They are carried by health physicists in nuclear power plants, hospitals, submarines, nuclear weapon stockpiles, and in all of the many industries that use radiation sources. Tens of thousands of them were built for use by civil defense radiation monitors in the event of nuclear war. The survey meter can see what we cannot, the presence of gamma rays, and it is essential for safety against radiation accidents or purposeful attacks.

The radiation survey meter gradually evolved from the earliest radiation experiments to the now familiar handheld unit over a few decades in the first half of the twentieth century. Bench-top radiation detectors had been cobbled together and used in physics labs for years, but serious development began to stir in the 1920s, when the popularity of medical radiation therapy and chest X-rays was blooming. Many instruments were built to find very expensive lost radium needles in cremation ashes and wadded up bandages, but these contraptions were heavy, and they often had to be plugged into the wall for power. Being an X-ray technician had become a hazardous profession, so similar instruments were being built to check for leaky shielding and accidental X-ray exposure to sensitive body parts.

The first radiation instrument that was easy to carry in one hand was the Victoreen condenser R-meter, introduced in 1928. John Austin Victoreen was a self-taught engineer from Ohio. He started the Victoreen Instrument Company on Hough Avenue in Cleveland, and the R-meter was his first product. It was a rectangular metal box, chromed all over and damascened to look pretty, with a microscope eyepiece sticking out of the top, a protruding radiation probe in front, a leather handle, and a chrome knob on the side. To use it to detect radiation, the operator spun the knob on the side. This action would build up a static electric charge by friction, loading electrons onto one plate of a capacitor in the radiation probe. The capacitor was simply two square sheets of brass, each 1 square centimeter, separated by 1 centimeter of air. By squinting into the microscope, the operator could read the voltage on the capacitor by seeing a magnified image of the edge of a tiny gold leaf, free to swing and held away from a second gold leaf, which was held stationary. The electron load on the capacitor was shared to the gold leaves by an insulated wire, and the negative charge on both leaves caused them to repel each other. The amount of movement of the free leaf indicated the voltage on the capacitor. Incoming gamma rays would knock the electrons off the capacitor plate, reducing the applied voltage.

The operator pushed a button on the side to light up the indicating leaf with a small lightbulb, powered by two penlight cells. The speed with which the voltage died on the capacitor indicated the radiation intensity, and the extent to which the leaf collapsed from its repelled state indicated the radiation dose on the capacitor. It was a brilliant product, and until after 1947, Victoreen was the only company in the United States that made radiation instruments.

Shortly after the introduction of the R-meter, in 1929 the stock market crashed, and a general collapse of the economy prevented people from splurging on an elegant Victoreen condenser R-meter. Victoreen kept building them, stacking them on shelves, developing better and more sophisticated instruments, and waiting for that special moment when everyone would need an R-meter.[203]

---

203 This type of instrument is now called a capacitive ion chamber. Victoreen was not the only manufacturer. The other was the National Bureau of Standards in Washington, DC. They built handheld instruments on a noncommercial basis. An excellent example with interchangeable radiation probes was the Lauriston Ion Chamber, developed by Lauriston Sale Taylor at the NBS. At least one Lauriston was supplied with the 29-inch cyclotron sold to Japan from Berkeley, California, right before World War II, and that instrument may have been their only handheld radiation meter until after the war. They were supposed to use it to confirm radiation produced by the cyclotron, instead of peering into the deuteron exit port. Reports of Japanese scientists poking around in

The final step in making the radiation survey meter truly portable and reversing the fortunes of the Victoreen Instrument Company was brought about because of a crisis condition that arose in 1944. The United States Army had intelligence intercepts indicating a possible new type of weapon being readied in Germany. It was a new and terrible way to kill soldiers. Today, we call it the radiological dispersal device, the weapon of mass disruption, or simply the dirty bomb.

Some of the most interesting intelligence fragments in World War II were coming from embedded agents in German-occupied France around 1943. The information was in technically garbled pieces, but if interpreted by knowledgeable specialists and assembled in just the right way, an ominous forecast was possible. Upon France's takeover by Germany, the Auer chemical company had confiscated from France tons of processed thorium-232, thorium ore, and uranium tailings from radium extraction and shipped it to Germany by rail. The Germans were also very interested in Jean Frédéric Joliot-Curie, Nobel Prize–winning nuclear physicist, married to Madame Marie Curie's daughter, Irène, and doing fission research at the Collège de France, down in the Latin Quarter in Paris. He had a cyclotron, and they were trying to think of a way to ship it to Berlin. What exactly were the Germans planning to do with a load of thorium, a cyclotron, and piles of uranium leftovers?

General Groves at the Manhattan Project headquarters in Oak Ridge, Tennessee, formed a committee of scientists, and they came up with an explanation. There was further information that the Germans had developed an anti-tank mine that had no metal parts, code-named Topfmine. It was made of compressed wood pulp, tar, cardboard, and glass. Containing 6 kilos of TNT explosive, it could easily blow the tracks off a Sherman tank that ran over it, and it was not metal, so it was invisible to the fancy electronic handheld mine detectors that the Allies were so proud of having developed. The Germans, not wishing to run over their own mines,

---

the Hiroshima debris with handheld Geiger counters as the smoke cleared are mythological, but they probably used the Lauriston to confirm the presence of radioactive ground contamination. It was called the Lauriston instead of the Taylor because Lauriston was a much more interesting word. Taylor, who worked in radiation research all his life, died at the age of 102 in 2004.

had developed a special detector to fit on the front of a tank. It was a Geiger counter.[204]

Putting the pieces together, the committee reasoned that the Germans were identifying their own mines with a radioactive tracer. It would have to be an active, high-energy gamma-emitting isotope so that the signal could be picked up through the dirt that buried the mine and was obviously stronger than the usual background radioactivity. The most logical radioactive isotope to be used was cobalt-60. It was very active, with a 5-year half-life, and it emitted a mega-electron-volt gamma ray. Cobalt-60 does not occur in nature. Where were they getting it? Cobalt-60 was made by neutron-activating natural cobalt, or by chemical extraction from spent nuclear reactor fuel. More important, where were they getting enough of it to tag a quarter of a million anti-tank mines?

A conclusion was that the Germans had confirmed self-sustained nuclear fission and had built reactors on an industrial scale. The production of isotopes to tag anti-tank mines was probably a by-product of a much larger intention, possibly to manufacture fissile plutonium-239, a concept that had grave implications. Moreover, the Germans' quick appropriation of all the thorium in France implied that they were approaching nuclear reactor design from an entirely different angle from the Manhattan Project. The Americans were using natural uranium, which contained less than 1% fissile material, uranium-235, and a critical mass capable of sustained fission had to be huge.

---

204 Information on the mine from intelligence sources probably referred to the Topfmine A, To.Mi.4531, also called the Pappmine (literally, cardboard mine). The handheld mine detector, used by German troops walking slowly ahead in front of the tanks, was the Stuttgart 43 detector, which had both a metal-detecting head and a simple Geiger counter on a 7-foot pole so that it could detect both German and Allied mines. The Topfmine had an additional advantage over steel or cast-iron mines. The Germans anticipated that after the war was won, all the hundreds of thousands of mines buried at 7-foot intervals on the Siegfried Line, which was 390 miles long, would have to be dug up and deactivated so that farming could resume. That would be an arduous, budget-killing task if the mines were the usual metal cans, but the Topfmines were biodegradable. They would last a couple of years in the field, which was the expected duration of World War II. Water would eventually seep in through a groove in the top of the case, eventually deactivating it, and the wood-pulp case would crumble into dirt, so it would not blow up a tractor tilling the soil. There was also a version fully waterproofed with tar for more permanent installations, the To.Mi.A4531. The radioactive tracer was incorporated into the black paint on the mine, called *Tarnsand*. The tank-mounted radiation detector that was assumed to exist by Allied analysts was not implemented.

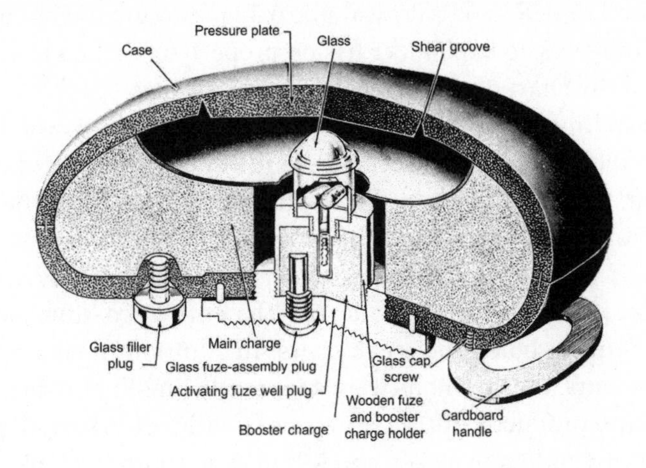

The Topfmine. It is made of wood fiber, cardboard, glass, and wood, and it is therefore invisible to a metal detector. It was painted black, and the gritty inclusions in the paint were radioactive. It was not meant to contaminate the tank crew but made it easy for the Germans to find their own mines.

What if the Germans were using thorium instead of uranium? Natural thorium occurred in only one isotope, thorium-232. It is not fissile, but if bombarded with neutrons from a powerful, cyclotron-driven neutron source, it would activate into protactinium-233, which quickly decays into long-lived uranium-233, which is just as fissile as uranium-235. There would be no inefficient, very bothersome isotope separation necessary. The product would be 100% uranium-233. Small, efficient nuclear reactors could be built using this fissile material, as well as bombs.

Another possibility was that the Germans had developed a subcritical reactor, using a smaller mass of fissile uranium-233 that could be driven to a simulated criticality or supercriticality using a cyclotron as an external neutron source. It would have all the advantages of a critical pile, making highly radioactive fission products, converting uranium-238 into plutonium-239, or even making more power output than it took to run the cyclotron. It would be easy to control, and it could be turned on and off with a switch. Just because the Manhattan

Project had zeroed in on a critical pile using uranium did not mean it was the only way to exploit the fission property of certain isotopes at the top of the chart of the elements.

The intelligence sources hinted at a problem even worse than nondetectable mines. The Germans were building the Atlantic Wall, fortifying the coastline from the north end of Norway to the south end of France, to prevent an Allied invasion of the continent. It was a typically grandiose German project, but there simply was not enough concrete in Europe to fortify the entire coastline, and the wall was full of holes and weak spots. In addition to concrete gun emplacements, the Germans were apparently making further use of their fission products, salting the beaches with powerful radioactive materials, intending to sicken and kill invaders with invisible, undetectable radiation. Furthermore, the Germans may have developed artillery rounds and aerial bombs filled with the dangerously radioactive material. These would be used to contaminate specific spots on the beach during an invasion, bombard the east coast of England, or give the inhabitants of London something that would terrorize them as mere explosives had not: a lingering, invisible source of delayed, certain death. Speculation as to German intentions approached the threshold of panic.

In 1943, Victoreen Instrument Company, anticipating a need for military radiation detection, developed an advanced, portable, gamma-sensitive survey meter: the Model 247. It was an ionization chamber, but unlike the antiquated capacitor meters, these instruments measured the ion drift between the two chamber plates caused by ionizing gamma rays ripping through the gas separating the upper and lower metal plates. The drift of the freed electrons moving toward the positively charged plate and the ionized atoms of gas drifting toward the negatively charged plate was measured as microamperes of electrical current. The level of the current was proportional to the number of gamma rays crossing through the ion chamber per second, and was read out as milliroentgens per hour (mR/h) on a small, round meter. The ion chamber was filled with 56 cubic inches of air and was enclosed in a rectangular box, 12 inches long by 9 inches wide and 10 inches high, with the meter, a leather handle, and six adjustment controls on the top face. It had four sensitivity ranges, from 2.5 to 2,500 mR/h. A special model X-247, with a range of up to 50 roentgens per day, was built on request for the army.

The 247 unit weighed 12 pounds, most of which was batteries. It used a 300-volt battery for charging the ion chamber, a 45-volt battery for the vacuum-tube DC amplifier with a 22.5-volt biasing battery, and two 1.5-volt flashlight cells for the filaments in the tubes. Cost to the government was $395 per survey meter, batteries not included. That is about $6,000 in 2016 money. They were not inexpensive. Grabbing the 300-volt battery across the terminals could drop a healthy horse stone-dead.

The army immediately bought twenty-four Model 247s and twenty-four Model X-247s for use in a special mission attached to Operation Overlord, the D-Day Normandy landing, named Operation Peppermint. There would be many more orders as the needs arose. General Groves of the Manhattan Project dispatched Major Arthur V. Peterson to England to brief General Dwight D. Eisenhower, commanding general of SHAEF (Supreme Headquarters Allied Expeditionary Force), on the purpose and importance of the special operation. The meeting was on April 8, 1944, and in further briefings, the British government was brought into the discussion. The Cavendish Laboratory at the University of Cambridge was tasked with providing close technical support, and eleven Victoreen 247 survey meters, one custom-built Geiger counter from the Met Lab in Chicago, and 1,500 photographic film dosimeters were shipped to England. Another twenty-five survey meters, five Geiger counters, and 1,500 dosimeters were kept in reserve, ready to be shipped to England by air at a moment's notice.

Major Peterson trained soldiers who were specialists in the Chemical Warfare Service in the use of the 247s, and all medical personnel for Operation Overlord were told to report any unusual fogging of X-ray film, which would indicate the presence of unusual radiation. Semi-scale rehearsals were staged to test the survey-meter operations, with radiation sources buried in beach sand, and aerial surveys were made, looking for radiation in recent bomb craters in England. There wasn't yet any evidence of a German radiation bomb having been used.

⚛

On the morning of June 6, 1944, the Operation Peppermint personnel were among the first off the landing crafts and onto the well-defended Utah and Omaha beaches. Under fire, they had to creep forward, setting the 247 down on the sand and waiting for the meter to settle down and show any radiation. It had seemed simple in the rehearsals, but on the beaches of Normandy, it was different. The 247s were not exactly watertight, and after being swum to the beach under heavy shelling and machine gun fire, the sensitive instruments, soaked with seawater while trying to amplify sub-microamperes, were unwilling to do any measuring. The twitchy, drift-prone vacuum-tubed amplifier would interpret the slightest hint of water vapor as a short circuit across the ion chamber. Somehow, with immediate projectile-induced death happening all around, everyone involved just wanted to get the hell off that beach as quickly as possible, and the concept of absorbing a radiation dose did not register a concern. Not a hint of radiation was found, and a brief review was sent back to Victoreen: Make it waterproof! And so was born the Model 247A.

The 247A would revolutionize the world of handheld radiation detection, but first there was the problem of the German Topfmines. A new program was organized to find the buried German mines, code-named Project Dinah.[205] Each mine was painted on top with something radioactive, the equivalent of two micrograms of radium, but not necessarily radium. It was basically anything at hand, such as ground up uranium or thorium oxide, or sometimes even very expensive radium. There seemed to be no fission products in use. A special tool was devised to find the mines, which were buried at a depth of 2 to 3 inches. Radiation survey meters were not adequate, as they were designed specifically to find dangerous levels of radiation. The mines were radioactive, but barely. There was just enough penetrating gamma radiation to register above background when the mine was buried. A Geiger counter, which is one hundred times more sensitive than an ion chamber device, was needed. Geophysical Laboratory in Houston, Texas, was contracted to build the AN/PRS-2. The Geiger probe was in

205 Rumor has it that the project director liked Dinah Shore, who was a very famous singer at the time. The British had a program with a similar goal, finding the hidden German anti-tank mines, but they used dogs trained to smell out the fumes from the TNT as it degraded in the leaky, cardboard mines.

a metal can 4 inches long and 2 inches in diameter, held at the end of a yard-long stick and cabled back to the power supply/amplifier box. The heavy box, mostly batteries, had an on/off switch and a metal handle with which to carry it. The operator wore a set of headphones and listened for clicks. One click meant that one gamma ray had struck the Geiger probe, and it was up to the operator to notice the presence of unusual radioactivity by the subtle pile-up of the random clicks.

Unlike the ion chamber, the Geiger counter produces a digital signal, literally counting the gamma rays. The ion chamber returns a voltage that is indirectly proportional to the extent to which the gas in the chamber is ionized, implying the density of the gamma rays ripping through it. The Geiger counter can register infinitesimally small radiation loads by simply returning one click every now and then. An ion chamber cannot do this, but for radiation safety checks, there is no need for such sensitivity, and an ion chamber can measure large, dangerous dose rates that would saturate a Geiger counter and render it useless. The Geiger counter is the right tool for tracking down trace radiation spills, hunting for uranium ore, or finding buried German mines.

The idea of marking buried mines with radioactivity was good, and an army program to do the same with the Allied mines was named Project Mamie.[206] Two types of mine markers, to be buried or placed atop mines, were produced for the army by the Massachusetts Institute of Technology: small glass ampules filled with cobalt-60-chloride dissolved in water, and fake rocks made of a porous ceramic material, saturated with an aqueous solution of cobalt-60-nitrate.

As the Allied army entered the European continent at Normandy and pushed east into Germany, the Alsos Mission, commanded by Colonel Boris Pash of the Manhattan Project, was tasked with finding the Germans' reactors and the extent of their atomic bomb development program, which by that time had obviously not reached the

---

206 Possibly named for General Eisenhower's wife, Mamie? She did not resemble Dinah Shore. The U.S. Army did not actually deploy any nonmetallic mines until 1955, when 1.5 million plastic antipersonnel "blast-type" mines, the MINE, APERS, MN, M14, were manufactured and stockpiled. It took ten years of experimentation to be convinced that we needed mines that could not be found with a metal detector. As a rule, all other US mines were made of steel.

production stage. The mission was disappointed at every turn, finding no indication that the Germans had achieved anything of importance toward atomic weaponry. There was no fission product, no plutonium, and no radiation poisoning weapon, which was what the most sapient Japanese scientists thought that we were developing. The Japanese suspected that bomb delivery of radioactive material was a foolish waste of effort, a feeling that was confirmed by their accumulated-dose measurements taken at the wrong bomb test at Alamogordo, New Mexico, in May 1945.

As it turned out, the Auer chemical company had shipped all the thorium and thorium ore stocks back to Germany from France in anticipation of a fad-driven need for radioactive cosmetics after the war was over. There was no scientific interest in thorium, and the Germans expected the conquest of Europe to be over and done with in a short time. Incredibly, radioactive cosmetics did have a small run in postwar Europe, ending quietly in about 1960. The popularity of thorium-laced toothpaste died in Germany along with the Third Reich.

The entire concept of a radiation weapon to stop or even slow down ground troops was an impractical plan dreamed up by scientific data analysts leading into Operation Peppermint. As we have observed in the previous chapter, it is possible to disable or kill an individual by external application of a radioactive nuclide, but to do so, a person must be exposed to a pure metal slug of the radioactive material, held almost touching the body, with a protracted exposure time. A soldier, trudging along, could walk right over a concentrated, powerful cobalt-60 gamma-ray source buried an inch in the ground below his feet and not be aware of it or notice any unusual symptoms. The fact that he was only close to it for a short period of time is enough to make it ineffective. Feet, as it turns out, are made of the tissue that is least affected by radiation, whereas the internal organs can be permanently harmed by a large enough gamma-ray source held close.

Clearly, the beach was not the place to bury sources of radioactivity. If they had meant to disable the Allied invasion force with radiation, then the place to put it would have been the point inland where the soldiers, exhausted from having struggled their way across the beach, up the hill, and through the concrete machine-gun bunkers, finally sat down to recover strength, preferably sitting down and leaning against a wall. Having radioactive sources embedded in such a wall, where soldiers would remain still for hours, could have ruined a lot of kidneys and started a panic against the invisible demons for which there was nothing to kill with a rifle. Just the vomiting would be worse than if the entire squad had put away a barrel of carelessly formulated moonshine. Even if a small percentage of soldiers were affected, there would still be the amplified fear created by an invisible enemy.

Trying to affect soldiers with radiation delivered by artillery shells or gravity bombs is probably a better plan than planted radioactive sources, but hazards are more severe for the attackers than for the attacked. It would seem logical that dangerous radiation could be spread economically using a military bursting explosive such as TNT, RDX, amatol, pentolite, picratol, or tetryl. Assume that a hollow, spherical charge of 22,000 curies, or 20 grams, of pure, metallic cobalt-60 is driven by a contained explosive charge in an air-delivered device, set to go off at an altitude of 1,000 feet above ground, and that the resulting radioactive aerosol is uniformly spread by a blast radius of 1,200 feet. Cobalt-60 is quite active, with a half-life of 5.28 years, and it generates a lot of heat just sitting there. A charge of 20 grams generates 400 watts. Imagine the heat generated by four 100-watt lightbulbs, closed up in a steel bomb case. Much more cobalt-60 than 20 grams would cook off the high-explosive charge, causing a premature explosion somewhere on the way from the factory to the aiming point.

Further assume the hollow sphere is to be quite thin, and consists of cobalt-60 powder glued to the surface of the explosive ball. The shell of cobalt-60 has so little depth, there is no self-shielding by the radioactive metal, and every gamma ray makes it to the bomb case. To be perfectly safe for the armorers, the bomb would have to be stored at the bottom of a 30-foot-deep pool of water. The dose rate at the bomb case is 21,000 rad per hour (210 sieverts per hour). Rub your hand on the bomb, and you will stagger away and die.

An area of 4.5 million square feet would be contaminated by the blast, at a density of 0.002 curies per square foot. This results in a gamma-ray exposure rate of 0.0047 rad per hour (0.47 sieverts per hour) to the vital organs for a person standing on that square foot, assuming that the cobalt-60 contamination is sitting on the surface of the ground. To absorb a dose that would result in nonfatal radiation sickness symptoms, a soldier would have to stand motionless on the spot for 10,000 hours. The dose rate would be "down in the noise" and barely detectable using a Victoreen 247 survey meter dialed to the lowest detection range.

It is ironic that the radiation bomb is extremely dangerous until it is exploded. Anyone near the cobalt-60 charge as the thing is being built, transported, loaded, and implemented is subject to agonizing death. Every action involving an unexploded radioactive munition has to be by remote control. Storage and transport must be done with the bomb contained in heavy lead shielding or, even better, uranium shielding. Uranium is even denser than lead, so it stops gamma rays with less volume of material put in the path of the radiation. Its half-life is so excruciatingly long, 5 billion years that it is only barely radioactive. A radiation survey meter cannot detect its presence. It takes a Geiger counter to recognize the radiation from uranium.

Exploding the radiological dispersal device, the dirty bomb, takes a fatal concentration of radioactive material and spreads it thin over a wide area. Aside from burying it in a deep hole, exploding it is the best way to render the device safe. Soldiers could walk through a freshly contaminated field with no medically detectable effects, and decontamination would consist of taking a shower, washing clothing, and hosing down the vehicles. There are, of course, long-term effects of lingering contamination on the ground, and in the special case of cobalt-60, it takes about fifty-three years for radioactive evidence of dirty bomb activity to disappear by radioactive decay. The radiation from cesium-127, another candidate for the active ingredient in a radiological weapon, would stay around for three hundred years.

This low-level radiation, while not immediately life threatening, threatens instead to eventually cause cancer in people living atop the contaminated soil. The magnitude of the probability is no matter. The possibility of radiation-induced cancer, no matter how small or how far into the future, is a psychological injury, and it has become recognized as a major detrimental effect of radiation-dispersal weapons. The other effect is being within range of the shrapnel and being killed by the chemical explosive itself. Death by radiation poisoning is not on the list.

The three factors that determine the dose effect of a concentrated radioactive source are time, distance, and shielding. Radiation damage to body tissues accumulates with time. To minimize the probability of damage near a radioactive source, go stand somewhere else. If your survey meter indicates a high radiation field, then run away from it, beat it off your clothing, or put it back in the block of lead just as quickly as you can. In the special case of radiation protection, speed does not kill. Speed sustains life. Brief exposure time means low accumulated dose, even from a source that is considered highly dangerous. Low accumulated dose means fewer damaged cells, less radiation burn, and less chance of nonfatal (cancer-inducing) DNA reproduction errors. The lower the dose, the better. The extent of radiation damage to your body is always probabilistic, and it is better to depend on brief exposure than on luck.

Distance makes a huge difference in radiation dosage. If you were an inch away from a 0.010-curie radiation source, in two hours you would be dosed by 29 rads (0.29 grays) of gamma radiation. That is a serious but nonfatal dose. You would recover from it, but memories of the vomiting experience would remain forever. If you had been 3 feet away from that same source for the same two hours, you would have an accumulated dose of only 0.026 rads (0.00026 grays). It would cause official concern, as it would show up on your dosimeter, but the physical effects would not be noticeable. If you had been 33 feet away from this source for the same amount of time, you would have an accumulated dose of only 0.00026 rads (0.0000026 grays), and your Geiger counter would not have noticed the source. Radiation blasts off from the source in no preferred direction, and it essentially is configured in a spherical cloud with the source at the center. The intensity of radiation and the resulting dose rate is inversely proportional to the square of the distance from the source.

Nuclear physicist Harry Daghlian, in a much-documented accident involving the "demon core" plutonium sphere on August 12, 1945, died of radiation poisoning after taking a 590-rad (5.9-sievert) dose to his internal organs. The security guard, Robert Hemmerly, was sitting 12 feet away from the radiation source when Daghlian was dosed. The only thing between him and the plutonium sphere was the air in the room. Hemmerly walked away from the accident, returned to active duty, retired, and died of old age. The difference between Hemmerly's outcome and Daghlian's outcome was determined by 12 feet of distance. Distance saves.

Shielding will also save you from radiation exposure. Anything solid between you and the concentrated source of radiation will interact with the gamma rays before they interact with you. They will be stopped, slowed down, or scattered off in another direction when they collide with a more or less stationary atom. The heavier the atom, the more effective the collision. There is a lot more stopping power in a plutonium atom than in a lithium atom. Lead makes a better shield than water. Water makes a better shield than wood. Wood is better than air. Anything is better than the vacuum of outer space, where gamma rays can cross intergalactic distances without hitting anything.

It can take several hits to stop a gamma ray and render it harmless, so the thickness of the shield material is as important as the size and mass of its individual atoms. Air is a lousy shield material, because the atoms making it up are so light and are so sparsely located that you can see through it; but you live at the bottom of a 90-mile-thick blanket of air, and as a radiation shield, it is thick enough to stop a constant, potentially deadly barrage of mixed cosmic radiation coming at Earth from all directions. Without the air blanket, Earth would be as sterile as Mars. A concrete wall between you and the radiation source is very good. Less so if the wall is concrete blocks. They are hollow, like Legos. It is ideal if the wall density is fortified with steel pellets, or "punchings," left over from the days when bridges and skyscrapers were made of steel beams, riveted together after having the rivet holes punched out. You are unlikely to be behind this wall in anything but a nuclear power plant. A chain-link fence is not going to be any help, but an oak tree to cower behind isn't bad.

Knowing the three ways to protect yourself against ionizing radiation (time, distance, and shielding) will keep you out of trouble in the event of a dirty bomb attack, as long as you have a radiation survey meter with you. Without some means of detecting the invisible danger, you have no idea of what to avoid, which way to run, or what to get behind, and that is the terror factor of the radiological dispersal device. Most people do not know to carry a gamma meter with them.

By now, international terrorists are aware of the advantages of a radiation weapon, and the deadly danger of implementing and delivering it are not major concerns. A dirty bomb exploded in the air or atop a tall building does not leave a crater. All it does is spread detectable radiation over a large area. It does not have to be deadly or even sickening radiation. It could easily panic a population unequipped to understand or measure the danger, and it would contaminate entire blocks of a large city. Decontamination of a city is different from cleaning up a battlefield. It would be a long and expensive process, as residents would want everything returned to the pre-bomb condition and would be unwilling to wait for the passage of ten half-lives of the active nuclide that was used.

By 1946, Victoreen had completely redesigned the 247 radiation survey meter, and the new product was the Model 247A. The new meter was a thing of technical beauty. The six knobs on top were reduced to two. The one on the left turned it on and checked the four batteries one at a time as it clicked clockwise. The knob on the right switched the sensitivity through four ranges, measuring gamma rays from 0.05 rads per hour up to 50 rads per hour. Pull the knob, and it became an adjustment, setting the round voltmeter, clearly marked as "R/hr," to read zero when there were no gamma rays.

The case was made of cast aluminum, one eighth of an inch thick, in two pieces held together with bolts, sealed with a rubber gasket, and finished in glossy gray paint. It could literally work under water. The Model 247A-Special, later renamed the 247B, was built for higher radiation fields and was distinguished by having the meter flange painted bright red. Eventually there was a 247C and a 247D for X-rays. It was precisely what the new, atom-bomb-equipped military needed, and it was designated the IM-3/PD. Many units were bought for the Operation Crossroads A-bomb tests at Bikini Atoll in 1946, and Victoreen followed with the development of multiple versions of the Model 263 portable Geiger counter, the Model 356 alpha radiation detector, the Model 488 neutron survey meter, the Vic-Tic Model 631 Geiger counter, made especially for uranium prospecting in 1954, and the wildly popular Cutie Pie Model 740 in 1955, made with health physicists in mind.

⚛

It is true that most people in the United States, despite the fact that 5,740,000 Civil Defense radiation detection instruments were made and distributed during the Cold War, have never held a Geiger counter, much less own or carry a survey meter. This was not always the case.

In 1949, the Atomic Energy Commission saw a lack of uranium mining in the United States as a crisis and decided to put the entire population to work hunting for yet unknown uranium ore deposits. There were newfound needs for fissionable material in the Department of Defense, as an atomic bomb stockpile was being amassed and a nuclear-powered submarine was being developed in secret. Every square inch of America needed to be searched for exploitable uranium-ore deposits, and even though there were only forty-eight states, this was a lot of ground to be explored. To this end, the U.S. Geological Survey published a booklet, *Prospecting for Uranium*. It was free for the asking. The December issue of *The Engineering and Mining Journal* announced that the AEC was going to jack up the price per ton of high-grade uranium ore. In March 1951, the AEC offered a bonus of $10,000 to anyone opening a productive uranium mine. The race was on.

Uranium is unusually easy to find because it broadcasts its presence by throwing gamma rays. You can find it underneath dirt or overgrowth without even seeing it. All you need is a small, handheld, battery-powered Geiger counter. Its probe, a small metal cylinder on the end of a cable, detects the gamma rays and causes the device to click or the needle on a meter to move. The more enthusiastic the clicks are, the more uranium is near the probe.

Within days, a new industry sprang to life, building consumer-grade Geiger counters. There were dozens of manufacturer start-ups, particularly out West, where the geology was usually dry and innocent of ground clutter. All the popular magazines, from *Collier's* to *Boys' Life*, devoted issues to uranium prospecting, and a stampede was encouraged on radio, television, newspapers, and at the movies. There was even a *Geiger Counter Magazine*, and toy manufacturers were quick to supply juniors with downsized examples, both pretend and working. Sears, Roebuck introduced its own brand of Geiger counter, Tower, as did Montgomery Ward. You could buy a Geiger counter at any gas station in the Southwest, where the craze was at its maximum fever pitch.

By 1955, if you did not have a Geiger counter, then you just were not participating in life. On television, every program, be it comedy or drama, had a Geiger counter in it, from the *Amos 'n Andy Show* to *I Love Lucy*.[207] In Colorado at the time, one was more likely to own a Geiger counter than an automobile. Even Donald Duck had a radiation detection instrument, although his was the far more sophisticated and expensive Precision Radiation Instruments (PRI) scintillation counter, a hundred times more sensitive to gamma rays than the best Geiger counter on the market.

A Geiger counter was good for finding uranium, but serious prospectors carried the PRI Model 111B De Luxe Scintillator as the tool of choice. It was a beautiful instrument, all chrome plated, with a solid-state, thallium-doped silver iodide crystal glowing every time a gamma ray hit it. The tiny flash was picked up by an RCA 6199 photomultiplier tube having ten cascading dynodes and running on 1,000 volts. It could detect a single photon, and the De Luxe Scintillator could find a source of radiation down to 0.00001 rads per hour. The detector head looked like a ray gun, and it connected to a battery cluster with a curled cable. It was expensive. A new one cost $495, which is about $1,500 in 2016 money. The demand was enormous, and PRI, on South La Brea Avenue in Los Angeles, California, became the "world's largest manufacturer of portable radiation instruments."

Large corporations could afford airborne surveys, plotting the radiation on the ground from scintillation measurements from low-flying aircraft. For this special application, PRI sold the Model 118 Royal Scintillator. The price could give a grizzled prospector the blind staggers: $1,995 (around $30,000 today). It had a very large three-inch sodium iodide crystal and was ten times more sensitive than the De Luxe Scintillator.

---

207 The episode in which Lucy, Ricky, Fred, and Ethel join Fred MacMurray to find uranium in Las Vegas, Nevada, "Lucy Hunts for Uranium," was actually on the *Lucy-Desi Comedy Hour* in 1958. *The Amos 'n Andy* episode "The Uranium Mine" was one of the last filmed in 1955. Jack Benny had a Geiger counter in "Jack Hunts Uranium," in 1955; Phil Silvers held his on *The Phil Silvers Show*, "Big Uranium Strike," in 1956; and George Burns and Gracie Allen had their turn on the *Burns and Allen Show*, "The Uranium Caper," in 1955. Everybody, it seemed, had a Geiger counter.

The uranium rush crashed in 1960, when the AEC found that it then had all of the uranium it could possibly use in the foreseeable future. The first significant strike had been made near Moab, Utah, on July 6, 1952, when Charlie Steen hit a huge deposit in the Big Indian Wash of Lisbon Valley. Further mine openings from Kern River, California, to Bone Valley, Florida, ensured that there would always be enough uranium. In the Four Corners area of the Southwest, there were four thousand mines where 225,000,000 tons of uranium ore were dug up, a lot of money was made, and, unfortunately, a lot of radon gas was breathed in by Native American mine workers. What remained at the ending was a Geiger counter in many closets, hundreds of failed companies, and a population that had been immersed up to the eyeballs in the atomic culture.

John Victoreen quit the radiation detection business, moved to Colorado Springs, Colorado, and worked as a consulting physicist at the Medical Center. His research efforts were devoted to hearing enhancement, and he started the Vicon Instruments Company, making hearing aids. PRI dropped making scintillation counters like a hot rock and became a record label. Their first release was PRI-3002, *David Pell Plays Harry James' Big Band Sounds.* For radiation instrument manufacturing, the big game was over.[208]

⚛

The US government has been studying and preparing for radiological dispersal device attacks on major cities inspired or instigated by Islamic terrorist groups at least since 2003. Medical supplies, decontamination

---

208 The popular market for radiation detection instruments in the 1950s was similar to the present combined demand for smart cellular telephones, iPads, Kindles, and laptop computers. It was a big market, and it was a shock to the system when it dried up in 1960. There was still a demand on a smaller scale for radiation detection equipment for nuclear power plants, hospitals, and laboratories. There was also a new market for mass-produced "Series CD V-700" Geiger counters, ionization chambers, and dosimeters for the Civil Defense Administration. Several manufacturers participated, including what was left of the Victoreen Instrument Company (it had been sold to General Electric), Universal Atomics in New York City, Bendix, Anton Electronics Labs, Chatham & International Pump and Machine Works, Electroneutronics Inc., Nuclear Measurements Corp., Nuclear Chicago Corp., Jordan Electronics, and Lionel, the maker of toy electric trains.

equipment, and radiation detectors are being stockpiled, quietly and without publicity. The Medical Preparedness and Response Sub-Group of the Department of Homeland Security Working Group on Radiological Dispersal Device (RDD) Preparedness has done a lot of work toward preparation, without publicly expressing terror at the thought of such an attack. Not to be left out, the Department of the Army and the U.S. Marine Corps have compiled Army Field Manual FM 3-14 (Marine Corps MCRP 3-37.1A), *Nuclear, Biological, and Chemical (NBC) Vulnerability Analysis*, still worrying about German radiation bomb or hidden battlefield radiation sources. It will not be a complete surprise if we are eventually attacked by a dirty bomb.

There is also a subspecies of dirty bomb to worry about: the smoky bomb. From a psychological standpoint, it may be more damaging than a dirty bomb, which is assumed to use a gamma-emitting nuclide as the active ingredient. The dirty bomb will scatter the pulverized radioactive material all over the streets, the sides of buildings, lampposts, sidewalks, roofs, automobiles, and people. It will definitely show up on radiation counters, and the stigma and inherent terror of embedded radiation will be difficult and extremely expensive to erase. For decades after an attack, no matter how the affected area is scrubbed and re-scrubbed, no matter how many buildings have been torn down and removed, an unusual background count of gamma radiation will be detectable. Even though no deaths will be scientifically attributed to it, the radiation dread will remain as long as it will make the Geiger counters click.

Another way to distribute radioactive mayhem is to use an alpha-emitting nuclide, such as polonium-210, the insidious alpha-particle poison that was used to assassinate Alexander Litvinenko in London on November 1, 2006. Litvinenko had been employed by the Russian Federal Security Service, and he was critical of the leadership in his former country. A very small amount of polonium-210 was secretly introduced into his system, and his internal organs were destroyed by alpha bombardment. He died in agony after three weeks of steady decline.

Alpha particles have an extremely short range, and this makes them difficult to detect. If a bomb were built to introduce polonium-210 internally, it would have the added terror effect of being invisible not only to human senses but to all but the most specialized radiation detection equipment. The internal dose would be administered by carrying polonium-210 on smoke particles, allowing it to be breathed. The smoky bomb can be set up by boiling out a large beaker of polonium-210 chloride dissolved in water in a wooden building. The steam condenses on everything and contaminates the entire interior of the building. Next, set fire to the building and let it burn down. The smoke will carry injuring doses of polonium-210 for miles. If you can smell the smoke, then you are infected with it. Firefighters and their equipment will be the first to be contaminated.

It doesn't take much polonium-210 to cause distress, but it has to be inside the body. Just having it on the skin doesn't do anything at all. The danger to firefighters or anyone with surface contamination is that when they peel off their respirators and their outer clothing, the polonium-210 rubs off on the fingers. Anything on the fingers eventually gets into the mouth, and there is the problem.

Announce the intention of the fire and its radioactive component on the Internet, and the trap snaps shut on the surrounding population. Panic will spread, and people who were not even in range of the smoke will immediately complain of radiation poisoning symptoms. Hospitals will be overrun.[209]

---

209 The model demonstration of this psychological phenomenon is the Goiânia Incident in Brazil, beginning on September 13, 1987. Metal thieves Roberto dos Santos Alves and Wagner Mota Pereira broke into an abandoned radiotherapy clinic in downtown Goiânia and found a big cancer teletherapy machine just sitting there. They found the valuable-looking core of the machine, removed it, and moved it to Alves's house. It contained a 1,375-curie cesium-137 gamma-ray source. Later that day, both men exhibited symptoms of radiation poisoning, but trips to the hospital did not diagnose it as such. They eventually managed to puncture the nested stainless steel source capsules and release the powdered ceisum-137-choride. It glowed blue, and on September 18, they were able to sell the exotic substance to a scrap yard, owned by Devair Alves Ferreira. Thinking it supernatural, he brought it home to show his wife, Gabriela Maria, and all his friends were invited over to see the magical glow. His six-year-old daughter, Leide das Neves, spread it over her body, causing her to glow like a fairy princess. On September 28, Gabriela Maria noticed that everyone around her was getting very sick, and it had something to do with that metal thing that her husband had brought home. She gathered up as much of the glowing powder as she could, put it in a plastic bag, and took it to the local hospital. The next morning a visiting physicist pointed his scintillator at it, and its rate meter slid off scale. By the end of the day, news of the stolen gamma-ray source was known internationally. The news media in Goiânia made certain that it was known locally, describing the symptoms of acute radiation poisoning. By that afternoon,

Fortunately, polonium-210 is hard to get, and so far, only a government has access to usable quantities.[210] It will take 100 grams of it to saturate the wood in the building and deliver doses that will sicken people. That is an entire year's production run in the only factory in the world that produces polonium-210 in industrial quantities. That facility is in Russia. You cannot build up a stockpile of polonium over several years by buying small, unnoticeable quantities, because it has a half-life of only 138 days. After sitting on a shelf for a little over a year, 87.5% of your polonium-210 stash has turned into non-radioactive lead-206.

In the history of warfare, there is not a case of a successful attack using radiation dispersal, but there have been at least two aborted attempts. The first occurred in November 1995 in the middle the First Chechen War of Independence. The Russian Republic of Chechnya, an Islamic state, had declared independence in November 1991 and renamed itself the Chechen Republic of Ichkeria. In 1994, Russian president Boris Yeltsin sent the Russian army to quell the rebellion and restore order. Chechen rebels, hoping to bring the war home to Russia, wrapped a cesium-137 radiation source in a high explosive and buried it at the Izmailovsky Park in Moscow. The bomb never went off, despite a rebel leader's boasting of it to the news media.

A second attempt by the same splinter group of Chechen rebels, occurring in December 1998, was similarly unsuccessful. It was in

---

130,000 people were nauseated and dizzy, fearful that they had radiation exposure, and were jamming every hospital emergency room. The government had to use the Olympic stadium as a medical facility. Eventually, 249 people were found to be contaminated, 20 people were treated for radiation sickness, and 5 people died. The first death was the daughter, Leida das Neves Ferreira, on September 28. At her burial, 2,000 people rioted and tried to block the cemetery road using bricks and stones, fearing that her corpse would poison the water supply. Contamination was eventually found on two scrap yards, a hospital, forty-two houses, fourteen automobiles, three buses, five pigs, and fifty thousand rolls of toilet paper.

210 The principal industrial use of polonium-210 in the United States is the StaticMaster antistatic brush, manufactured by Amstat Industries, Inc. It is a soft brush with a 0.00025-curie polonium-210 alpha-particle source at the base of the bristles. The heavy alpha flux from a recently purchased StaticMaster will vigorously neutralize the free electrons (static electricity) on a smooth surface. This device was very popular when photo enlargements were made from film negatives, as it was the only sure way to sweep the dust off the negatives. They are still used in high-tech industries in which any hint of dust must be eliminated. A replacement polonium cartridge sells for $134.

the interwar period, between the First Chechen War and the War of Dagestan. Two Russian brigades had been left in place to enforce the peace, Wahhabism was epidemic, and the chief economic driver in Chechnya was kidnapping. The Chechen Security Service announced having found a container of radioactive materials wrapped around a land mine near the railroad track in Argun, a suburb of Grozny, the capital of Chechnya.

There have been several thefts and underhanded purchases of industrial radioactive sources from the oil-drilling industry in the Middle East, probably gamma-producing iridium-192, but none have shown up yet attached to explosives. Bear in mind that the Nuclear Regulatory Commission estimates that in the United States an average of one industrial radiation source is stolen, lost, or abandoned per day. There has been one arrest in the United States of a person suspected of planning to detonate a dirty bomb, on May 8, 2002. The charge was dropped.

The Department of Defense has put a lot of effort into anticipating a terrorist attack using radiation, from the formulation of triage procedures of irradiated personnel to printing a handy pocket guide, *Terrorism with Ionizing Radiation General Guidance*. Despite the title, it is not instructions on how to terrorize using radiation but is a handy summary of radiation sickness symptoms. Basically, in cases of sickness in which no treatment is necessary (subclinical, up to 200 rads acute dose), vomiting starts 3 to 6 hours after exposure. There are two sublethal ranges requiring immediate care, with nausea onset in 2 to 4 hours and 1 to 2 hours. For lethal acute doses over 600 rads, expect vomiting to start in less than an hour. For the triage of radiation sickness victims, first-contact medical personnel are encouraged to estimate the elapsed time between radiation exposure and the onset of nausea and write it on the tag attached to the patient. It is a reliable indicator of the treatment that a patient will require. Dust contamination is confirmed with radiation counts on mucosa swabbings, and dosimetry is accomplished by blood count.

If you are in range of a terrorist explosion, move quickly away from the suspected center of the detonation and breathe through a handkerchief or a cloth garment. Anything you can do to avoid taking in smoke and dust will help you survive. Keep your mouth closed on the way out, of course. The last thing you want is radioactive dust in your gastrointestinal tract.

If you ever suspect that you have been dosed by a radiological dispersal device, your most important action is to get a shower as soon as possible. Do not worry about contaminating the sewage system with radioactivity. That is a long-term cleanup consideration, and diluting the radiation into the water supply may be the best thing you can do with it. Your first concern is to get away from it with all possible speed. Remember: radiation protection comes down to time, distance, and shielding. Throw away the clothes you were wearing, stuff them into a plastic bag or, if you prefer, wash them. Discard your shoes. Shower until your skin wrinkles, and shampoo your hair repeatedly. Do not use a conditioner, as it will only seal radioactive contaminants into your hair, and it will have to be shaved off. Hair is bad about collecting and saving radioactive dust. If you have access to a gamma spectrometer, a scintillation counter, or a Geiger counter, check yourself carefully for any lingering contamination. If these instruments will not operate because of count saturation and an ion chamber is necessary to evaluate the radiation remaining on your skin, then you are in trouble. Get in line at the triage tent while you can still walk.

A CD V-700 Geiger counter, which is often available on eBay, is a good, usable low-level radiation detection device.[211] Hundreds of thousands of them were manufactured for the Cold War. It runs on flashlight batteries. Learn to use it, and test it for functionality using the radiation source pasted to the side of the instrument.

211 The CD V-700 is the only Civil Defense radiation instrument that is actually a Geiger counter. There are many other V-7XX units for sale, but they are all ionization chambers. The Geiger counter is one hundred times more sensitive to gamma and beta rays than any ionization chamber, and it is appropriate for finding residual radiological contamination. Professional-grade Geiger counters, such as the Eberline E-120, are available as surplus, but they are usually sold without the probe assembly, which is what wears out when in constant use in a nuclear power facility. The probe is the expensive part, and installation of a new one will require calibration. The "Smart Geiger" sold for use on an iPhone or Android by FTLABS is interesting, but it is not a Geiger probe. It is a PIN diode. While it is sensitive to gamma radiation on a linear scale and it makes use of the telephone's computer to integrate a dosage, its sensitivity is inferior to an ion chamber. Smart Geiger plugs into the earphone jack. If you get a phone call while using the FTLABS app, the ring will feed back through the microphone contact, and gamma counts will pile up rapidly, appearing that someone has just slipped a cobalt-60 source into your pocket.

The drug therapy for cesium-137 ingestion, which is a likely result of being close to a dirty bomb explosion, is Prussian Blue. Prussian blue, or ferric hexacyanoferrate, was the first synthetic dye made, and it is now used in everything from blue crayons to ballpoint pens. It is manufactured in 10,000-ton quantities yearly. It is also nontoxic, and it binds with cesium or thallium, prevents these elements from being absorbed in the gastrointestinal tract, and leads to fecal excretion. Take two pills three times a day, and it will turn the commode contents a beautiful blue. Its effectiveness was fully tested in the Goiânia Incident in Brazil in 1987. It is available only from Heyl Chemisch-Pharmazeutische Fabrik GmbH, Berlin, Germany, and is $15.80 per 30-capsule bottle when ordered in 100,000-bottle quantities.

Potassium iodide has received a lot of unjustified publicity as an "anti-radiation" drug, as it is in first aid kits in nuclear power plants and is often given out to families living near such facilities. In reality, it only protects the thyroid gland against iodine-131 contamination, and it must be taken just *before* the bomb goes off. It is very unlikely to be used in a radiation dispersal attack, as the half-life of the nuclide, which is an easily escaping fission product, is only 8 days.

Trisodium calcium diethylenetriaminepentaacetate, or Ca-DTPA, is the drug of choice if you have swallowed plutonium, americium, californium, or curium. It is made by Heyl, is distributed by the Oak Ridge Associated Universities, and a single-dose ampule costs $1.15 when ordered in 100,000-unit blocks. There are 5 ampules per unit. It is administered intravenously within 6 hours of exposure, and is a chelating agent used all over the world. It promotes renal excretion. Although plutonium, californium, or curium are unlikely to be used in a dirty bomb attack, americium is not out of the question, as americium-242 is widely used in smoke detectors and has a 141-year half-life. Plutonium ingestion is possible if the attack involves a low-grade nuclear explosive.

If we are attacked with a uranium-based atomic bomb, which is not impossible, the best drug to take is sodium bicarbonate, or baking soda. Swallow 4 grams to start with, and follow up with 2 grams every four hours until your urine pH is 8 or 9, which should happen in about 24 hours. The Chic-4 nuclear weapon design, so popular in the Middle East, is a uranium weapon. A gram is about the weight of one raisin.

Amifostine, or phosphorylated aminothiol (WR-2721), is an FDA-approved radioprotectant. Take it before the bomb hits and your cells are protected against radiation damage. It is a free-radical scavenger that protects cell membranes and macromolecules such as DNA from radiation-induced free radicals. It prevents cell death, the development of cancer due to radiation, radiation toxicity, and bone-marrow depression. It costs about $1,400 per intravenous dose, to be administered 1 hour before radiation exposure. It seems the wonder drug for a dirty bomb attack, but its use would require pre-knowledge of the explosion. If you knew that it was going off in an hour, would you not have left town?

The Department of Homeland Security may have stockpiled this drug for use by medical personnel. While they may not have been exposed to the initial blast of contamination from the bomb, they would be exposed to radiation every time they examine a victim. First responders could accumulate doses hundreds of times as fresh, pre-decontaminated patients are brought in. Amifostine may therefore be administered to medical personnel before they are deployed in the field.

Do not be afraid of a dirty bomb attack in the United States, but be conscious of the possibility. Know what the danger of it is, and know what the danger is not. A dirty bomb does not kill people. A dirty bomb causes people to panic. Panic kills. In this world and this era, we do not really know what is going to happen next.

CHAPTER 10

A Bridge to the Stars

"Where are they?"

—Enrico Fermi's response to a statement that there must be billions of highly advanced civilizations in the universe having developed interstellar travel and visiting Earth, while walking to lunch with Edward Teller, Emil Konopinski, and Herbert York, 1950

THERE ARE TWO POSSIBILITIES. EITHER the infinite universe, with 200 billion observable galaxies and 200 billion stars per galaxy, has a percentage of those stars with planets capable of supporting life, which has evolved into sentient beings, and a percentage of those beings are far enough along to wonder if there is anyone else in the infinite universe, or it does not. Either outer space is teeming with life and, given that our civilization is probably about average, some of it is coming for us; or we are the only living things in the universe. Although the second possibility makes the vast universe seem rather pointless, there is no law or principle in physics indicating that the universe cannot be pointless. Either possibility is slightly terrifying upon contemplation.

The great Italian nuclear physicist Dr. Enrico Fermi wondered aloud why if there are so many advanced civilizations out there they are not all over us, buzzing in and out and landing all over the place. The great British astronomer Sir Fred Hoyle countered, stating that they *are* buzzing all around. We call them "insects," a class of life for which he could see no evolutionary reason for existing, so they must have come here from outer space.[212]

Finding, confirming, and making contact with extraterrestrial civilizations may well be mankind's ultimate quest, and modest steps are currently underway. First, we must discover whether planets, gravitationally gathered globular clumps of supernova debris that have by chance fallen into orbit around a star, are a novelty in our solar system or are common. We had long ago determined that stars are common and more recently found that they are all fusion-powered energy sources, warming up planets against the near-zero cold of outer space and melting the ice on those planets lucky enough to be at just the right distance from the fusion. In just the last decade, improved optical telescope technology has taken us a step further, discovering planets orbiting faraway stars and indicating that they are not a local fluke.[213] In the past few years, 2,111 "exoplanets" have been discovered in 1,354 planetary systems. It is now estimated that each star in the Milky Way Galaxy has an average of 1.6 planets orbiting around it. In 2011, the first Earth-sized planets were found orbiting a star, Kepler-

---

212 Fred Hoyle died in 2001 at the age of eighty-six. He was also a science fiction writer, and his book *The Black Cloud*, released in 1957, was critically acclaimed. One of his better quotes is "It's better to be interesting and wrong than boring and right." Enrico Fermi was fifty-three years old when he died in 1954, a year before his patent for the nuclear reactor was declassified. A street in Rome, Italy, is named for him.

213 The first discovery of an exoplanet was actually made using a radio telescope. In 1992, Alexander Wolszczan and Dale Frail discovered something large orbiting pulsar PSR 1257+12 and periodically getting in the way of the radio signal coming from the pulsar. A pulsar is actually a ruined star that has blown up in a supernova and left an energetic remnant. The radio-occluding object must be what is left of a giant, rocky planet that was sterilized in the supernova event. Since then, Earth-type planetary searches are carried out by the Kepler Space Observatory telescope, launched into solar orbit by NASA in 2009.

20, which is about the size of our Sun. Our Sun, when compared to all the other stars, is somewhat ordinary. It is not too big and not too small, and it is an average kind of star, a GV2 main-sequence type, commonly called a "yellow dwarf."

About 20% of these G-type stars have an Earth-sized (class M) planet orbiting in the habitable zone, a distance from the star that paints the planet with enough energy to allow complex organic molecules to exist without freezing or boiling away. The nearest such planet is probably about twelve light years away. That is the distance that light travels through the vacuum of deep space in twelve years, or 72 trillion miles. Light clocks out at 186,000 miles per second or 300 million meters per second. As Dr. Albert Einstein pointed out in 1904 in his special relativity theories, nothing can travel faster than light, and it takes an impossible amount of energy to accelerate to any speed approaching that of light.

Before worrying about how we are going to travel twelve light years to meet our neighbors, the next step in extraterrestrial life determination is to search for characteristic light-absorption spectra for organic gases in the atmospheres of class M planets orbiting G-type stars in the habitable zone, looking for scant evidence as their orbits pass between our advanced telescopes and their suns. This will not be easy. Close surface analysis of our own partner class-M planet which is sort of within the habitable zone, Mars, has yet to discover evidence that life ever existed on it. If only we could find some fossilized remnant of even the most primitive life, it would bolster the quest for interstellar life. If life could have once existed on Mars, back when it had lakes of water and a heavier atmosphere, then life would show a tendency to come about under less-than-perfect conditions.

Anticipating success in this step, there have been concerted efforts to find radio beacons or obviously intelligent radio communications signals originating somewhere in the cosmos. Methods for scanning the sky for extraterrestrial radio signals came about with the introduction of radio astronomy. In 1932, Karl Jansky, working for Bell Telephone Laboratories in New Jersey, built a large, rotating shortwave antenna array to find out where radio static was coming from. He found that it was coming from the sky, and most of it seemed to be received with the antenna pointed toward the center of the Milky

Way galaxy, straight at the Sagittarius constellation. From this one observation was born radio astronomy, and in the second half of the twentieth century many large antennae were built and pointed up into the cosmos. Bright objects in the universe, as it turns out, emit radiation over the entire electromagnetic spectrum, from gamma rays to radio waves, and we had only been seeing them using a tiny slice of this spectrum, the visible light.[214]

On November 28, 1967, at the United Kingdom Mullard Radio Astronomy Observatory on the outskirts of Cambridge, a strange signal was received at 19 hours 19 minutes right ascension and plus 21 degrees declination, in the constellation Vulpecula. It was a continuous, precisely pulsing radio signal, with a period of 1.2272 seconds and a pulse width of 0.04 seconds. Jocelyn Bell Burnell and Antony Hewish were the first to hear it. It did not sound like the usual static from outer space. It sounded as if it were a purposeful, artificially constructed alarm, like a radio beacon. They named it LGM-1, meaning "Little Green Men," and cautious speculation as to its source began at once.

Astronomers Sir Fred Hoyle and Thomas Gold immediately interpreted the signal as radio frequency radiation from a very rapidly spinning neutron star, the small but extremely heavy remnant from a supernova explosion. The radio astronomers had not discovered a beacon from another race of beings. They had discovered a new type of thing in the cosmos, the pulsar, and it was designated PSR B1919+21. Hewish, at least, won the Nobel Prize in physics for it in 1974. Hoyle protested loudly that Burnell, as joint discoverer, should have been included in the prize.

By 1971, NASA was interested in searching for extraterrestrial radio communications, and Bernard M. Oliver and John Billingham started

---

214 Electromagnetic energy is released into the nuclear-powered universe in a very wide spectrum of frequencies and energies, from infrared light to extremely high-powered gamma rays. This spectrum of available photons from the sky has a peak right in the middle of the visible light spectrum, right at the color yellow-green, and falls off to lesser intensities as the frequency trends higher and lower on either side of the peak. Yellow-green happens to be the most sensitive part of the narrow visible light spectrum for our eyes. Human vision seems to have evolved for us to see as much of the universe as possible without the use of electronic instrumentation. Just look up on a dark night and see structures that are thousands of light years away, without the use of a 305-meter radio telescope antenna.

Project Cyclops, and *A Design Study for a System of Detecting Extraterrestrial Intelligent Life* was published by the Ames Research Center in Silicon Valley, California. It made an impact on the large-antenna community of radio astronomers, and it was read with interest. Billingham went on to found SETI, the Search for Extra Terrestrial Intelligence, in Mountain View, California.

Six years passed, and SETI scientists discovered a lot more pulsars. On August 15, 1977, Jerry R. Ehman was working on a SETI project at the Big Ear radio telescope at the Perkins Observatory in Delaware, Ohio. The antenna was pointed northwest of the globular cluster M55, near the Chi Sagittarii star group in the constellation Sagittarius. The radio signal was computer processed and displayed graphically on a video terminal, and it always looked about the same: mostly blank spaces, with clusters of ones, some twos, threes, and an occasional four, all indicating a radio signal intensity. Out of the darkness came an unusual signal at 10:16 that evening, lasting for 72 seconds. It was a tight cluster on the monitor screen, making the alphanumeric sequence "6EQUJ5." Ehman had never seen a signal quite like it. He printed it out, circled the cluster, and wrote "Wow!" next to it.

There was no information coded on the signal. It was simply a continuous radio wave, and the alphanumeric readout indicates only the strength of the signal as it gradually rose to a peak and then sank back to background level. It looked as if the fixed signal source moved into and out of the sensitive focus of the fixed antenna as Earth slowly rotated. The Very Large Array radio telescope in New Mexico was also listening that evening, and even though it is much more sensitive than the Big Ear, it did not register a similar signal. Decades of looking for a repeat of the signal reception have yielded nothing. The Wow! signal was as close as SETI has ever gotten to a radio signal from extraterrestrial intelligence.

On the 35th anniversary of the Wow! signal in 2012, the Arecibo Observatory radio telescope in Puerto Rico transmitted a response into that same piece of night sky, consisting of ten thousand Twitter messages. The problem with that gesture is that the M55 globular cluster, from which the signal may have come, is 17,500 light years away. That means that when that signal was transmitted, humankind was in the Paleolithic age, they had just domesticated dogs, mastodons were dying out, and a long ice age had descended from the north. The response signal will arrive in another 17,500 years, and by that time at the very least we will have long forgotten what "twitter" means. Over interstellar or intergalactic distances, radio communications become completely unusable. If we were to transmit "Hi there!" to an Earth-like planet in a habitable orbit twelve light years away, it would take twenty-four years to get back a response of "What?" Using the Project Cyclops model, SETI may have been looking for a signal from extraterrestrial intelligence for the past forty-five years using the wrong technology.

Nothing can travel faster than light, and that includes any other electromagnetic radiation besides light, including radio waves, infrared radiation, ultraviolet radiation, X-rays, gamma rays, spaceships, cannonballs, thrown bricks, microscopic dust, loose neutrons, electrons, protons, ionized molecules, and basically anything that displays a physical presence. But, what about something as abstract as information? It has no mass and none of the physical constraints of solid objects. Can information travel faster than light? There is a possibility.

To understand and be convinced of the existence of superluminal communications (information transfer faster than the speed of light) requires a long, scary drop into the darkness at the core of nuclear physics, a place where rational, right-thinking scientists must abandon common sense and connection to the real world. It is a place where scientists must accept blind faith, at least in mathematics, as a guide to concepts that confound and amaze but turn out to be unwaveringly true. It is quantum mechanics. The explanation will involve spontaneous parametric down-conversion, quantum vacuum fluctuation, nonlinear photonic crystals, the three-polarizer experiment, quantum erasure, photon entanglement, and Rick Steenblik. Please hang on.[215]

---

215 I shall scrupulously avoid the use of "Schrödinger's Cat" as a method of dipping the laymen into the quantum world. Dr. Erwin Schrödinger, a revered quantum theorist who invented wave mechanics, concocted the cat example as a gedanken experiment to illustrate the superposition of quantum states, and it was a clever way to bring the phenomenon into the macroscopic world where it could be effectively touched and seen, but I and others have found that it has a semi-violent counter-effect on people who may not have majored in physics. They tend to zoom in on the poor, half-dead/half-alive cat and miss the abstract point. Just as I try desperately to explain that Schrödinger did not actually put a cat in danger, I fear that the angry villagers are starting to come up the driveway carrying pitchforks and torches. The cat experiment is covered specifically in many books, but not here.

Physics is fundamentally based on the work of Sir Isaac Newton in the seventeenth century, and it is named "classical mechanics." In general, classical mechanics is a set of mathematical rules that can predict the position, speed, acceleration, and trajectory of physical objects, from galaxies to dust particles, with accuracy as fine as can be measured. Using Newtonian classical mechanics to meticulously plan a trip, mankind can blast a rocket off the surface of the Earth, fly 50 million miles through outer space, and land softly on a target the size of a soccer field on the planet Mars.

It is impossible to beat the accuracy of classical mechanics until the sizes of objects to be moved are too small to be seen, as is the atomic nucleus. At this incredibly tiny scale, a lot of what we know about classical mechanics goes out the window. It turns out that many physical attributes, such as momentum, the brightness of light, or the volume of water, cannot be divided down forever. If a glass of water is cut by half over and over, eventually the volume gets so invisibly small that it cannot be further divided. Eventually all that is left is a single molecule of water, and when that is divided, it is no longer water. What is left are the components of a water molecule, which are an oxygen atom and two hydrogen atoms locked together as a hydrogen molecule. The water molecule is therefore the *quantum* of water, or the smallest possible quantity of water. Dividing the hydrogen molecule, which is the quantum of hydrogen gas, yields two hydrogen atoms, each of which is the quantum of the element hydrogen. Dividing a hydrogen atom yields a proton and an electron, which are subatomic particles. Dividing the oxygen atom yields an oxygen nucleus plus eight loose electrons, and dividing the nucleus yields protons and neutrons, which are subnuclear particles, or *nucleons*. Subnuclear particles may be further divided into component parts. It is at this quantum scale where certain parts of classical mechanics no longer apply.

Some classical specifications are modified to fit the quantum scale, such as the conservation of energy. Classical mechanics says that energy can be neither created nor destroyed, but quantum mechanics allows for energy to be converted to matter and for matter to be converted to energy, and the larger energy-matter sum is on most occasions conserved. Matter can be destroyed, as occurs in nuclear fission, but it shows up as an energy equivalent. Momentum is also conserved at the quantum level.

Electromagnetic radiation, which is observable by telescopes of all kinds, has some interesting, weird characteristics at the quantum level. Its physical presence has a dual-mode, or a *superposition of states*. It can be either a photon or a wave, depending on how you look at it. Photons are discrete puffs of energy, shot out of a light source like machine-gun bullets. The wave mode is a continuous, alternating magnetic and electric field, vibrating at a specific frequency and wavelength. In both modes, electromagnetic radiation travels at the speed of light. Measure it expecting a photon and it will be a photon. Try to find out that it is a wave and it will be a wave. It cannot be both at the same time, and until a sentient being turns on his instrument and points it at the light, it is neither and it is both. Until it is measured, its physical state is undetermined.

Information can be coded into electromagnetic radiation, which it is used for everything from local radio to fiber-optic digital service. Across the surface of the earth, a time lag due to the finite speed of electromagnetic radiation is hardly noticeable. For astronomical distances that consider the spacing between stars, the time lag kills the idea of using it for communications or data transmission.

Another property of electromagnetic radiation is polarization. The direction of vibration is either horizontal or vertical. In the wave mode, looking straight at the light source, the wave either undulates up and down or side to side as it approaches. In the photon mode, the particle either hops up and down or slides to and fro as it races forward. As is the case of the photon-wave traveling mode, the polarization (the direction of vibration) is undetermined until someone who knows what he is looking at tries to determine the state of polarization.[216] The two polarization states are superimposed until observed. It does not

---

216 How do we define horizontal and vertical? If, for example, the observer is floating in space, there is no up and down and no vertical and horizontal. In this case, hold a polarizing filter in front of the eyes and observe a polarized light beam. (The light has already been through a polarizer and has been assigned a discrete polarization.) Turn the filter until the beam is blocked by the filter and the light is no longer observable. Name it "vertical." Turn the filter 90 degrees in either direction. Define that position as "horizontal," and observe as all the light comes through the filter. The polarization state of the incoming light is, by definition, horizontal. The only difference between horizontal and vertical polarization is that 90-degree angle.

matter how long the light has been traveling or over what distance it has traveled. The light from the M55 globular cluster, which is 17,500 light years away, has an undetermined polarization state until someone holds up a polarizing filter and gazes at the light through it.[217]

The most intriguing quantum property of light is the *entangled state.* The mathematical equations that describe photons seem to allow two photons, traveling in different directions and even having different vibrational frequencies, to share a common quantum state, such as polarization. Moreover, that polarization is undetermined for both photons, no matter how far apart they are or how long they have been traveling, until someone sticks a Polaroid filter in front of one of them. At that instant, the polarization angle of both photons snaps into being. The equations indicate a zero delay between establishing the polarization of the photon traveling through the Polaroid filter and the other photon that is traveling alone. Time and distance are not part of the equation. There is no speed-of-light lag time. If the two photons are separated by 6 trillion miles, the length of a light year, it does not matter. Both assert the same polarization angle simultaneously.[218]

---

217 If you have ever worn a pair of polarizing sunglasses, then you have used polarizing filters. The filters that make up these glasses are set up to transmit vertically polarized photons. The usual sunlit scene as viewed through the glasses is shaded to one half the available photons. As they bounce off the scenery and into your eyes, the photons are randomly polarized, so exactly half of them turn out to be vertically polarized and make it through the filters into your eyes. However, the photons that reflect off the hood of your car are horizontally polarized, due to the fact that the hood is basically a flat, horizontal surface laid out in front of you. Those photons do not make it through the filters, so the polarizing glasses do the neat trick of saving your eyes from the glare of the sun on the hood. Edwin H. Land invented this type of filter in 1929, using iodoquinine sulphate crystals bound in a transparent polymer film. This polarizing filter, the J-sheet, was improved in 1938 by embedding pure iodine crystals in a polyvinyl alcohol. This formula was named the H-sheet. Land went on to develop the highly successful Polaroid instant camera in 1948. He made a fortune and then lost it all in 1981 after having developed the Polaroid instant movie camera. The home video cameras using VHS tape landed right on top of his instant movie camera introduction, and it was a solid wipeout for Polaroid.

218 As is the case of some simplifications used to make these explanations understandable, this is almost true. There are actually two modes of shared polarization between entangled photos: Type I and Type II. In Type I polarization correlation, the two separate photons have exactly the same direction of polarization, and they are deemed parallel. In Type II polarization, the two photons always become polarized in perpendicular directions. The angles are exactly 90 degrees out of phase. The polarization mode depends on what type of crystal is used to generate the entangled pairs.

This weird prediction from Niels Bohr's otherwise productive "Copenhagen Interpretation" of the physical world came to light in 1935, when the Gang of Three—Albert Einstein, Boris Podolsky, and Nathan Rosen (EPR)—jumped on this ridiculous entangled photon indication as a proof that quantum mechanics was at least incomplete and did not necessarily correspond to reality. Einstein had a word for entanglement: spooky! He immediately saw it as an implication that information coded into light polarization could travel faster than light, an action that was probably forbidden by his theory of special relativity. This presented as a paradox, at the least, and was seen as an indication that quantum mechanics was full of big, ragged holes. The EPR gang carried a lot of weight, and for decades, the existence of entangled photons was argued and debated, knowing that the credibility of the quantum mechanical world-model depended on consistent, accurate predictions.

The elegant EPR work from 1935 was blown out of the water in the early 1980s, while a lecturer at the École Normale Supérieure de Cachan near Paris, Alain Aspect, was working on his doctorate in physics. He managed to build a series of physical experiments in which two entangled photons, separated by an arbitrary distance, obviously assumed the same state of polarization when one of them was measured. Aspect's demonstrations proved that the weird Copenhagen version of reality is correct, while the orthodox reasoning of the EPR arguments is not. Reality had been redefined.

Discovering that two photons can be entangled was a big step, but finding a way to produce continuous streams of entangled photons, making them useful on a more practical level, would be another leap. By 1995, Dr. Paul Kwiat at the University of Illinois at Urbana-Champaign and others had built and tested optical setups that produce a pair of steerable beams of light in which a photon in one beam has a doppelganger in the other beam. Each pair of matched photons has an entangled, undetermined state of polarization, waiting to be purposefully snapped into either horizontal or vertical orientation as they sail off in opposite directions at the speed of light.

The technique used to make entangled photon pairs is *parametric down conversion*. Down conversion has been used in electronic work for decades to change the frequency of an incoming radio signal to a lower value. It is commonly used in cable and satellite television to reduce an extremely high-frequency microwave signal down to something that a television can handle, and it is used in home data systems to down-convert electromagnetic signals carried by cable and fiber-optical systems to make them readable by a Wi-Fi or Ethernet receiver operating at a lower frequency. For down-converting light, which involves frequencies far removed from the microwaves used in electronic systems, exotic methods are necessary.

An optical down-converter uses a powerful laser producing a tightly controlled beam of ultraviolet light. This beam is directed into a nonlinear photonic crystal, which absorbs each ultraviolet photon and instantly reemits it as a lower-frequency photon. A prime example of a down-converter crystal is the beta barium borate (BBO). Others include potassium dihydrogen phosphate (KDP) and lithium niobate (LN).

A photon's frequency is directly proportional to its energy, with Planck's constant as the scaling factor. If the photon is converted to one having a lower frequency, then the conservation of energy and momentum at the quantum level demands that the missing energy be accounted for. In a cable television the down conversion is justified by dumping the excess energy and momentum into the heat sink on the integrated circuit in the cable box. In the optical down-conversion crystal, the only way to even up the energy accounting is to make another photon. An ultraviolet photon thus turns into two red photons coming out the other end of the down-converter crystal. The two new photons retain the quantum features of the original ultraviolet photon. Every characteristic, including the energy, still adds up to that one, original photon, and yet it is now divided into two.

If an ultraviolet laser puts out a beam of photons vibrating at 1×10^{15} Hz (a wavelength of 300 nanometers), then each has an energy of 6.6×10^{-19} joules. If this incoming ultraviolet photon breaks evenly into two photons while blasting through a BBO crystal, that gives each new photon 3.3×10^{-19} joules, or a frequency of 0.5×10^{15} Hz (a wavelength of 600 nanometers, which corresponds to the color red). The wavelength or the frequency of visible light manifests as color.

Turning one photon into two lesser photons may seem simple, but making two entangled photons is a bit more complicated. Entangled conversion is a rare event, happening in only one out of a trillion down conversions in the appropriate crystal. It is a perfectly random event. It happens only if at the exact location in the crystal where a down conversion is in progress there happens to be a spontaneous vacuum fluctuation.

In the vast, empty vacuum that separates the subatomic particles making up the beta barium borate molecules and everything else that we consider solid, there exists a fundamental energy. In the quantum world, even absolute nothing has energy, and for a very short period of time the energy can become mass, for no other apparent reason than that it is mathematically allowed to. This vacuum energy and its spontaneous fluctuations may have been vitally important in the creation of the universe. It could also explain the current expansion of the universe and answer questions that we have yet to ask. For producing entangled photons, it is essential.

Fortunately, we can predict exactly where the two streams of entangled photons will exit the down-converter crystal, and the rest of the laser beam and its down-converted-but-not-entangled-photons are discarded. A new photon resulting from a down conversion of the ultraviolet laser beam is constrained to exit the crystal in a cone-shaped region, slightly off the axis from the input beam. Ultraviolet photons that do not participate in down conversion continue straight through the crystal. There are two converted photons per conversion, almost all of which are not entangled. Half of them are horizontally polarized, and half are vertically polarized.[219] The two down-converted photons are separated into two exit cones out of the crystal, with the vertical polarized ones slightly off the laser axis on top and the horizontally polarized ones slightly skewed on the bottom.

In circular cross section, the two exit cones intersect. They overlap, like a Venn diagram demonstrating two sets intersecting. The two outgoing lines where the two cones intersect define the trajectories of those down-converted red photons that happen to be entangled. One

219 The fact that these photons are sorted into vertical and horizontal polarizations upon emerging from the crystal means that their polarization has been measured and they are locked into their polarization states. The situation in which the scientist notes the two red cones coming out of the crystal, illuminating dust particles floating in the air of his supposedly dust-free lab, is enough to make it so.

goes left and one goes right, and these slightly angled paths can be diverted prismatically without measuring the polarization and spoiling the entanglement. The entangled red photons, lacking any defined polarization, hasten away in opposite directions at the speed of light.

By the time Paul Kwiat was publishing papers about making continuous streams of entangled photons, Rick Steenblik, recovering from our adventure with cold fusion, had resigned his research scientist position at the Georgia Tech Research Institute and formed a company, Chromatek, Inc., to exploit his patented 3-D optical system. Commerce rolled, and Steenblik enjoyed success.

Forever seeking a new technical challenge, Steenblik spontaneously decided to use the new, exciting work that described making entangled photons to develop a superluminal communications system. He formed a new company, Ansible, Inc. The name derived from a fictional machine that could send and receive messages instantly over interstellar distances, the ansible, appearing in Ursula K. Le Guin's science fiction novel from 1966, *Rocannon's World*. By 1997, he was working on a design and anticipating a patent application. The project was a deep industrial secret, code-named TAXI.[220]

One February morning, he came to my office at Georgia Tech, bubbling with quantum mechanical jargon, and produced a very large, heavy green notebook that came crashing down on my desk. Would I please review this document for technical accuracy? He was asking a lot, considering the nonexistence of spare time, but he had tweaked

---

220 In a recent light-speed communication with Rick Steenblik, I finally found out where the code word TAXI comes from. I quote him: "The TAXI name (not an acronym) arose as a code word. Shortly after I started doing serious research into the published methods of creating an entangled photon source, I had to go on a business trip for Chromatek. I was reading published papers on the flight and then wanted to talk with Mark about what I had learned. I called him from the taxi, en route from the airport to my destination, and (in the interest of maintaining the security of the ideas) I explained the entangled photon sources in terms of the taxi engine, drive shaft, transmission, wheels, and so on. Mark immediately understood the analogy and we had an excellent brainstorming session about generating entangled photons without ever uttering a word about photons, entanglement, lasers, or nonlinear crystals. Naturally, we had to code-name the project TAXI." Although the ansible concept first appeared in Le Guin's novel in 1966, Steenblik had never heard of it. He picked up ansible from Orson Scott Card's *Ender's Game*, which was introduced in book form in 1985. Mark Hurt was Steenblik's business partner.

that tangle of nerves that cannot help but respond to something new and weird.

Imagine a trip to Mars. It could take over a year, and the onboard personnel want to be able to have phone conversations with people back on Earth every day as they travel toward the red planet. Using focused radio beams, daily communications are certainly possible, but in a few days the delay due to the speed of radio signals (light) as the distance between the ship and Earth lengthens puts an irritating interruption into the conversations. It is like trying to talk to someone in Europe over an undersea cable. Just a 1-second delay is enough to make the two who are conversing trip over each other every time there's a pause, waiting for a response. By the time the Mars ship is only a fifth of the way to the destination, the delay has become an agonizing minute.

Now, imagine a trip to Mars with a small, automated communications-relay ship following the main ship, traveling at exactly half-speed. It is always behind the ship, halfway between it and Earth. On the relay are a pair of ultraviolet lasers, running continuously on a thermoelectric power source, like a Russian lighthouse. One is for ship-to-Earth transmission, and one is for Earth-to-ship transmission. Each laser assembly is sending out two beams of entangled red photons, with one aimed at the ship and one aimed back to Earth. Send/receive telescopes on the Mars ship and back on Earth keep the photon beams in sight. The ship-to-Earth laser is spaced forward, toward the ship, and the Earth-to-ship laser is spaced back, lagging behind and just a little bit closer to Earth than to the ship.

The telephone signal in the Mars ship is translated into digital ones and zeros, and is fed to an electronically stimulated polarization filter placed at the focal point of the ship's transmitter telescope. The stream of red photons coming in at the telescope, not quite yet having decided on a quantum polarization state, are snapped into a modulated pattern of polarizations at the filter position, with a vertical polarization being a one and a horizontal polarization being a zero. The transmitter telescope is slightly closer to the photon source than the receiver back on Earth, so it is the transmitter that establishes a polarization state in the photons.

Immediately after having been given a polarization by the electrically flipped filter, the entangled photons are picked up on Earth by its receiver telescope and are focused onto a fixed, vertical polarization filter. The red signal from the relay ship blinks on and off, as the now polarized photons either slide through the filter or are stopped by it. A blink on means a one, and a blink off means a zero. Communication between the Mars ship and Earth is established, and there is no speed-of-light delay. The entangled photons respond instantly to the measurement of polarization using previously established beams of light. It doesn't matter that it may have taken several minutes for the photons to travel. The fact that an entangled pair exists simultaneously at the transmitter and the receiver end makes all the difference.

To send reply signals back to the Mars ship from Earth, the other photon source, the one that is permanently positioned slightly closer to Earth than to the ship, is used in the same way. If that is insufficiently dramatic, then imagine a robot mission to a star that is twelve light years away. Using advanced ion drive, the automated craft accelerates to a respectable speed, but it is still going to take twenty years or so to make it to the distant planetary system and start mapping the area. What makes the mission possible is that there are no humans on board who have to be kept alive for forty years inside an enlarged telephone booth. Back on Earth, the ship is being flown like a drone, complete with a television camera and even hellfire missiles, if hostility is expected. Having an entangled photon relay ship trailing behind is the key to interstellar exploration.

Steenblik walked outside one clear night and picked a star out of the heavens. He held up a Polaroid filter and flipped it vertically and then horizontally several times. He wanted to be the first person on Earth to send an entangled photon signal to a receiver somewhere out there, using the supposition that the extreme, continuous photon load out of a star produces, by quantum probability, a number of entangled photons per second that are bound to hit Earth right before the entangled ones are intercepted by a planet that is slightly farther away on the other side of the star. If someone on that planet were searching for a digital polarization with a telescope, then a communication had been achieved.

An intriguing idea, is it not? If only it would work. There is one fatal flaw in the setup, and it is a big one. When a Polaroid filter is presented to an entangled photon with an undetermined state of polarization, it does not cause it to be polarized in the direction of the filter orientation. It *finds* the polarization, and it can be either horizontal or vertical. The quantum polarization of a photon is utterly random, as only a quantum property can be. (Although things can certainly *seem* random, nothing outside a quantum property actually *is* random.) Upon being impelled to reveal its polarization, a photon of undetermined state effectively tosses a coin. Heads it is horizontal; tails it is vertical. This is why a Polaroid filter blocks out half the light, and it makes a good sunglasses lens. On average, 50% of all photons coming directly from the light source are vertically polarized, and 50% are horizontally polarized. When you look through polarizing glasses at a light source, only half the available photons make it through the filter.

The sender and the receiver in the spaceship communications system both have the exact same sequence of random numbers, available at both ends with zero speed-of-light delay, for what it is worth.[221]

Maintaining focus, Steenblik continued the investigation with research into current developments, experimentation, and contemplation. Flipping through the pages of my green, not-to-be-reproduced notebook I found an inserted envelope labeled "3 POLARIZERS." Inside were three Polaroid filters, cut from cinematic 3-D glasses. He had marked the direction of polarization on each filter with a pen. Steenblik had run across one of the simplest, most mysterious, and least understood puzzles in all of optical physics: the three-polarizer experiment.

Buy or borrow three plastic Polaroid filters. Hold one of them up to a light source, like a desk lamp, while looking through it. The light will look a bit dim. That's because, now that you are measuring them, the photons are at randomly selected polarization angles. No matter how you hold the filter, half the photons will make it through the

---

221 A good technologist can make a silk purse out of a sow's ear, making use of the fact that one system's fatal flaw is another's feature. The simultaneous, continuous stream of digital random numbers at either end of the communication is being used for extremely high-speed banking solutions in which random numbers are needed for encryption. Using fiber optics and a centrally located source of entangled photons, two banks separated by hundreds of kilometers can have the same random encryption key appear at both ends, instantly. The numbers are completely random, and are not generated by a crackable algorithm. It is the first practical use of entangled photons. Einstein would find it irritating.

filter at an angle that it defines as "vertical" while the other half tries to go through 90 degrees out of phase and is captured by the filter, never making the crossing to your eyes. Now, take another filter, and put it in front of the first one. Rotate it 180 degrees, or one half of a full turn. You will watch as the second filter goes from transparent to opaque. At the opaque angle, the filter is set to the horizontal position, so you have one vertical and one horizontal filter in series, and nothing gets through. The first filter sets the previously undetermined polarization angle, and the half that are vertical make it through the filter. Everything coming through is therefore set to vertical, so there is no way it can make it through a horizontal polarization angle and the stack of two filters goes black. (Depending on the quality of your filters, the blackness may not be absolute.)

So far, everything is quite logical and simple. Now, take the third filter and slip it in between the two previously positioned filters, at a 45-degree angle. Presto! The filter pack goes clear, and the light source is now perfectly visible through the stack of three filters. The first one is vertical, the second one is at 45 degrees, or at an angle halfway between vertical and horizontal, and the third on is horizontal. You are actually looking at a light source dimmed to one quarter or 25% of its original brightness. Turn the third filter, in the middle of the stack, 180 degrees and watch as the light fades out to blackness and returns. Do it in front of a physicist. If he or she is not familiar with the experiment, the scientist will find it hard to believe what is happening and may scramble desperately for an explanation.

It turns out that photons have two independent polarization modes. There is the familiar vertical/horizontal mode and there is the lesser-known +45-degree/-45-degree mode. Wherever vertical is defined, there is another, completely different polarization with the photon vibration angle turned clockwise and set at the midpoint between vertical and horizontal. It is 45 degrees out of phase with vertical as one orientation, and perpendicular to 45 degrees as the other possibility, with the polarization axis turned counterclockwise and put halfway between vertical and horizontal.

A photon can be set by measurement to be either vertical/horizontal or +45 degree/-45 degree. It *cannot* be both at the same time.

When those photons make it through the first Polaroid filter, their polarization angles have been frozen in place by the act of measurement. All the photons striking the first filter are defined as either vertical or horizontal, and half of them make it through. The intensity of the light source is 50%. Next, the vertically defined photons hit a filter that is canted at 45 degrees. This filter is neither vertical nor horizontal. Passage through the filter is not defined, because there is no way for the photon to be in both polarization states at the same time.

Not having a logical path through or not through the filter, the quantum state of polarization is *erased*. The state of the photon's quantum polarization is reset to its premeasured value, and we start over from scratch. The photons proceed to the third filter, but there are still only 50% of them. Half of them were irretrievably lost in passage through the first, now canceled, measurement.

The photons with newly undefined polarization states proceed through the horizontal filter, and they are redefined as either horizontal or vertically polarized. Half of them make it through the filter. The result is a clear view of the light source, cut down to 25% of its original intensity. Odd as it may seem, the light will go through a stack of three filters, but not through two of them. It does not matter whether the third polarizer, in the middle, is set for +45 degrees or -45 degrees. The effect is the same either way.

This three-polarizer phenomenon has been proven to give the same paradoxical results using entangled photons as it does with single photons in repeated experiments in multiple quantum-optics laboratories.

Steenblik's inspired solution to the problem of not being able to send a meaningful series of ones and zeros faster than light using vertical/horizontal polarization was to use the quantum erasure phenomenon. Instead of flipping a vertical/horizontal filter between the two orthogonal states to make ones and zeros, this modified scheme uses a fixed vertical polarizer at the source of entangled photons pointed at the receiver, presetting both the receiver and the transmitter photons to have either horizontal or vertical polarization. At the faraway location of the transmitter, set slightly closer to the source than the receiver, ones and zeros are made by introducing a 45-degree polarizer in and out of the beam right in front. With the 45-degree filter in the beam, the quantum polarization is erased and a one is being sent. With the filter out of the beam, it remains defined as vertical/horizontal, and a zero is being sent.

At the receiver end, the beam is viewed through a vertical Polaroid filter. If the 45-degree filter is out of the transmitter circuit, then each photon has a 50% chance of making it through the filter. If the 45-degree filter is inserted into the transmitter circuit, then each photon has a 25% chance of making it through the filter. The difference in probabilities is interpreted as ones and zeros, the binary numbers.

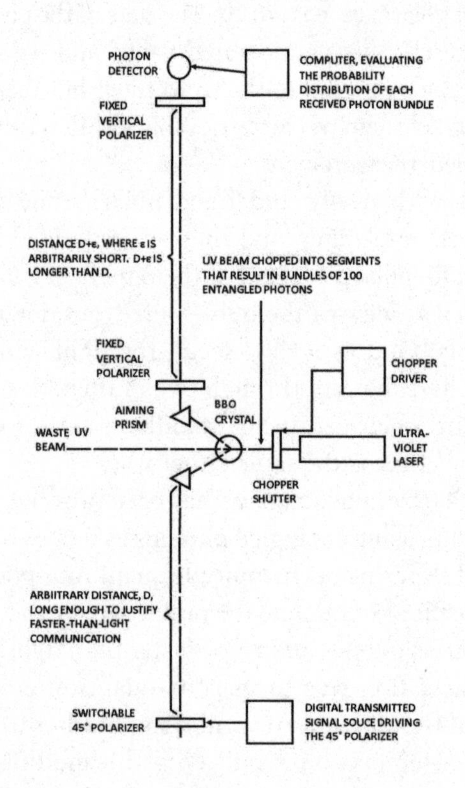

This schematic diagram shows basically how the faster-than-light communication system works using entangled photons. The problem of random polarizations from entanglement may be solved using quantum erasure. The signal is sent with the beam to the transmitter and the beam to the receiver both vertically polarized. To send a binary "1," the vertical polarization is erased by switching on a 45-degree polarization into the transmitter beam. The received beam responds instantly to the erasure, regardless of the distance.

The enhanced technique to Steenblik's design is to evaluate packets of multiple-sent photons, instead of evaluating one photon at a time. The beam of entangled photons is chopped into bundles of many photons per bundle. The number of photons per bundle is counted at the faraway receiver end. In a standardized bundle of 100 photons, if approximately 25 make it through the receiver filter, then that is a one. If 50 photons are counted in a bundled interval, then that is a zero. Send a powerful stream of photon bundles across twelve light years of space and before long you have downloaded the latest Alvin and the Chipmunks movie at a speed that defies relativity.[222]

It is true that this method may cause instant agreement between the sender and the receiver as to the states of the photons, regardless of the distance separating the two, but the communication is not exactly instantaneous. There is a slight delay as the photons in a packet bundle are counted, deciding whether a one or a zero has been sent. There is also a speed-of-light delay introduced by the fact that the transmitter must affect the photon stream first, before it hits the receiver. The receiver stream must be longer than the transmitter stream. For interstellar distances, precise positioning of the entangled photon source becomes problematic, and uncertainty can introduce a large length offset between the transmitter and the receiver. This offset length adds a receive delay equal to the length of time it takes light to traverse the offset. It will be grand if such practical considerations are the only thing standing between humanity and superluminal communications.

This design for a superluminal communications system is a handsome piece of logic, but will it work? Perhaps. The microscopic

---

222 I have described the first of four modes of superluminal communication claimed in Richard A. Steenblik's US patent number US 6,473,719 B1, *Method and Apparatus for Selectively Controlling the Quantum State Probability Distribution of Entangled Quantum Objects,* applied for on January 11, 1999, and awarded on October 22, 2002. Steenblik's earlier patent of the same name, US 6,057,541, is referenced in Mark John Lofts's US patent application, US 2006/0226418 A1, filed on November 10, 2004, *Method and System for Binary Signaling via Quantum Non-Locality.* Lofts admits that Steenblik's method will work, but he claims that his is less unwieldy. The abstract of this patent is difficult to comprehend. Be warned: because a scientific concept is granted a US patent does not necessarily mean that it works. Remember, T. G. Hieronymus was granted US patent number 2,482,773 for his crazy material analyzer (see Introduction). Having been granted a patent does, however, mean that the invention does not use or imply the use of cold fusion.

causality postulate of axiomatic quantum field theory seems to imply that sending data faster than light is impossible, but such orthodox quantum theories have been wrong before. Before he could absolutely prove the concept with elaborate experimental setups, Steenblik had to lay Ansible aside and put food on the table with more immediately lucrative optical inventions. Faster-than-light communications is a fundamentally important concept, but that does not mean that it is profitable. Meanwhile, other scientists have begun to dip a toe into the ocean of entangled-photon transmission and reception techniques.

Dr. Anton Zeilinger, an Austrian quantum physicist at the University of Vienna, has conducted some large-scale related experiments, such as achieving quantum teleportation over 144 kilometers separating two Canary Islands. He was the first to implement quantum cryptography with entangled photons in 1999. He has demonstrated communications by entangled photons using both free-space transmission and fiber optics across the Danube River, and later between those same Canary Islands. His dream is to put a source of entangled photons into orbit in an Earth satellite and prove that the superluminal methods obviously communicate faster than light.

In the past few years, the SETI Project has branched out its message-receiving modes to include optical telescopes, looking for signals encoded into laser beams.

In parallel with the problem of interstellar communications, as beliefs are trampled, antennae are aimed, and nonorthodox quantum mechanics is plumbed, there was no way to prevent the next step from having started. The effort may be premature, but design studies of interstellar transportation devices have been conducted since the beginning of the Space Age.

The American physicist Robert W. Bussard first found work in 1955, when he moved to Los Alamos, New Mexico, and became one of the Rover Boys, designing a series of nuclear-fission-powered rocket engines. This team eventually designed and tested the remarkable Kiwi engine, which was intended for use sending humans to Mars. In 1960, just as the Kiwi design was beginning to look practical, Bussard branched off on his own to design a nuclear-fusion engine. His designs were so interesting, he was not ordered to stop to focus on the Kiwi project.

The Bussard engine is intended for interstellar travel, powered by hydrogen fusion and using fuel scavenged from the immense vacuum of outer space. Hydrogen is the most common substance in the universe, and a large percentage of it exists as ions (naked protons) just floating between stars. Bussard's spacecraft concept scoops it up with a large, forward-looking funnel. Hydrogen caught in the funnel is directed to a collection hole at the center by magnetism and is conducted to a continuously operating fusion power reactor. The highly energetic exhaust of the reactor is used as a rocket engine, directed out the back to accelerate the ship forward. This now famous design is called the Bussard Interstellar Ramjet.

As seems the case with all interstellar transportation system designs so far, there is more than one intractable problem with the Bussard ramjet vehicle. Plain hydrogen does not fuse easily into heavier elements, which is the process by which fusion creates energy. In fact, the fusion cross section of this reaction (the probability that it will happen) is immeasurably small. The only reason the proton-proton fusion idea works at all in a reactor the size of a star is that the reactor is typically hundreds of times wider than Earth. There is no hope of fusing two hydrogen ions together in a vehicle that is built to weigh as little as possible.

Confronted with this criticism, Bussard countered by saying his reactor would use the carbon-nitrogen-oxygen cycle to fuse hydrogen into helium. This "CNO cycle," as proposed by Dr. Hans Bethe of Manhattan Project fame, uses the three heavier elements as reusable catalysts in a complex star-based reaction. Again, the mass required to create minimum environmental conditions for such a reactor is of stellar proportions. CNO cycle fusion has never been accomplished in an Earthbound experiment.

The estimated density of loose protons in interstellar space has more recently been downgraded, and it will take a much larger collection funnel to fuel the reactor than was originally estimated. Assuming that these protons are standing still in space, they will crash into the collector and must be accelerated to match the speed of the ship. The calculated drag produced by having to continuously bring the fuel up to speed may exceed the possible thrust from a fusion engine, if such an engine were possible. Some protons in space are traveling in random directions at near relativistic speeds, in which case they will go clean through the collector funnel and out the other end of the spacecraft without participating in propulsion. Critiques have not even approached the question of how a ramjet that can only produce power speeding forward is supposed to slow down and stop when it reaches the objective. In interstellar space, the only way to stop is to accelerate in the opposite direction with enthusiasm equal to that which achieved forward speed. The ramjet cannot do that.

Bussard passed away in 2007 in the middle of efforts to gain further funding for his Bussard Polywell fusion power plant, which had been in development under a U.S. Navy contract since 1994. Work on this reactor continues to this day at the EMC2 Fusion Development Corporation, which is soliciting for tax-deductible donations. Net power has yet to come forth from the Bussard Polywell fusor.

In 1933, Philip E. Cleator founded the British Interplanetary Society (BIS) in Liverpool, England, to promote, research, and speculate about ways to get people off the planet. Although its creation was preceded by American, German, and Soviet societies of similar bent, it is the only one that was not absorbed into government stewardship and survives to this day as a nonprofit, privately held organization. Finding itself unable to do physical experiments with rockets like the rest of the societies due to the British Explosives act of 1875, the group was free to imagine atmosphere-escaping rocket ships with no practical constraints or dangerous chemicals. The concept of noisy, disrupting rockets was frowned upon in England before World War II, and British rocketry research literally did not get off the ground.

That lack of experimentation did not prevent detailed design work and mission planning. By 1938, the BIS had put together a Moon-landing trip, complete with a parachute landing back on Earth. The boost vehicle was a multistage affair, to be built using thousands of small solid rockets, each having approximately the energy density of a road flare. The Moon-exploring individuals were protected from the Sun bearing down in a zero-atmosphere environment by a tarp-like outer garment resembling a rain poncho. It was a bold, exciting exercise, designed to spur interest in outer space exploration.

By 1973, the BIS had moved beyond the solar system and began planning for an unmanned interstellar probe. The effort was named Project Daedalus. A criterion was that it must be able to reach the system surrounding Barnard's Star, 5.9 light years away, within a human lifetime. At that time, the only known or feasible way to move in outer space was using a reaction engine, blowing a high-speed exhaust out the back of the vehicle. The design team, led by Alan Bond, reasoned that it would have to be a fusion-driven rocket. Given the maximum attainable speed using rockets as calculated by the team, 12% of the speed of light, it would take 50 years to reach Bernard's Star.

The star-probe would have to be assembled in orbit, as it would weigh 54,000 metric tons, with 50,000 metric tons of that being fusion fuel consisting of small, cryogenically frozen pellets of a deuterium and helium-3 mixture. The first-stage engine burns for 2 years, accelerating the probe to 7.1% of the speed of light. The first stage is cut free, and the second stage starts up, firing for 1.8 years and increasing the speed to 12% of the speed of light. After second-stage-engine shutdown, the probe coasts for 46 years, then blows past the objective, Bernard's Star, doing 22,320 miles per second. There is no fuel left for braking action, but telescopes on board will automatically survey the region, quickly, and radio the recorded data back to Earth. The signal will travel 5.9 years.

The design was completed in 1978. There are some problems. The engine, Friedwardt Winterberg's inertial-confinement-fusion drive, uses multiple, focused electron beams directed at deuterium-helium-3 fuel pellets, one at a time. The deuterium-helium-3 fusion requires more effort and more ultrahigh temperature to initiate than does a deuterium-tritium fusion, but it yields 18.354 MeV per fusion, as opposed to 17.571 MeV for the easier reaction.

This type of energy-producing fusion using inertial confinement has never proven to be possible, despite forty-five years of intense, expensive research and experimentation at the Lawrence Livermore National Laboratory in California and other laboratories in Europe. Billions of dollars have been spent so far trying to make a tiny pellet of frozen deuterium-tritium fuse and produce net energy by aiming a few hundred trillion watts of focused beams at it. The deuterium-tritium fusion is the least difficult configuration to fuse. No joy yet.

A further complication is the fuel for the fusion engine. About 25 million kilograms of helium-3 is required. Helium-3 is rare and expensive. Industrial consumption and production of it in the United States is about 8 kilograms per year. That much helium-3 costs somewhere between $6 million and $120 million, depending on demand. It is collected from decaying stockpiles of tritium, which is still used in nuclear weapons.

Seeing this as a problem, the BIS team suggested mining it in the upper atmosphere of Jupiter using robotic factories supported by very large hot-air balloons. It would take about twenty years of mining to accumulate the fuel-load for Daedalus. The engineering problems to be solved for this plan plus the expense are mind-boggling.

Independent calculations using the maximum theoretical thrust of the Winterberg engine indicate that it would take 100 years to accelerate to the desired coasting speed, and not 3.8 years as planned, if only it were possible to achieve 12% of light speed using any type of rocket. These engineering problems, plus the small dividend of information resulting from a project that would require the undivided economic attention of the entire world, suggest that implementation of Daedalus is not likely. It is, however, a start, and the BIS team deserves recognition for having the nerve to propose it.

After thirty years, seeing that most of the Daedalus engineers had died or retired, BIS decided to reengineer the interstellar probe concept and announced a new initiative, Project Icarus, on April 4, 2009, at the United Kingdom Space Conference at Charterhouse, Surrey. The specifications for Icarus are similar to those of Daedalus, but the mission length has been drawn out to one hundred years, and fusion engines using something besides helium-3 are allowed.

Back in the United States, a NASA project was awarded to the U.S. Naval Academy in 1987 for a concept design of an unmanned interstellar probe. It was named Project Longshot. Its mission was to fly to and assume orbit around the nearest star, Alpha Centauri B. The Alpha Centauri system, which consists of three stars orbiting around one another, is only 4.37 light years away. The Longshot specifications are somewhat more practical than those for Daedalus, with a mission length of one hundred years and a maximum speed of 4.5% of the speed of light, but it too relies on the inertial-confinement-fusion engine.

A workable inertial-confinement-fusion reactor seemed right around the corner in the last decades of the twentieth century, if one read the glowing reports from Lawrence Livermore, but the idea of causing self-sustained, pulsed hydrogen fusion by making very tiny H-bombs seemed to run into a series of brick walls, and success no longer seems as guaranteed as it once did. The Longshot vehicle, unlike Daedalus, was to have an onboard fission power plant making 250 kilowatts to run the inertial-fusion lasers and the telemetry transmitter aimed back at Earth. The total mass of the vehicle, 396 metric tons, seemed reasonable, but the fuel-load would include 132 metric tons of helium-3. That was better than the Daedalus design, but helium-3 was still an impossibly exotic substance.

Project Breakthrough Starshot, a new initiative to send an unmanned probe to the Alpha Centauri system, was announced in New York City on April 12, 2016, by physicist and venture capitalist Yuri Milner, British cosmologist Stephen Hawking, and Facebook CEO Mark Zuckerberg. This project is true, outside-the-box thinking. It is, in fact, not in the same building with "the box," but for interstellar transportation ideas free thinking is appropriate.

The Starshot probe is to consist of an unmanned rocket boosted into high-earth orbit. The orbiting vehicle then releases one thousand "nano-spacecraft," designated StarChips, one at a time. Each StarChip weighs only a few grams and occupies only about 1 cubic centimeter of space. It automatically deploys a featherlight, 4-by-4-meter sail and awaits launch into interstellar space from an external influence.

The launch influence is several ground-based 100-gigawatt lasers, all aimed straight at the sail and giving it a jolt of a few billion joules of light energy for 10 minutes, which will accelerate it to interstellar coasting speed and send it on its way. A StarChip should make it to Alpha Centauri in twenty or thirty years, hastening along at 15% of the speed of light. Hazards include space-dust collision wipeout, which is why one thousand of them are launched sequentially. Statistically, at least one should make it. Each StarChip is equipped with a plutonium power source, computer, camera, sail deployment mechanism, and laser communications system with sufficient power to send continuous data back to Earth.

Design challenges will begin with making a plutonium thermoelectric generator that stays within the weight specification of a few grams. To make 10 watts of heat requires 20 grams of plutonium-238. To turn that heat into electricity requires a silicon-germanium thermocouple and an integrated circuit voltage converter. The heat-to-electricity efficiency of the best possible thermocouple generator is 7%, so a 10-watt plutonium-238 generator gives 700 milliwatts of electricity. Assuming that a 700-milliwatt laser beam, which is an expensive green laser pointer, will make it back to Earth from 4.37 light years away, the weight budget is already blown.

The deployment of several 100-gigawatt ground-based propulsion lasers raises a power issue. A large nuclear power plant running at full capacity will generate one gigawatt of electrical power. The largest solar power plant in the world, the SEGS in California, generates 0.354 gigawatts at high noon. The biggest wind farm in the world, the Gansu in China, supposedly makes 6 gigawatts on a windy day. Supplying power to these lasers will be expensive.

The ground-based lasers to be used in this project have unusually high power specifications. Lasers with beams in the trillions of watts have been built for use in inertial fusion experiments, but these devices can deliver a 1-nanosecond pulse exactly once. The propulsion lasers must deliver continuous power for 10 minutes. The most powerful industrial lasers capable of semicontinuous duty currently in use have power ratings as high as 25 kilowatts. To produce a sustained laser beam of 100 gigawatts, which is about 4 million times greater, will require active development. Note that the StarChip sail must have a 100% reflectivity of these laser beams, lest it absorb a few billion watts and quickly reduce to a superheated puff of individual atoms. Any laser-weapon engineer will be glad to complain loudly about the impossibility of aiming a high-energy light source into outer space through the 80 miles of turbulent air surrounding the Earth to hit something as small as an intercontinental ballistic missile. There is a lot of challenging work to be done for implementing Starshot.

In general, given the current state of physics, cosmology, technology, and space exploration budgets, no subject of interstellar travel being discussed now is remotely possible, including antimatter rockets (Project Valkyrie), wormholes, quark matter, or the Alcubierre drive. This only means that the new and future generations of scientists and engineers are in for an exciting ride.[223] If constructive communications with other civilizations in the interstellar community is established, then priorities and goals may snap into focus. You have it to look forward to.

Rick Steenblik now lives the good life on an island in Hawaii. I write books.

223 Bear in mind, when I was an undergraduate, the terabyte hard drive was utterly impossible.

CHAPTER 11

Conclusions

"The historian's mission: to bring out of chaos—more chaos."

—Robert Harris, *Fatherland*, 1992

IN THE SEVENTH GRADE, I read a lot of science fiction. I had a subscription to the pulp magazine read by scientists at the Los Alamos nuclear lab, *Analog: Science Fiction–Science Fact*, edited by John W. Campbell. I read all the classics of the genre in the school library, from writers Robert Heinlein to Alan E. Nourse. My imagination grew expansive, and I daydreamed of space exploration and encounters with faraway civilizations.

At the time, I developed a parallel interest in what were termed UFOs (Unidentified Flying Objects) or flying saucers. The flying saucer spacecraft had become a science fiction staple by then, being used for interstellar travel in such movies as *The Day the Earth Stood Still* and *Forbidden Planet*, but the configuration, a dinner plate with another plate inverted on top and an inverted cup topping it off, was not the invention of screenwriters. In this case, art seemed to be copying life. There had been hundreds of sightings of round, flat flying machines that were beyond our ability to construct. Beginning in the late 1940s, newspaper and magazine accounts of flying saucers were appearing often. I read every book on the subject of flying saucer sightings, recovered crashes, and human abductions in the Maud M. Burris Library in downtown Decatur, Georgia. They were cataloged in the basement, under Dewey Decimal 629.133. According to everything I read, we seemed to be on the brink of a solid contact with people from a distant star system, from civilizations more technically advanced than ours.

It turned out that people had been documenting strange sightings for a long time, possibly starting in the Old Testament, where the prophet Ezekiel describes a noisy but soft landing of a manned spacecraft, seeming to touch Earth on four, wheel-equipped struts. Through the centuries, the described machines always seem to be just beyond our ability to duplicate what they were capable of doing. In the nineteenth century, for example, "sky-ships" were seen only at night, from Boston to San Francisco, and they looked like elongated balloons with engine-driven propellers, always with a gimbal-mounted searchlight scanning the ground and blinding Earthbound observers. It was a technology that could be imagined in the nineteenth century, but it was just beyond an ability to own it.

By the 1950s, an unidentifiable flying machine could outrun our fastest jet pursuit planes, turn instantly, hover with a slight wobble, and shut down a car engine by apparently suppressing the flow of electrons in the ignition system. The interstellar explorers in the ship could paralyze a human being, collect data and fluid samples, and almost wipe out his or her memory of having been probed. As in the previous century, they seemed to come at night with their running lights on.

There were obvious hoaxes. In Atlanta in 1953, police officer Sherley Brown happened across an odd-looking burned place in the middle of the street. Getting out of his prowl car to have a look, he met three traumatized witnesses pointing to a body on the side of the road. They claimed that a flying saucer had landed there, aliens piled out, and they hit one with their truck. They could not stop fast enough. The others jumped back into the spaceship and blasted off, leaving that burn on the street surface. Officer Brown could still smell the smoldering tar. The victim looked very small, green, and dead.

The three witnesses, barbers Edward Waters and Tom Wilson and butcher Arnold Payne, together had all the expertise needed to make a dead rhesus monkey look like a Martian. Payne cut off the tail, Waters and Wilson removed the hair using a great deal of depilatory, and a soaking in green food coloring completed the setup. Multiple residents confirmed that a flying saucer had been buzzing around, and a veterinarian confirmed that the creature was not of this Earth. The U.S. Air Force was called to the scene.

The story fell apart quickly when two Emory University doctors, Herman Jones and Marion Hines, took a glance at the corpse. "If it came from Mars, they have monkeys on Mars," was the last word.[224]

The most famous and prolific hoaxer was George Adamski in Palomar Mountain, California. He made personal contact with an alien from Venus named Orthon on November 20, 1952, and he had a photograph of Orthon's spaceship to prove it. Close examination of Adamski's picture indicated that it was a chicken brooder. The three bright orbs underneath the ship that Adamski identified as landing gear turned out to be three Westinghouse lightbulbs used to warm young poultry. Still, Adamski sold 200,000 copies of his three books, *Flying Saucers Have Landed, Inside the Space Ships*, and *Flying Saucers Farewell*.

False sightings and mistakes by inexperienced observers were inevitable, but there were many credible sightings by Air Force pilots, military personnel, entire crowds, and even scientists. Clyde Tombaugh, the

---

224 This clumsy attempt at hoax was heard around the world, and the model for the flying saucer pilot became the small, green-colored biped that exists to this day in a jar of formaldehyde on a shelf in the Georgia Bureau of Investigation Laboratory. The signal from the first discovery of a pulsar was named LGM-1 from a first impression that it may have come from "little green men." The stereotype remained in place until the early 1980s, when it was replaced by "the gray," a thin, bald guy with large, black eyes and delicate hands.

astronomer who discovered the once-planet Pluto, gave a detailed report of having seen something strange in the desert sky one evening. Everyone could not be wrong.

The Roswell Incident of June 1947 was the mother of all flying-saucer stories. It had everything: a crashed spacecraft, dead aliens, secret transport of the remains to Hangar 18 at an airbase, and an obvious, massive government conspiracy to cover it up and deny that anything had ever happened. On July 8, 1947, the story hit the newswires. The Air Force had taken possession of a crashed flying disc on a remote ranch near Roswell, New Mexico.[225] The "flying disc" had been forwarded to the 8th Air Force at the Fort Worth Army Air Field in Texas.

The next day, the Air Force retracted the story, trying to suppress speculation that they had found an alien spacecraft. A modified press release identified it as a weather balloon that got caught in a downdraft and hit the ground on the Foster homestead. A ranch foreman, Mac Brazel, had noticed a field of odd-looking stuff while on his rounds. It looked like strips of rubber, aluminum foil, sticks, and paper. News of the find got back to the Roswell Army Air Field, home of the 509th Bomber Group that had dropped the A-bombs on Japan. A team was sent to clean up the debris and erase evidence of its having been on the ranch. The story of a crashed flying saucer was cooked up, perhaps too hastily, to hide whatever the weather balloon was doing. A tight lid was pulled

---

225 The term "flying saucer" was twelve days old when it was used in the *Roswell Daily Record* top story headline on July 8, 1947: RAAF CAPTURES FLYING SAUCER ON RANCH IN ROSWELL REGION. It was first used on June 26, 1947, in the page-two headline: SUPERSONIC FLYING SAUCERS SIGHTED BY IDAHO PILOT; "Speed estimated at 1,200 miles an hour when seen 10,000 feet up near Mt. Rainier," in the *Chicago Sun*. On June 24, 1947, Kenneth Arnold, a fire-extinguisher salesman from Idaho, was flying to Yakima, Washington, in a CallAir A-2 airplane, when he saw out the left window a string of nine shiny things flying around Mt. Rainier. They looked like "saucers skipping on water" glinting in the sunlight, and they seemed as if they were tied together on a string, moving like the tail on a Chinese kite. Seeing them go behind the mountain, he estimated the distance at 25 miles and saw enough detail to say that they were "crescent-moon-shaped," they had to be bigger than DC-4 airliners, or over 100 feet wide. Arnold thought they might be experimental military aircraft or possibly visitors from another world. Thus was born the flying-saucer phenomenon.

over it, and no further information was available from the government authorities.[226]

After a couple of years of intense research into eyewitness accounts of dramatic flying saucer sightings and encounters, it started to dawn on me, dimly at first, that science is not a democracy. It doesn't really matter how many people have reported observations if the observations make no physical sense. We don't take a vote to see if flying saucer attributes, such as antigravity, force fields, and speeds of thousands of miles per hour in the lower atmosphere, are technically achievable. Such features either exist or they do not, and they are judged on compatibility with existing proofs and the results of confirmatory research. There are many strange and even unbelievable phenomena that have been proven and re-proven to exist in the nuclear physics model of the universe, but the capabilities of your common flying saucer are not in this set.

It was possible for everyone who witnessed a flying saucer incident to be simply wrong. In fact, if every person on the planet Earth believes that we have been visited by aliens, that does not make it a correct assertion. Absolutely everyone can be wrong. I was at peace with this scrap of philosophy and was even willing to be proven

---

226 The weather-balloon cover story was weak, but it was the last word on the Roswell flying-saucer crash until 1980, when Charles Berlitz and William Moore rounded up ninety witnesses who had been there, interviewed them all, and wrote *The Roswell Incident*. In their research, the authors found that the debris field reported by the Air Force was only a touch-down point for the disabled spacecraft. It actually came to a stop miles from there, digging a trench and coming to rest against a hill. One of the crew was found hanging out of the opened hatch, and every one inside was dead on impact. This imaginative account was never mentioned in any story from 1947, but it bears a resemblance to another story from the Land of Enchantment in 1948 that was widely published. In March of that year, a large, disc-shaped spaceship, 99 feet in diameter, unsuccessfully attempted a controlled landing in Hart Canyon, New Mexico. The nearest town was Aztec. The Air Force quickly gained control of the crash site and transported the craft plus the sixteen dead humanoids found inside to Hangar 18 at Wright-Patterson Air Force Base in Ohio for analysis. A strange detail was that the crew was dressed in Edwardian livery, as if they had expected to land in 1890 and didn't want to stand out. Frank Scully wrote about the incident in his book, *Behind the Flying Saucers*, in 1950. A few years later, the incident was revealed as a hoax cooked up by Silas M. Newton and Leo A. Gebauer. The scheme was to sell a device that could detect oil, gas, and gold, based on technology found aboard the spacecraft that had crashed near Aztec. Both men began to serve time for fraud in 1953. The Air Force claimed to be unaware of a spacecraft crash, anywhere, ever.

wrong, but somewhere, deeply buried, there was a tiny detail. It was an unanswered question: what was so utterly secret about a weather balloon that the Air Force would fire off that crazy press release at the Roswell Air Field about a crashed flying saucer just to smother its actual mission?

On Wednesday, April 30, 1989, I was slumped over my desk at the Georgia Tech Research Institute, still taking heat in the slow cooldown from the cold fusion fiasco. People were calling all the time to discuss cold fusion, to propose another experiment, to debate the existence of cold fusion, to float a theory past me, to condemn the rigidity of science in general, or to convince me that I was wrong in my negative findings. I had allowed myself to grow weary of both cold fusion and telephonic communications.

In the midst of these tribulations, a visitor came to the Electronics Research Building, wanting to see me. As it turned out, he and I had sipped from the same fire hose. We had each earned a bachelor of science in physics at Georgia Tech. I replied to the receptionist's call saying that I would be glad to talk with him. I would be in the conference room. Please send him up.

He introduced himself as Clarence H. "Judge" Ellison, a retired industrialist and physicist, living in suburban Atlanta. He was tall, fair-haired, and getting on in age. That certain rearrangement of the nerve pathways from having done time at Georgia Tech was still there, and he had a twinkle in his eye as he unwound his story as can only a son of the South. When he escaped from school, just as World War II was ending, he had no trouble finding work. The Air Force had new types of airplanes, new types of weapons, and new enemies, and it needed new technical talent fresh out of school. He was hired to work in a top-secret project named "Mogul." The goal of Mogul was to detect Soviet A-bomb tests from a great distance. The Soviet Union was thought to be involved in a crash program to come to parity with the United States in nuclear weapons, and the fate of western civilization was at stake. Knowing when secret nuclear airbursts occurred was part of the counterstrategy.

It was vitally important that the Soviets not know that we were developing methods to secretly monitor their progress toward making atomic bombs. They had to think that their secrets were still secret, and unfortunately, it had been proven during the last war that our most sensitive activities were riddled with Soviet spies and an impressive intelligence-gathering system. Mogul would have to be as contained and secure as could be managed. All personnel would be thoroughly vetted and monitored twenty-four hours a day, and there would be cover stories on top of cover stories.

Detonating an atomic bomb in the air produces an enormous air-pressure spike, radiating spherically with the speed of sound, with the leading edge reflecting off the ground and causing a secondary shock wave. The effect is similar to that of dropping a depth charge in the ocean to wreck an enemy submarine, and in several decades the navy had developed special, low-frequency microphones, or "hydrophones," to detect such explosions and sounds caused by submarines. By seeding the ocean surface with a matrix pattern of sonobuoys equipped with hydrophones, connected by radio signals to a surface ship, the source of these underwater disturbances could be triangulated and the exact time and position of otherwise unseen activity could be derived. This developed surveillance system is SOFAR, SOund Fixing And Ranging.

The distances over which these sounds can travel is remarkable. In the ocean, which is a fluid having complex characteristics of temperature and pressure, sound channeling develops, in which low-frequency components of a shock wave can travel 3,000 miles. In Project Mogul it was reasoned that the atmosphere is also an Earth-spanning ocean, only it is made of air instead of water. Theory had it that there were also sound channels in the atmosphere, and it should be possible to pick up the low-frequency components of an airburst from 3,000 miles away. Instead of dropping floating sonobuoys on the top of the sea, they would send special microphones to the top of the atmosphere. Like the sonobuoys, they would be linked to a central receiver by radio signaling. The vehicles would be helium balloons, and the launching point would be Alamogordo, New Mexico, the forbidden place where in 1945 the first atomic bomb had been tested.

Although the principle was simple and it might even work, there were a few technical details to be worked out. Weather balloons, which were by 1948 being used to find the temperature, air pressure, and humidity in the unmeasured regions of the upper atmosphere, would typically rise until the balloon burst. The below-freezing temperatures would make the rubber balloon brittle, and in the low air pressure the helium filling the balloon would expand until the cold rubber reached the end of its elasticity. The Mogul balloons would have to have some sort of automatic altitude control mechanism.

To triangulate the position of a bomb detonation, multiple balloons in different locations would have to detect the blast simultaneously, and the position of each balloon would have to be known when an explosion signal was received from its radio transmitter. There was no Global Positioning System (GPS) in 1948, but the Air Force owned long-range radar systems, able to track airplanes hundreds of miles away by sending out pulses of tightly directed microwave radio waves and receiving back reflections off metal surfaces on the targets. The directional position of a target airplane was derived from the aimed direction of the radar antenna, and the distance was found by timing the returned pulse. Each balloon would therefore have one or more corner-reflector radar targets hanging from a cord, lightweight, made of aluminum-coated paper and sticks, and named "the kite."

The kite was three flat squares, interleaved at right angles to form a sort of inside-out box. However you looked at the kite, from the side, top, or bottom, you were seeing an inside corner, which will always send an incoming radar signal back exactly where it came from. Think about throwing a basketball against the wall in a rectangular room. Just hitting a flat wall, the basketball will not necessarily fly back into your hands, but if you aim for a corner, be it a ceiling corner or a floor corner, the ball will shoot right back into your cupped hands. That paper kite, which was about 4 feet tall, looked like a battleship on the radar display. A great deal of the microwave pulse sent out by the radar transmitter would be bounced back to the radar receiver, looking like a very big object.

The electronics package to be lofted by the balloons consisted of a navy ERSB (expendable radio sonobuoy), model AN/CRT-1. It was a cardboard tube, 4 inches in diameter and 30 inches long, with a 39-inch telescoping quarter-wave rod antenna sticking out the top. Inside the tube was a lightly constructed aluminum chassis on which was built a low-frequency audio amplifier driving an FM radio transmitter operating at a frequency selectable between 67 and 72 megahertz. An airplane orbiting below the approximate position of the Mogul balloon could pick up and record the signal from the AN/CRT-1 transmitter from 35 miles away, using an AN/ARR-3 receiver screwed into its air-transport rack. A radio operator would sit in front of it to twist the knobs and wear the headphones.

Judge Ellison's job was to figure out how to reengineer a hydrophone and make it sensitive to "infrasound" signals transmitted in thin air at frequencies below the threshold of human hearing. Together, Ellison's microphone, the AN/CRT-1, and its batteries weighed 17.5 pounds. In 1947, there were no transistors and certainly no integrated circuits. Today, an FM radio transmitter can be built on a tiny wafer of silicon, but in those days it required five vacuum tubes, a transformer, four coil assemblies, seventeen capacitors, seventeen resistors, and two very heavy carbon-zinc batteries, 1.5 and 135 volts, all connected together with insulated copper wires. It was a masterpiece of optimized design, and it could operate in the near vacuum at 100,000 feet above ground in subfreezing temperatures.

If a balloon goes up, then it must eventually come down, dropping the AN/CRT-1 and its microphone with it. If anyone were to discover it lying on the ground, pick it up, and say, "What's this?" they would have to shoot him. Any other part of the Mogul package, from the top balloon to the radar corner reflector, could be explained away as a weather-sampling expedition, but the fact that there was a sonobuoy up in the air gave away the entire Mogul project. That is why the tests were conducted in rural New Mexico, where the population density was similar to that of the lunar surface.

The test vehicle was not just a simple balloon with a sonobuoy attached underneath. It was a string of objects, 596 feet long, consisting of three radar corner reflectors, twenty-three polyethylene helium balloons (spaced at 20-foot intervals), four recovery parachutes, a radiosonde weather data transmitter operating at 74.5 megahertz, three automatic cutoff devices set to separate various balloons or ballast at set altitudes, eight aluminum tubes filled with sand ballast, the AN/CRT-1 sonobuoy with special microphone, and a couple of 3-inch metal rings for holding the thing down before letting it go.[227]

On the morning of June 38, 1947, the Project Mogul contractors, twenty-three men from New York University and the Watson Laboratories at Columbia University, arrived at Alamogordo to test two innovations for the balloon string. They had replaced the troublesome neoprene balloons with much stronger polyethylene units, and the plastic tubes containing the sand ballast had been replaced with aluminum. As is characteristic of such research, they learned something every time they sent up a balloon string and it failed, and the performance was improving incrementally. On June 4, 1947, they released test flight no. 4 and quickly lost sight of it as it climbed away.[228]

On June 7, events began to unravel at the Roswell Army Air Field, 60 miles from the launch point at Alamogordo. It was reported that the balloon string had come down on the plains east of the Sacramento

---

227 Ellison's infrasound microphone looked like a snare drum, about 2 feet in diameter, and it was supported with its axis vertical by three cords, right above the AN/CRT-1, which was mounted upside down with its antenna pointed earthward. It was almost the last object in the 300-foot string. Below it was the sand ballast, supplied with a barometric cutoff that would drop it if the balloon string dropped to a specific altitude. Certain balloons in the string would be automatically cut away at 40,000, 42,500, and 45,000 feet, indicating that the optimum operating altitude to catch the sound channel was between 40,000 and 45,000 feet. The batteries supplying power to the AN/CRT-1, four size D flashlight cells plus two 67.5 volt batteries, would only last for four hours. If Mogul wanted to continuously monitor the Soviet A-bomb testing, then at least two balloons would have to be released every four hours for several years. As soon as they had figured out how to keep a balloon up for more than four hours, the next problem was going to be extended battery life.

228 On May 29, 1947, flight no. 3 was released. It was a configuration duplicate of flight no. 4. The balloons apparently ran into high-speed wind at altitude and disappeared from radar heading north. Flight no. 3 was never recovered. Kenneth Arnold's sighting twenty-six days later, on June 24, 1947, of a string of nine shiny objects, flying as if they were tied together "like a Chinese kite tail," may be a description of the top segment of the Mogul balloon vehicle, blown all the way up to Mt. Rainier in the high-altitude jet stream. The top segment was separated from the rest of the string by an electrically activated explosive squib as it reached 45,000 feet. That portion of the vehicle string would have been nine polyethylene balloons, looking highly reflective and metallic at a distance. The top two would have been big, 1-kilogram balloons, separated by 36 feet of braided nylon rope ("lobster twine"). After another 79 feet of rope, a series of seven smaller, 350-gram balloons would have followed, each spaced at 20-foot intervals. If so, then Arnold's description of the speed and distance of the objects would have been misjudged.

Mountains. The balloons, still having some lift capability, had dragged the string for miles, shredding the fragile radar corner reflectors and spreading debris over a long stretch of ground. Personnel from Roswell were sent immediately to recover the sonobuoy and especially the "flying disc," which was the code name of Ellison's special microphone, before anybody stumbled over it.

I was stunned. Judge Ellison had just filled in the last piece of the puzzle of the Roswell Incident and closed the lid for me. But that was not what he came to talk about.

Project Mogul dried up in 1949, having pioneered polyethylene balloons, altitude maintenance by aerostat, and the infrasound microphone.[229] Its mission of bomb-test detection was taken over by high-flying surveillance aircraft having dust collectors to pick up fission debris released into the upper atmosphere by a test explosion. After working on high-vacuum systems at the American Instrument Company in Silver Spring, Maryland, in the 1950s, Ellison moved to Atlanta, developed a unique product, and started a fledgling industry. His company, which flourished back in the 1960s, manufactured cold-fusion neutron generators exploiting the odd properties of palladium. As I absorbed his tale, my head started to hurt. I had fallen not only into the maelstrom of flying-saucer conspiracies but into another cold-fusion wormhole as well. No one could say that working at Georgia Tech Research Institute was not interesting.

Ellison's patent, number 3,283,193, applied for in 1962, was for an "Ion Source Having Electrodes of Catalytic Material." His device, named the Activitron, would produce an impressive 2×10^{11} neutrons per second. He produced an eight-by-ten glossy studio portrait of this baby. It looked like an anti-aircraft weapon from a Flash

229 A remainder of the Mogul infrasound microphones exists today in the ground-based geophysical MASINT, Measurement And Signal INTelligence system, which is installed worldwide. On February 15, 2013, a meteor, coming in at a shallow angle above Chelyabinsk Oblast, Russia, exploded 97,400 feet overhead. About 1,500 people were injured and 7,200 buildings in six cities around Chelyabinsk were damaged by the shock wave. It was the equivalent of setting off a 500-kiloton atomic bomb high above the city. The blast was picked up by MASINT microphones all over the world, and the data from these measurements were used to calculate the size of the energy spike, which was exactly what the Mogul balloons were supposed to do if the Soviets tested a nuclear device after World War II.

Gordon episode. Set to roll on casters was a metal tower, about 4 feet tall, tapering from bottom to top, with an access door hinged on the front. Atop the tower was balanced a horizontal, chest-high ray gun, complete with a barrel made of polished stainless steel and a bulbous blunt end, ending in a half-sphere. The business end, where one was advised not to stand, was a water-cooled blank metal cap from which high-energy neutrons would stream when you threw the switch into the on position.

Ellison built a factory on Roswell Road, in what is now the service department for a Buick dealership, where he manufactured and sold twenty-three of these devices for use in identifying trace elements using neutron activation. Using Ellison's Activitron, it required no chemistry to analyze ore samples, bomb debris, things hidden in lead-lined boxes, or any unknown elements obscured in ways making identification impossible. The neutrons streaming out the end of the Activitron were captured by radiologically inert atoms in the sample, becoming radioactive species of the same elements, each readily identifiable by the energies of the resulting gamma radiation. The multichannel analyzers, developed in the 1950s, made it possible to identify gamma rays across their entire energy spectrum using energy-sensitive scintillation detectors. Even better resolution of individual energies could be found using the new lithium-drifted germanium crystal detectors perfected in the early 1960s. An Activitron could count the number of gold atoms in a shovelful of dirt or turn an aluminum oxide crystal into a blue sapphire, and it did not require a nuclear reactor to make the neutrons.

At the zenith of his machine's popularity, Ellison's company was bought by the Technical Measurements Corporation. He produced a photo of himself and a small crew, standing at the open tailgate of a Jeep station wagon. The hot end of an Activitron was pointing at the ground, where they were testing for traces of gold for the state geologist in, of all places, the Dawson Forest.

He sold these things all over the world. One wound up in Hungary in 1966, through a trade-restriction loophole in Austria, eventually to an obscure university called Lajos Kossuth. In March 1989, students there had needed a piece of palladium to try the cold fusion experiment, and they took one from the ion generator of the old Activitron, found languishing in a storage room. They used it to log the world's first confirmation of cold fusion. There was no follow-up publicity following this announcement, so I had assumed that fault was found with their experiment.

I was in an extremely cautious mood, but I had to ask the question. "What was special about the palladium in your Activitron?" I was not ready to grant the Hungarians a legitimate confirmation, but it did seem interesting.

"It was alloyed with a little bit of silver," he replied. "It kept the palladium from sputtering in the ion bombardment." Pons and Fleischmann, he thought, were going about loading the palladium the wrong way. In his machine, he would shoot deuterons toward tritium, using 103,000 volts of electricity, and neutrons would result from deuterium-tritium fusions.

To be struck by flying deuterons, the tritium atoms had to hold still at the end of the accelerator run. To accomplish this, the tritium atoms were embedded in the surface of a thin titanium disk. His description of the machine sounded like another version of the well-studied Cockcroft-Walton accelerator, except that the accelerating voltage seemed too low to induce fusion and thus neutrons.[230]

What made it interesting was the fact that he discovered that his machine would make neutrons with a reduced acceleration potential of 25,000 volts, which seemed impossibly low, and they streamed out at the tritium target end. The palladium cap on the ion generator was used as a controlled leak to dole out deuterons one at a time into the accelerated beam. Although the electrically driven particle-accelerator was sealed tightly and kept at a hard vacuum using pumps, a window at the front end of the long, thin chamber was made of Ellison's special

---

230 When Ellison delivered an Activitron to a middleman buyer in Austria, it was strictly forbidden to export an object containing radioactivity. The Activitron was completely inert except for the deuteron target, which was a 1-inch disk of titanium saturated with tritium, a radioactive isotope of hydrogen with a 12-year half-life. When the Austrians asked advice on how and where they could get the 10-curie tritium source for use in the new Activitron, to their shock Ellison pulled it out of his shirt pocket and laid it on the table. Yes, it was an extremely active source of radiation, but the beta rays produced by the tritium are too weak to make it out of a paper envelope. The radiation from the tritium-infused titanium was not only perfectly safe to handle, it could only be detected by dissolving it in a scintillator fluid. He had walked through the airport terminals with no fear of being stopped and searched for radioactivity.

palladium-silver alloy. Hydrogen atoms, including the deuterium variety supplied from a pressurized tank, were able to migrate through this metal window freely, as if it were not there. Ellison theorized that the individualized, nascent deuterium atoms in the gently accelerated beam improved the deuterium-tritium fusion cross section over the usual deuterium molecules, made of two atoms stuck together.

I asked where I might find one of these machines, that I could try it and see for myself. "I gave two of them to Georgia Tech," he replied. Efforts to find these instruments in Georgia Tech inventory were deeply disappointing.[231] Accounts of his Activitron's workings were very interesting, but it ultimately suffered from the usual fusion-reactor problem. The power required to run the vacuum pumps and make the high-voltage current far exceeded any energy recoverable from the neutron output. It made no net power.

My path through life intersected with Judge Ellison again in 1994. The Soviet Union had fallen apart, and scientists in Russia who had once worked in defense ministry research were scrambling for something to do. A team consisting of A.B. Karabut, Y.R. Kucherov, and I.B. Savvatimova had discovered cold fusion, submitting a paper to the refereed journal *Fusion Technology*, "The Investigation of Deuterium Nuclear Diffusion at Glow Discharge Cathode." Judge had closely studied the experiment and found it very similar to the effect he had observed back in the 1960s on his Activitron. He wanted to duplicate the Russian experiment. I could not help but agree to participate.

I built the reactor in my home shop, machining the parts for the ion chamber and the vacuum system on my metal lathe and supplying the vacuum pumps, high-voltage power supply, and the instrument rack from my personal hoard. I borrowed the neutron detection equipment from the Georgia Tech Nuclear Research Center, and Nancy Wood loaned me a brand new BF3 tube.

---

231 I finally found evidence of an existing Activitron in Sydney, Australia, and I made a trip there to find it, giving cold-fusion lectures at the University of New South Wales in Sydney and the University of Western Australia in Perth. The Activitron was not available for inspection. One Australian scientist got up and angrily claimed that an Activitron could not possibly work, and he had refused to sit through a demonstration of it. Another said it was being considered for use to irradiate luggage at the airport, detecting barium-based explosives by neutron activation.

The electrodes, spaced 5.0 millimeters apart in the reactor, were machined from pure molybdenum, with water-cooled stainless steel support structures. Ellison paid to have an insulating cap made of aluminum oxide, threaded to screw into the cathode structure and holding a thin disc of palladium against the molybdenum block. From that disk, a streaming out of neutrons would confirm that deuterium was fusing. A thermocouple was embedded in the cathode and wired back to a readout in the equipment rack so that we could monitor the temperature. Kucherov had indirectly specified 44 degrees centigrade.

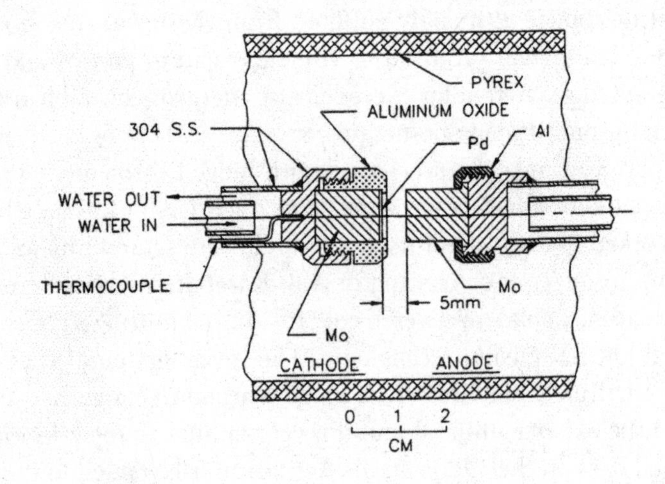

A cross-sectional view of our Kucherov glow-discharge cold-fusion reactor.

Ellison was right in the middle of assembling the system when one afternoon his front doorbell sounded. He was in the basement, working on something, and he had to put it down, climb the steps, and open the door. There stood a large man saying, "My car is stalled can I use your telephone?"

Ellison lived in Brookhaven, right west of Peachtree Road, in an elegant older house, and this fellow was not from around here. His request would not have been less believable if he claimed to be selling Girl Scout cookies. Ellison hesitated a second, and the stranded motorist took the opportunity to produce a pistol, point it at Judge's center of mass, and force open the door. His intent was to steal all he could carry and leave in his victim's car.

Judge was furious, but he had the presence of mind not to wrestle with the robust-looking individual over the gun. He was told to lead him to the bedroom, where he was tied up, hands and feet, using neckties, and left face down on the floor while the robber stuffed his pockets with everything shiny in the drawers. His father's Patek Philippe pocket watch went into one pocket, and the car keys off the top of the dresser went into another.

Ellison heard him leave the room and open the door to the garage; after a few seconds, the garage door motor started grinding. The miscreant wasn't very good at tying knots. Ellison ripped his hands free, stomped his feet out of their binding, grabbed his .22 caliber rifle out of the closet, loaded it, and ran to the sound. The robber was in the driver's seat of Judge's black Mercedes-Benz 300SD sedan, figuring out how to start the diesel engine. He looked up in time to see Ellison going for the down button for the garage door, and yelled, "Don't!"

"Too late," replied Ellison. The door motor backed up and the wooden garage door started down. The robber gunned the Benz in reverse, it smashed handily through the door in an explosion of splinters, shifted into drive, and screeched down the driveway to the street. Ellison emptied his rifle into the back of his car as it sped away, causing several problems with the paint.

Ellison's car was found hours later in the train-station parking lot on Peachtree. The robber was never identified. Ellison's wife, traumatized by the event, had the last word. "When you are through with that cold-fusion thing, we are moving somewhere else."

We changed plans. Instead of housing the experiment in my basement, we set up the equipment in Ellison's basement so that his wife would never be left alone in the house while he was doing cold fusion. We took our time and made sure that everything was correctly configured. Ellison did the copper plumbing for the water-cooled electrodes, and he built concrete-block shielding in case the thing was successful, started throwing neutrons, and activated his basement into radioactivity. It was an excellent exercise in scientific experimentation, and we did everything correctly, complete with the control experiment, of course, using a pressure tank of radiologically inert hydrogen gas as the inactive ingredient.

The radiation-counting background in Ellison's basement was established at about 28 events per hour, over a period of nine months, and we simulated an expected neutron flux out of the palladium anode using a 52.5 microcurie californium-252 neutron source. Finally, one night we ran the experiment. We used the best DC-704 silicon diffusion pump oil, liquid nitrogen in the trap, and a molecular sieve in the foreline to get as clean a high vacuum as we could. After an hour of pumping, the pressure was down to a simulation of deep space, at 5×10^{-8} Torr. Ellison carefully opened the valve on the pressure tank full of deuterium gas, adjusted the injection pressure, and addressed the control panel, slowly turning up the voltage across the electrodes to 600 volts, as prescribed by Kucherov. Monitoring the neutron count and keeping an eye on the electrode temperature, we awaited the commencement of fusion. Over a period of eight months, we ran the experiment with some variations nineteen times.

We submitted a detailed paper describing the experiment to *Fusion Technology*. It was refereed and published in the January 1996 edition: "An Investigation of Reports of Fusion Reactions Occurring at the Cathode in Glow Discharges." The last sentence in in the paper reads, "Therefore, if copious quantities of fast neutrons can be produced at the cathode of a glow discharge, it is concluded that these mechanisms and any others that are similarly independent of loading, are not responsible."

The ion chamber made a haunting blue glow, due only to the 600 volts across the electrodes, but in many hours of operation we were unable to find a single neutron over background. The Kucherov cold-fusion idea slid away into obscurity, but I predict that cold fusion by palladium hydride will be rediscovered sometime in the next fifty years. It never goes completely away, but it is possible for everyone who witnesses a cold-fusion incident to be simply wrong.

The Ellisons moved to Cashiers, North Carolina. I, as usual, soldiered on at the Georgia Tech Research Institute.

There are still many more atomic adventures that the world has yet to experience. There is most likely at least one forming up right now, probably but not necessarily on or near the surface of the Earth. Let us hope that it is a beneficial adventure, and not one of the destructive, unpleasant ones. Whatever it turns out to be, it will be deeply interesting, very technical, and out on the scientific shock wave. Be prepared.

Bibliography

Author's Note: Stories Told at Night around the Glow of the Reactor

1. Garrett, Franklin M., *Atlanta and Environs: A Chronicle of Its People and Events, Volume II*. Athens Ga.: University of Georgia Press, 1954.
2. Graham, W. W. III, and D. M. Walker. *Safety Analysis Report for the 5 MW Georgia Tech Research Reactor*. Atlanta Ga.: Georgia Institute of Technology, 1967.
3. Kalush, William, and Larry Sloman. *The Secret Life of Houdini: The Making of America's First Superhero*. New York: ATRIA Books, 2006.
4. *Atlanta Constitution*, April 18, 1915, April 20, 1915, April 22, 1915, Dec. 11, 1911, and Dec. 14, 1911.

Introduction: The Curious Case of the N-Rays, a Dead End for All Times

1. Blondlot, R., *N Rays: A Collection of Papers Communicated to the Academy of Sciences with Additional Notes and Instructions for the Construction of Phosphorescent Screens; Trans. J. Garcin*. London: Longman, Green and Co., 1905.
2. Campbell, John W. Jr., "Correction and Further Data on the Hieronymus Machine," *Astounding Science Fiction*, Aug. 1956.
3. Campbell, John W. Jr., "Psionic Machine—Type One." *Astounding Science Fiction*, Jun. 1956; Goodavage, Joseph, "An Interview with T. Galen Hieronymus." *Analog Science Fiction*, Jan. 1977.
4. Hieronymus, T. G., *Patent no. 2,482,773, Detection of Emanations from Materials and Measurement of the Volume Thereof*. Washington, DC: US Patent Office, Sept. 27, 1949.
5. Lagemann, Robert T., "New Light on Old Rays: N Rays." *American Journal of Physics*, March 1977.
6. Wood, Robert W., "The N-Rays." *Nature*, Sept 29, 1904.

Chapter 1: Cry for Me, Argentina

1. Adelberger, E. G., et al. "Solar Fusion Cross Sections." *Review of Modern Physics*, Oct. 1998.
2. "Argentina Lacks Development Resources even if in Theory its Atomic Tests are Possible." *New York Times*, April 1, 1951.
3. Balserio, Dr. José A. *Report About the Experiments of Dr. R. Richter, According to What Was Witnessed by Me During the Visit Made to the Atomic Energy Plant at Isla Huemul, from 5 to 8 September 1952*. Buenos Aires: National Atomic Energy Commission, 1952.
4. Balserio, Dr. José A. *Report of Dr. José Antonio Balseiro Referring to the Inspection Carried out in the Isla Huemul in September 1952*. Buenos Aires: National Atomic Energy Commission, 1988.
5. Cabral, Regis. "The Latin American Nuclear Debate." *Science Studies*, Vol. 4, 1991.
6. Goñi, Uki. *The Real Odessa.* Cambridge, England: 2002.
7. Hahn, Paul-J. and Rainer Karlsch. *Scharlatan oder Visionär? Ronald Richter und die Anfänge der Fusionsforschung.* Willstätt, Germany: Dr. Paul-Jürgen Hahn, 2016.
8. Hansen, James R. "Secretly Going Nuclear." *American Heritage of Invention & Technology*, Vol. 7 No. 4, Spring 1992.
9. Mahaffey, J. A. *Fusion*. New York: Facts on File, 2012.
10. Mariscotti, Mario J. *El secreto atómico de Huemul*. Buenos Aires: Editorial Planeta, 1985.
11. Mayo, Santos. "More on the value of Ronald Richter's work." *Physics Today*, March 2004.
12. Miley, G. H. *Fusion Cross Sections and Reactivities*. Urbana: University of Illinois, 1974.
13. Nestor, Lt. Col. Virgil N. (Air Attaché, Buenos Aires). *Dr. Ronald Richter.* Confidential report to the Director of Intelligence, USAF, Washington, DC, July 3, 1956.
14. Nestor, Lt. Col. Virgil N. (Air Attaché, Buenos Aires). *Status of Dr. Ronald W. Richter.* Confidential report to the Director of Intelligence, USAF, Washington, DC, April 1, 1957.
15. Oliphant, M.L.E., P. Harteck, and Lord Rutherford. "Transmutation Effects Observed with Heavy Hydrogen." *Nature* 133, 1934.
16. "Peron Is Irked By Atom 'Find' Disbelievers: Scientist Says He Knows H-bomb Secret, Told Not To Make It." *Washington Post*, March 25, 1951.
17. "Peron's Atom." *Time*, April 2, 1951.
18. Richter, Dr. Ronald. "Argentina Has No Atomic Bomb." *United Nations World*, July 1951.
19. Richter, Dr. Ronald. "Ball Lightning" (unpublished letter to the editor). *Scientific American*, June 5, 1963.
20. Richter, Dr. Ronald. *Improved Ram Jet Engine.* Air Intelligence Information Report IR-79-56, USAF, 1956 (declassified April 26, 1999).
21. Richter, Dr. Ronald. *Separate Sheets nos. 1 through 13.* USAF, 1956. These are believed to be Richter's biographical sketch detailing his work up until 1956, probably attached to IR-79-56 (declassified April 26, 1999).

22. Richter, Dr. Ronald. *Thermonuclear Exponential Experiments*. Air Intelligence Information Report IR-80-56, USAF, 1956 (declassified April 26, 1999).
23. Richter, Dr. Ronald. *Thermonuclear Propulsion System*. Air Intelligence Information Report AR-145-56. USAF, 1956 (declassified April 26, 1999).
24. Stong, C. L. "How to Build a Machine to Produce Low-Energy Protons and Deuterons." *Scientific American*, Aug. 1971.
25. "U.S. Nuclear Experts Doubt Peron Claim of New Way of Controlling Atom Energy." *Washington Post*, March 25, 1951.
26. Von Ardenne, Manfred. *Ein glückliches Leben für Technik and Forschung*. Zurich: Kinder Verlag, 1972.
27. Winterberg, Friedwardt. "Ronald Richter, Genius or Nut?" *Physics Today*, August 2003.

Chapter 2: AFP-67 in the Dawson Forest

1. Broad, William J. *Teller's War: The Top-Secret Story Behind the Star Wars Deception*. New York: Simon & Schuster, 1992.
2. Goodchild, Peter. *Edward Teller: The Real Dr. Strangelove*. London: Weidenfeld & Nicolson, 2004.
3. Baucom, Donald R. *The Origins of SDI: 1944-1983*. Lawrence: University Press of Kansas, 1992.
4. McCallie, Samuel Washington. *Gold Deposits of Georgia*. Atlanta, Ga.: Franklin Printing and Publishing, 1897.
5. Jones, S. P. *Second Report on the Gold Deposits of Georgia: Bulletin No. 19*. Atlanta, Ga.: Chas. P. Byrd, 1909.
6. Schneer, George H. and Lt. Colonel F. E. Redding. *"Project Little Eva": Operational Engineering Service Test of Radio Set AN/ARC-34*. Dayton, Ohio: WADC, April 1954.
7. Gantz, Lt. Colonel Kenneth F. *Nuclear Flight: The United States Air Force Programs for Atomic Jets, Missiles, and Rockets*. New York: Duell, Sloan, and Pearce, 1960.
8. Penner, S. S. *Advanced Propulsion Techniques*. New York: Pergamon Press, 1961.
9. Cottrell, W. B., H. E. Hungerford, J. K. Leslie, and J. L. Meem. *Operation of the Aircraft Reactor Experiment*. Oak Ridge, Tenn.: ORNL, December 6, 1955.
10. Roberts, C. T. *Nuclear Familiarization*. Marietta, Ga.: Lockheed Aircraft Corporation, 1957.
11. Anon. *Capabilities for Determining Bio-Medical Effects of Nuclear Radiation*. Marietta, Ga.: Lockheed Nuclear Products, 1957.
12. Butz, J. S. Jr. "Tests Simulate Nuclear Flight Environment." *Aviation Week*, 1959 (reprint).
13. Anon. "Soviets Flight Testing Nuclear bomber: Atomic powerplants producing 70,000 lb. thrust are combined with turbojets for initial operations." *Aviation Week*, Dec. 1, 1958.
14. Anon. *Review of Manned Aircraft Nuclear Propulsion Program: Report to the Congress of the United States*. Washington, DC: GAO, Feb. 1963.
15. Anon. *Nuclear Meteorology*. Atlanta, Ga.: Lockheed Nuclear Products, 1970.
16. Poore, H. L. *Decommissioning of Radiation Effects Reactor License R-86, Docket No. 50-172*. Marietta, Ga.: Lockheed-Georgia Company, April 12, 1971.

17. Greene, Ralph E. and Philip S. Baker. *Proceedings of Information Meeting on Irradiated Wood-Plastic Materials: Held at Conrad Hilton Hotel, Chicago, Illinois, September 15, 1965.* Oak Ridge, Tenn.: ORNL, April 1966.
18. Patterson, Reese. *Gamma Industries, Inc.: Home Grown Service for Isotope Radiography and Related Fields.* October 23, 2006.
19. Anon. *A nuclear-propelled airplane carrying a missile which can air-launch has a military punch similar to that of the missile-carrying, nuclear-powered submarine.* Marietta, Ga.: Lockheed-Georgia Division, March 6, 1959.
20. Stewart, J. W. *Infiltration and Permeability of Weathered Crystalline Rocks: Georgia Nuclear Laboratory, Dawson County, Georgia.* Washington, DC: Government Printing Office, U.S. Department of the Interior, 1964.
21. Anon. *Components Irradiation Test No. 7: 2N834 Transistors, 1N540 and 1N649 Diodes, S1752A Zener Diodes.* Marietta, Ga.: GNL, Sept. 28, 1964.
22. MacKallor, Jules A. *Aeroradioactivty Survey and Areal Geology of the Georgia Nuclear Laboratory Area, Northern Georgia (ARMS-I).* Washington, DC: USAEC and U.S. Geological Survey, January, 1962.
23. Selby, J. M., C. A. Willis, B. M. Bowen, and J. H. Edgerton. *Radiation Effects of a 10-Megawatt Reactor on Environs of AFP 67.* Marietta, Ga.: Lockheed Nuclear Products, 1959.
24. Boyd, R. M. *Air Tracer Experiments at Georgia Nuclear Laboratories.* Marietta, Ga.: Lockheed Nuclear Products, Feb. 1960.

Chapter 3: Inside Cold Fusion and Chapter 4: Good News and Bad News

1. Anon. "The Reported Conversion of Hydrogen into Helium." *Nature*, October 9, 1926.
2. Close, F. *Too Hot To Handle.* Princeton, N.J.: Princeton University Press, 1991.
3. Peat, F. D. *Cold Fusion; The Making of a Scientific Controversy.* Chicago: Contemporary Books, 1989.
4. Huizenga, J. R. *Cold Fusion; The Scientific Fiasco of the Century.* New York: Oxford University Press, 1992.
5. Taubes, G. *Bad Science; The Short Life and Weird Times of Cold Fusion.* New York: Random House, 1993.
6. Jones, S. E, E. P. Palmer, J. B. Czirr, D. L. Decker, G. L. Jensen, J. M. Thorne, S. F. Taylor, and J. Rafelski. *Observation of Cold Nuclear Fusion in Condensed Matter.* Provo, Utah: Brigham Young University, March 23, 1989.
7. Rafelski, J. and S. E. Jones. "Cold Nuclear Fusion." *Scientific American*, July 1987.
8. Mahaffey, J. A. *Georgia Institute of Technology Research Notebook No.930.* Atlanta, Ga.: GIT, 1989.
9. Wood, N. *Radiation Detectors.* Chicago: N. Wood Counter Laboratory, Inc., 1989.
10. Toon, John. "New 3-D Technique Overcomes Old Problems." Atlanta, Ga.: *Research Horizons; A quarterly publication of the Georgia Institute of Technology*, Summer 1992.
11. Beebe, G. *GTRI/GTRR COLD FUSION CONFIRMATION EXPERIMENT; Data Collection Form.* Atlanta, Ga.: GIT, April 8, 1989, 1632-2305.
12. Steenblik, R. *GTRI/GTRR COLD FUSION CONFIRMATION EXPERIMENT; Data Collection Form.* Atlanta, Ga.: GIT, April 9, 1989, 0000-0700.

13. Mahaffey, J. *GTRI/GTRR COLD FUSION CONFIRMATION EXPERIMENT; Data Collection Form*. Atlanta, Ga.: GIT, April 9, 1989, 0810-1529.
14. Acree, D. *GTRI/GTRR COLD FUSION CONFIRMATION EXPERIMENT; Data Collection Form*. Atlanta, Ga.: GIT, April 9, 1989, 1630-2245.
15. Beebe, G. *GTRI/GTRR COLD FUSION CONFIRMATION EXPERIMENT; Data Collection Form*. Atlanta, Ga.: GIT, April 8, 1989, 1632-2305.
16. Livesay, B. *GTRI/GTRR COLD FUSION CONFIRMATION EXPERIMENT; Data Collection Form*. Atlanta, Ga.: GIT, April 9, 1989, 2350-0708.
17. Steenblik, R. *GTRI/GTRR COLD FUSION CONFIRMATION EXPERIMENT; Data Collection Form*. Atlanta, Ga.: GIT, April 10, 1989, 0809-1530.
18. Steenblik, R. *GTRI/GTRR COLD FUSION CONFIRMATION EXPERIMENT; Data Collection Form*. Atlanta, Ga.: GIT, April 10, 1989, 1621-2300.
19. Mahaffey, J. *Possible temperature effect in a BF3 tube*. Atlanta, Ga.: GIT, April 18, 1989.
20. Anon. *Transcription of Cold Fusion News Conference at Georgia Tech 4-13-89*. Atlanta, Ga.: GIT, April 14, 1989.
21. Anon. *"Cold Fusion" News Conference Georgia Tech—April 25, 1989*. Atlanta, Ga.: GIT, April 25, 1989.
22. Mahaffey, J., D. Acree, and R. Steenblik. *Interim Report; STGC Project E904-031; "Cold Fusion Confirmation Experiments."* Atlanta, Ga.: GIT April 30, 1989.
23. Mahaffey, J., D. Acree, and R. Steenblik. *Interim Report 2; STGC Project E904-031; "Cold Fusion Confirmation Project."* Atlanta, Ga.: GIT, May 2, 1989.
24. Mahaffey, J. A. *Application for Principal Investigator Status for Acquisition and Use of Radioactive Materials; Procedure 9501, Revision 01*. Atlanta, Ga.: Neely Nuclear Research Center, April 14, 1989.
25. Ellison, D. H. *Ion Source Having Electrodes of Catalytic Material*. Washington, DC: United States Patent Office, November 1, 1966.
26. Spauschus, H. O. *FY89 STGC Budget Review*. Atlanta, Ga.: GIT, April 28, 1989.
27. Champlin, C. *Tech Researchers Renounce Cold Fusion Results as Error*. Atlanta, Ga.: The Technique, April 21, 1989.
28. Baker, A. "Cold Fusion: Red-Hot Scientific Brouhaha Is Cooling as Experiments Fizzle." *Fort Worth Star-Telegram*, August 12, 1989.
29. Toon, J., and G. Pinholster. *Georgia Tech Researchers Use Special Preparation Techniques to Reproduce Cold Fusion Experiments*. Atlanta, Ga.: Research Communications Office, April 10, 1989.
30. Browne, M. W. "Claim of Achieving Fusion in Jar Gains Support in 2 Experiments." *New York Times*, April 11, 1989.
31. Williams, J. "2 Groups: Fusion Test Duplicated." *USA Today*, April 11, 1989.
32. Parker, W. "Georgia Tech Researchers Confirm Fusion Experiment." *Gwinnett Daily News*, April 11, 1989.
33. Straus, H., and S. Sternberg. "Tech Reports Cold Fusion Re-Created: Tech Apparently Backs Controversial Finding." *Atlanta Constitution*, April 11, 1989.
34. Anon. "Cold Fusion' Duplicated; Texas A&M, Georgia Tech Scientists Report Success." *Asheville Citizen*, April 11, 1989.
35. Bishop, J. E. "Fusion Test Matched, But Mystery Persists." *Wall Street Journal*, April 11, 1989.

36. Anon. "Chemist Defends Claim of Cold Fusion Before Skeptical Colleagues." *Atlanta Journal and Constitution*, April 13, 1989.
37. Holtz, R. E. "Tech Scientists Retract Fusion Claim, Must Repeat Experiment." *Atlanta Constitution*, April 14, 1989.
38. Straus, H. "Tech Learns Risk of Shouting 'Eureka.'" *Atlanta Journal and Constitution*, April 16, 1989.
39. Browne, M. W. "Scrutiny of Fusion Experiment Produces Few Believers Among Physicists." *New York Times*, April 16, 1989.
40. Budiansky, S., and W. J. Cook. "In Hot Pursuit of Cold Fusion." *U.S. News & World Report*, April 24, 1989.
41. Anon, "New Findings Support Cold Fusion." *Chemical Week*, April 19, 1989.
42. Stipp, D. "Competing Theories About Fusion Devised by Two Scientists at MIT." *Wall Street Journal*, April 17, 1989.
43. Lemonick, M. D., Kane, J. J., and Thompson, S. "Fusion Fever is on the Rise." *Time*, April 24, 1989.
44. Anon. "Table-Top Fusion Looks Like a Parlor Trick." *Business Week*, April 24, 1989.
45. Straus, H. "Fires, Blowups: Unpredictability of 'Cold Fusion' Raises Red Flag." *Atlanta Journal and Constitution*, April 18, 1989.
46. Straus, H. "Scientists Skip, Stumble Along In the Chase After Cold Fusion." *Atlanta Journal and Constitution*, April 24, 1989.
47. Roundtree, P. "Tech Researchers Repeating Experiment On Cold Fusion After Qualifying Claim." *Georgia Tech Whistle*, April 24, 1989.
48. Stipp, D. "Georgia Group Outlines Errors That Led To Withdrawal of 'Cold Fusion' Claims." *Wall Street Journal*, April 26, 1989.
49. Broad, W. J. "Test to Confirm Fusion Was Flawed." *International Herald Tribune*, April 15–16, 1989.
50. Straus, H. "Tech Scientists Confirm Cold-Fusion Test Flawed." *Atlanta Journal*, April 25, 1989.
51. Straus, H. "Tech Scientists Confirm Error on Cold Fusion." *Atlanta Journal and Constitution*, April 26, 1989.
52. Lemonick, M. D. "Fusion Illusion? Two obscure chemists stir up a fascinating controversy in the lab, but new tests challenge their hopes of creating limitless energy." *Time*, May 8, 1989.
53. Hudson, R. L. "Pons and Fleischmann Withdraw Fusion Paper from U.K. Journal." *Wall Street Journal*, April 20, 1989.
54. Naj, A. K. "Corporate Scientists Seeking to Confirm 'Cold Fusion' Report No Successes So Far." *Wall Street Journal*, April 20, 1989.
55. Yoder, S. K. "Japan's Scientists join Race to Verify Fusion Experiment at Utah University." *Wall Street Journal*, April 18, 1989.
56. Begley, S., Harry Hurt III, and A. Murr. "The Race for Fusion." *Newsweek*, May 8, 1989.
57. Carey, J., W. E. Marbach, N. Gross, M. Maremont, and W. Symonds. "Fusion in a Bottle: Miracle or Mistake?; Frustration is building as proof remains out of reach." *Business Week*, May 8, 1989.
58. Broad, W. J. "Fusion Researchers Seek $25 Million From U.S." *New York Times*, April 26, 1989.

59. Anon. "Cold-Fusion Experiment Gives Tech Cold Shoulder." *Tech Topics*, Summer 1989.
60. Pollack, A. "Beating a Path to Fusion's Door; Companies in Race To Be in on Creating New Energy Source." *New York Times*, April, 28, 1989.
61. Champlin, C. "Cold Fusion Claims Melt After Heated Arguments; Neutron counters proved ineffective for precision of measurement needed." *The Technique*, April 29, 1989.
62. Ellison, C. H., and James A. Mahaffey. "An Investigation of Reports of Fusion Reactions Occurring at the Cathode in Glow Discharges." *Fusion Technology*, Volume 29, Number 1, 178–187, January 1996.

Chapter 5: The Lost Expedition to Mars

1. Angelo, Joseph A. Jr. *Rockets*. New York: Facts On File, 2008.
2. Dewar, James A. *To the End of the Solar System: The Story of the Nuclear Rocket*. Ontario, Canada: Apogee Books, 2007.
3. Finseth, J. L. *Overview of Rover Engine Tests: Final Report*. Huntsville, Ala.: Sverdrup Technology, February 1991.
4. Grant, Col. Kenneth F. *Nuclear Flight: The United States Air Force Programs for Atomic Jets, Missiles, and Rockets*. New York: Duell, Sloan, and Pearce, 1961.
5. NA. *America in Space: A Pictorial Review*. Washington, DC: U.S. Government Printing Office, 1964.
6. NA. *NASA Project Names*. Washington, DC: U.S. Government Printing Office, 1962.
7. Timmerhaus, K. D. *Advances in Cryogenic Engineering: Proceedings of the 1962 Cryogenic Engineering Conference*. New York: Plenum Press, 1963.

Chapter 6: The Chic-4 Revolution

1. Anon. *Restricted Data Classification Decisions 1946 to the Present (RDD-7)*. Washington, DC: U.S. Department of Energy Office of Declassification, January 1, 2001.
2. Irving, David. *The German Atomic Bomb: The History of Nuclear Research in Nazi Germany*. New York: Simon and Schuster, 1967.
3. Khan, Feroz Hassan. *Eating Grass: The Making of the Pakistani Bomb*. Stanford, Calif.: Stanford University Press, 2012,
4. Reed, Thomas C., and Denny B. Stillman. *The Nuclear Express: A Political History of the Bomb and Its Proliferation*. Minneapolis, Minn.: Zenith Press, 2009.
5. Sublett, Carey. *Nuclear Weapons Frequently Asked Questions: Section 3.0 Matter, Energy, and Radiation Hydrodynamics; Section 4.0 Engineering and Design of Nuclear Weapons; Section 6 Nuclear Materials*. http://nuclearweaponarchive.org/Nwfaq/Nfaq0.html, August 9, 2001.
6. Veda, V. N. *Pakistan's Nuclear Weapons*. New Delhi, India: KW Publishers Pvt Ltd, 2012.

Chapter 7: Japan's Atomic Bomb Project

1. Anon. *Disposition of Uranium Oxide Impounded by SCAP: Report by the State-War-Navy Coordinating Subcommittee for the Far East*. Declassified from TOP SECRET Jan. 12, 2000.

2. Anon. *Report on Interrogation of the crew of U-234 Which Surrender to the USS Sutton on 14 May, 1945, in Position 47°-07'N - 42°-27'W.* Portsmouth, N.H. (Op-16-Z), June 27, 1945. Declassified from Confidential Jan. 6, 1983.
3. Benke, Richard. "New evidence tracks Japan's efforts to create atomic bomb." *San Diego Union-Tribune*, June 1, 1997.
4. Clarke, Brig. Gen. Carter W. *"MAGIC"—Far East Summary: German-Japanese Liaison on Jet Propulsion.* TOP SECRET ULTRA. Washington, DC: War Department, June 13, 1945.
5. Ferguson, P. H. "Scientist Describes Japan's World War II Quest for Atomic Bomb." AP News Archive, July 19, 1995. http://www.apnewsarchive.com/1995/Scientist-Describes-Japan-s-World-War-II-Quest-for-Atomic-Bomb/id-9b2cd43fe9e33262ca8195c38c7412da
6. Fukushima, Shingo. "Scientists mining new trove of information on Japan's wartime A-bomb program." *Asahi Shimbun*, August 5, 2015.
7. Hiroki, Kato, "The Radios that Started and Ended World War II in the Pacific; An examination of the radios used during the attack on Pearl Harbor and the bombing of Hiroshima." *QST*. April 2016.
8. Parrott, Lindesay. "Five Cyclotrons Wrecked in Japan." *New York Times*, Nov. 23, 1945.
9. Rider, Dwight R. *Tsetusuo Wakabayashi, Revealed.* Fredericksburg, Va., 2014.
10. Scalia, Joseph Mark. *Germany's Last Mission to Japan: The Failed Voyage of U-234.* Annapolis, Md.: Naval Institute Press, 2000.
11. Snell, David. "Japan Developed Atom Bomb; Russia Grabbed Scientists." *Atlanta Constitution*, Oct. 3, 1946.
12. Sublett, Carey. *Nuclear Weapons Frequently Asked Questions: Section 4.1 Elements of Fission Weapon Design; Section 6.0 Nuclear Materials; Section 10.0 Chronology for the Origin of Atomic Weapons.* http://nuclearweaponarchive.org/Nwfaq/Nfaq0.html, August 9, 2001.
13. Wagner, D. Ing. H. *KERNPHYSIK: Techn. Stand und Anwendungsmöglichkeiten.* Berlin, Germany: HENSCHEL Flugzeuweke AG, August 5, 1941.
14. Wilcox, Robert K. *Japan's Secret War: Japan's Race Against Time to Build Its Own Atomic Bomb.* New York: Marlowe & Company, 1995.

Chapter 8: The Criminal Use of Nuclear Disintegration

1. Anon. *Criminal Dies Stealing Radioactive Material. NTI.* September 14, 1999.
2. Anon. *Crocker v. State.* Houston, Texas: Court of Criminal Appeals of Texas, Panel No. 2, May 10, 1978.
3. Anon. *Health Effects of Project SHAD Chemical Agent: Phosphorus-32 [Radiotoxic Effects].* Silver Spring, Md.: The Center for Research Information, Inc., Spring 2004.
4. Anon. *NRV Vindicates NIH Response to Contamination Incident.* Bethesda, Md.: NIH, October 21, 1997.
5. Anon. *Radiation Sickness or Death Caused by Surreptitious Administration of Ionizing Radiation to an Individual.* The Molecular Biology Working Group to The Biomedical Intelligence Subcommittee of The Scientific Intelligence Committee of UDIB, Report No. 4, August 27, 1969.

6. Anon. "Recent Events Involving Potential Loss of Control of Licensed Material." North Bethesda, Md.: USNRC. *NMSS Licensee Newsletter* March 1996–April 1996. NUREG/BR-0117; N0. 96-1.
7. Anon. "Report Clears Scientist of Radiation Poisoning." *Gazette*, September 24, 1997.
8. Anon. *The Radiological Accident in Tammiku*. Vienna, Austria: IAEA, October 1998.
9. Bailey, Edgar D., and Martin C. Wukasch. "A case of felonious use of radioactive materials." *IRPA Proceedings 4, Vol. 3*, April 1977, pp. 987–990.
10. Calio, Jim. "A Father's Monstrous Crime Against His Son Has Caused Kirk Crocker a Life of Pain." *People*, Vol. 23, No. 14, August 8, 1985.
11. Chao, J. H., C. L. Tseng, W. A. Hsieh, D. Z. Hung, and W. P. Chang, 2001, "Dose estimation for repeated phosphorus-32 ingestion in human subjects," *Applied Radiation and Isotopes*, 54:123–129.
12. Dunn, Wesley M. *Cobalt-60 Incident Update*. Houston, Texas: Texas Department of Health, March 11, 1996.
13. Ilyin, L. A., V. Yu. Soloviev, A. E. Baranov, A. K. Guskova, N. M. Nadezhina, and I. A. Gusev, May 2004, "Early medical consequences of radiation incidents in the former USSR territory," 11th International Congress of IRPA.
14. Streeper, Charles, Marcie Lombardi, and Lee Cantrell. "Nefarious uses of radioactive materials." Institute of Nuclear Materials Management 48th Annual Meeting, July 2007.

Chapter 9: The Threat of the Dirty Bomb

1. Anon. *Army Field Manual 3-14; Nuclear, Biological, and Chemical (NBC) Vulnerability Analysis*. Washington, DC: Department of the Army, November 1997.
2. Anon. *Department of the Army Technical Manual TN 0-1900, Department of the Air Force Technical Order TO 11A-1-20, Ammunition General*, Washington, DC: Departments of the Army and the Air Force, June 1956.
3. Anon. *Instruction and Service Manual, Countmaster Geiger Counter-Assayer*. Los Angeles: Hoffman Laboratories, 1954.
4. Anon. *Model 111B Scintillator ad Model 117 Special Scintillator Operation and Maintenance Manual*. Los Angeles: Precision Radiation Instruments, 1954.
5. Anon. *Precision Model 118B Royal Scintillator Operation and Maintenance Manual*. Los Angeles: Precision Radiation Instruments, 1954.
6. Anon. *Terrorism with Ionizing Radiation General Guidance; Pocket Guide*. Washington, DC: Department of Defense, October 2001.
7. http://national-radiation-instrument-catalog.com/new_page_4.htm
8. Koenig, Kristi L. et al. *Department of Homeland Security Working Group on Radiological Dispersal Device (RDD) Preparedness; Medical Preparedness and Response Sub-Group*. Washington, DC: Department of Health and Human Services, May 2003.
9. MacKinney, John. "Why Has There Been No Dirty Bomb Attack?" *Homeland Security Today*, February/March 2016.
10. Nelson, R. A. *Initial Management of Irradiated or Radioactively Contaminated Personnel*. Washington, DC: Department of the Navy, Bureau of Medicine and Surgery, December 1998.

11. Nininger, Robert D. *Minerals for Atomic Energy.* New York: D. Van Nostrand Company, 1954.
12. Norwow, L. *Quality Instruments for Locating Uranium and Metals for Industrial, Laboratory and Medical Use.* Los Angeles: Precision Radiation Instruments, 1954.
13. Reeves, Col. Glen I., *Triage of Irradiated Personnel*, Bethesda, Md.: Armed Forces Radiobiology Research Institute, September 1997.
14. Zoellner, Tom. *Uranium; War, Energy, and the Rock That Shaped the World.* New York: Penguin Group, 2009.

Chapter 10: A Bridge to the Stars

1. Baggot, Jim. *The Meaning of Quantum Theory.* Oxford, England: Oxford University Press, 1992.
2. Chiao, Raymond Y., Paul G. Kwiat, and Aephraim M. Steinberg. *Quantum Nonlocality in Two-Photon Experiments at Berkeley.* Ithaca, N.Y.: Cornell University Library, quant-phy, Jan. 18, 1995.
3. Haji-Hassan, T., A. J. Duncan, W. Perrie, and H. Kleinpoppen. "Polarization Correlation Analysis of the Radiation from a Two-Photon Deuterium Source Using Three Polarizers: A Test of Quantum Mechanics versus Local Realism." *Physical Review Letters*, Volume 62, Number 3, Jan. 16, 1989.
4. Herzof, Thomas J., Paul G. Kwiat, Harald Weinfurter, and Anton Zeilinger. "Complementarity and the Quantum Eraser." *Physical Review Letters*, Volume 75, Number 17, Oct. 25, 1995.
5. Herzog, Thomas J., Paul G. Kwiat, Harald Weinfurter, and Anton Zeilinger. "Complementarity and the Quantum Eraser." *Physical Review Letters*, Volume 75, Number 17, Oct. 23, 1995.
6. Kwiat, Paul G., Klaus Mattle, Harald Weinfurter, and Anton Zeilinger. "New High-Intensity Source of Polarization-Entangled Photon Pairs." *Physical Review Letters*, Volume 74, Number 24, Dec. 11, 1995.
7. Lofts, Mark John. *Method and System for Binary Signaling via Quantum Non-Locality.* Washington, DC: United States Patent Office, Pub. No. US 2006/0226418 A1, Oct. 12, 2006.
8. Steenblik, Richard A. *Method and Apparatus for Selectively Controlling the Quantum State Probability Distribution of Correlated Quantum Objects.* Washington, DC: United States Patent Office, patent no. 6,057,541, May 2, 2000.
9. Steenblik, Richard A. *Method and Apparatus for Selectively Controlling the Quantum State Probability Distribution of Correlated Quantum Objects.* Washington, DC: United States Patent Office, patent no. 6,473,719 B1, Oct. 29, 2002.
10. Steenblik, Richard A. *TAXI.* Alpharetta, Ga.: Chromatec, Inc., May 15, 1996.

Chapter 11: Conclusions

1. Anon. *Naval Sonar.* Washington, DC: Bureau of Naval Personnel, 1953.
2. Berlitz, Charles, and William L. Moore. *The Roswell Incident.* New York: Grosset & Dunlap, 1980.

3. Ellison, C. H., and J. A. Mahaffey. "An Investigation of Reports of Fusion Reactions Occurring at the Cathode in Glow Discharges." *Fusion Technology; a Journal of the American Nuclear Society*, Vol 29, No. 1, January 1996.
4. McAndrew, Capt. James. *The Roswell Report; Case Closed.* Washington, DC: U.S. Government Printing Office, 1997.
5. Weaver, Col. Richard L., Lt. James McAndrew. *The Rowell Report; Fact vs Fiction in the New Mexico Desert.* Washington, DC: U.S. Government Printing Office, 1996.

Illustration Credits

Illustration credits are listed in the order the images appear in the art inserts. The following eleven illustrations are courtesy of:

Dr. Paul-Jürgen Hahn, Administrator of Dr. Ronald Richters scientific legacy 77731 Willstätt, Germany. Email: paul-j.hahn@t-online.de http://www.p-j-hahn.de.

See also:

Paul-J. Hahn, Rainer Karlsch: "Ronald Richter und die Anfänge der Fusionsforschung" in R. Karlsch, H. Petermann (Ed.): "Für und Wider ›Hitlers Bombe‹" Cottbuser Studien zur Geschichte von Technik, Arbeit und Umwelt Waxmann-Verlag 2007,

Dr. P.-J. Hahn: "Das Richter-Experiment," http://www.p-j-hahn.de/richter.html,

Dr. P.-J. Hahn: "Dr. Richter: Der Auftakt der Fusionsforschung" http://www.p-j-hahn.de/pdf/Richter2.pdf,

Dr. W. Ehrenberg: "Die argentinischen Kernfusionsversuche in neuem Licht," ATOMPRAXIS, 4. Jahrgang April 1958 (http://www.p-j-hahn.de/atompraxis1.html),

Mario della Janna: "Ein Zeitzeuge berichtet," http://www.p-j-hahn.de/pdf/janna_1991.pdf,

Dr. P.-J. Hahn, (della Janna): "Die Janna-Skizze," http://www.p-j-hahn.de/pdf/Janna-Skizze.pdf, *and*

Dr. P.-J. Hahn: "Ball-Lightning: A HF Cavity Resonator?" http://www.p-j-hahn.de/resonator.html.
Young Ronald Richter in his lab (Courtesy of Dr. Hahn)
Richter's lab at Cordoba (Courtesy of Dr. Hahn)
Ronald, Ilse, and Monica at Bariloche (Courtesy of Dr. Hahn)
Fusion reactor under construction (Courtesy of Dr. Hahn)
First fusion reactor inside the lab (Courtesy of Dr. Hahn)
The large electromagnet (Courtesy of Dr. Hahn)
Richter at the reactor console desk (Courtesy of Dr. Hahn)
Two lesser magnets forming a gap (Courtesy of Dr. Hahn)
Another angle of the control desk (Courtesy of Dr. Hahn)
Equipment behind the console desk (Courtesy of Dr. Hahn)
Richter and Epsilon (Courtesy of Dr. Hahn)
The lab complex as it now exists (Courtesy of Guy Walters)
Inside a lab building on Huemul Island (Courtesy of Guy Walters)
Reactor being installed (Courtesy of the Georgia Department of Natural Resources)
Mel Dewar at the CEF (Courtesy of the Georgia Department of Natural Resources)
Existing ground-level portal (Courtesy of the author)
Billy Statham, reactor operator (Courtesy of the Georgia Department of Natural Resources)
East end of the hot cells (Courtesy of the author)
Palladium electrode (Courtesy of the author)
The team in RM127 (Courtesy of the author)
Preparing to remove the hydrogen from the palladium (Courtesy of the author)
The cold-fusion cell (Courtesy of the author)
The cold-fusion cell being lowered into the graphite block (Courtesy of the author)
Working on the neutron detector setup (Courtesy of the author)
KIWI-B4-A setup at Test Cell A (Courtesy of NASA)
The clamshell vacuum chamber for the NRX test (Courtesy of NASA)
The 27-inch cyclotron at RIKEN (Courtesy of Lawrence Berkeley National Laboratory)
Cyclotron accelerator chamber (Courtesy of Lawrence Berkeley National Laboratory)
RIKEN laboratory before the war (Courtesy of Wikimedia Commons)
Dr. Yoshio Nishina (Courtesy of Wikimedia Commons)
Cyclotron being discarded (Courtesy of Getty Images)
Soviet laser pistol (Courtesy of the Peter the Great Military Academy)
The IK17 Szhatie Soviet laser weapon (Courtesy of Vitaly V. Kuzmin)
A cluster of four Soviet radio-thermal electrical generators (Courtesy of Wikimedia Commons)
Vicotreen Model 247 (Courtesy of the UCLA Library)
Militarized Victoreen Model 247A (Courtesy of ORAU)
Hoffman Countmaster (Courtesy of the author)
Blossom Barton at her bath (Courtesy of UP)

PRI Model 118 Royal Scintillator (Courtesy of the author)
Entangled photon generator (Courtesy of Wikimedia Commons)
Bussard interstellar ramjet (Courtesy of NASA)
Daedalus probe (Courtesy of Adrian Mann)
The Activitron (Courtesy of the author)
Glow-discharge reactor on display (Courtesy of the author)

Interior illustration credits are listed in the order the images appear.

The Georgia Tech Research Reactor control room (Courtesy of the author)
The author sitting at his experimental apparatus (Courtesy of the author)
The author atop the reactor core (Courtesy of the author)
Proton-proton fusion (Courtesy of Infobase Learning)
Deuterium-tritium fusion (Courtesy of Infobase Learning)
Richter's diagram of his fusion reactor (Courtesy of Dr. Hahn)
Map of GNAL (Courtesy of the Georgia Department of Natural Resources)
Arial view of GNAL (Courtesy of the Georgia Department of Natural Resources)
Plan of the control room (Courtesy of the Georgia Department of Natural Resources)
Plan of the control room (Courtesy of the Georgia Department of Natural Resources)
Tupolev Tu-95LAL representation (Courtesy of Dreamtime.com)
The team gathered in RM127 (Courtesy of the author)
The team gathered in RM127 (Courtesy of the author)
Lab notebook page (Courtesy of the author)
Second lab notebook page (Courtesy of the author)
KIWI-A Prime nuclear rocket engine (Courtesy of NASA)
Data chart from the KIWI-B-1A test (Courtesy of NASA)
Relative sizes of nuclear rocket engines (Courtesy of NASA)
NRX nuclear rocket engine test setup (Courtesy of NASA)
Cyclotron diagram (Courtesy of Radio-Craft)
Thermal gas diffusion column (Courtesy *Physics Review Letters*)
Plans for the centrifuge (Courtesy of afloimages.com)
The Topfmine (Courtesy of Wikimedia Commons)
Faster-than-light communication (Courtesy of the author)
Kucherov glow-discharge reactor (Courtesy of the author)

Acknowledgments

Unfortunately, two of my very knowledgeable nuclear physics friends, Monte Davis and Don Harmer, passed away before they were able to read my latest manuscript, and I missed their nuclear-grade fact-checking skills. This left a burden on my third physicist friend, Doug Wrege, but he was able do a fine job trying to keep me in line. Harvey V. Lankford MD, Endocrinology and Nuclear Thyroidology, gave it a hard, clinical look from a different angle, and this action tuned the technical issues even further. Rick Steenblik's memories of our cold-fusion work were extremely helpful, and his very active, inventive mind is ultimately responsible for most of chapter 10. Robert Boyd's extensive collection of material and his detailed memories from the Georgia Nuclear Aircraft Laboratory, as well as reminiscences from the late Billy Staham, were essential for writing chapter 2. My English-language skills were given the usual hard scrubbing by my wife, Carolyn, her cousin, Alma Kiss, and Monte's widow, Nancy Davis, who together edited the living hell out of the manuscript. I am grateful to Suzie Tibor for finding a wealth of never-published photos from the Huemul Island project, and to Paul Jurgen Hahn and Monica Richter for providing them. I am further indebted to Jessica Case, deputy publisher of Pegasus Books; my literary agent, George Lucas of Inkwell Management; and to Lance, who is always here.

Index